T0313837

Current Practice in Forensic Medicine

Current Practice in Forensic Medicine, Volume 3

Edited by

John A.M. Gall and J. Jason Payne-James

This edition first published 2022

Registered Offices
111 River Street, Hoboken, NJ 07030, USA
The Atrium, Southern Gate, Chichester, West Sussex, PO19 8SQ, UK

Editorial Office
The Atrium, Southern Gate, Chichester, West Sussex, PO19 8SQ, UK

For details of our global editorial offices, customer services, and more information about Wiley products visit us at www.wiley.com.

Wiley also publishes its books in a variety of electronic formats and by print-on-demand. Some content that appears in standard print versions of this book may not be available in other formats.

Library of Congress Cataloging-in-Publication Data applied for
[ISBN: 9781119684091 (hardback)]

Cover Design: Wiley
Cover Image: © Virrage Images/Shutterstock

Set in 10/12pt SabonLTStd by Straive, Pondicherry, India
Printed and bound by CPI Group (UK) Ltd, Croydon, CR0 4YY

C9781119684091_270722

Contents

List of Contributors

Peter G. Blain CBE, BMedSci, MB, BS, PhD (Neurotoxicology), FBTS, FRBS, FFOM, FRCP(Lon), FRCP(Edin).
Professor Peter Blain is Emeritus Professor in the Translational and Clinical Research Institute at Newcastle University UK. He is a clinical professor and hospital physician with over 30 years' experience in the application of medical knowledge to intelligence, security, and operational issues, including high value asset protection for UK Government. He is a recognized international expert in clinical medicine and medical research, and provides high-level expert advice in CBRN medicine, related sciences, and emergency response medicine to both UK and US Governments and major international bodies. He received more than £11M in funding over the last decade for research into toxic mechanisms, diagnostic biomarkers, role of pluripotent stem cells in health protection, CBRN medical countermeasures and novel medical interventions. He was awarded a CBE in 2002 for services to medicine and defence.

Anthony Bleetman PhD, FRCSEd, FRCEM, DipIMC, RCSEd
Dr Bleetman is in full-time active clinical practice as a consultant in emergency medicine, formerly serving as the clinical director of urgent care at Kettering General Hospital NHS Foundation Trust. Prior to taking this post, he was the lead consultant in emergency medicine at the North West London Hospitals NHS Trust. He holds the position Honorary Clinical Associate Professor at the University of Warwick Medical School. Tony holds a part-time contract as a senior emergency physician in Beilinson Hospital, Israel. He served in the Israeli Defence Forces in a number of roles between 1981 and 1991. He completed medical school in 1989. He trained on a surgical rotation in Glasgow and received the FRCSEd in 1993. He commenced higher specialist training in accident and emergency medicine in 1994 and was appointed consultant in accident and emergency medicine at Birmingham Heartlands Hospital in 1996. The hospitals evolved into a Foundation Trust incorporating three hospitals, and Tony served as the clinical lead for emergency medicine at Good Hope Hospital until May 2010 prior to moving to London to assume the lead for emergency medicine at North West London Hospitals NHS Trust.

Tony received a PhD in occupational health from the University of Birmingham in 2000. He directs Advanced Trauma Life Support courses and regularly instruct on other accredited life support and resuscitation courses. He served as clinical director for HEMS for West Midlands Ambulance Service and continued to fly on air ambulances providing emergency medical and trauma services until 2013. In 1992, Tony was awarded the Diploma in Immediate Medical Care by the Royal College of Surgeons of Edinburgh. He was awarded the Queens Golden and Diamond Jubilee Medals for his pre-hospital emergency work.

Tony has written and exercised multi-agency major incident plans. He sat on government advisory committees for disaster and emergency planning. He is a medical advisor to the Ministry of Defence serving on SACMILL (Scientific Advisory Committee on Less Lethal Weapons). His PhD was for work on developing body armour for the police. This arose from his development work for the Home Office and the Police Federation on officer safety programmes, addressing protection from knives and bullets. Tony continues to work for the police on these programmes and is the first doctor to qualify as a police instructor for unarmed defensive tactics, safe prisoner restraint, handcuffing, tactical communication skills, incapacitant sprays, and knife defence. Through this interest, Tony has been able to offer opinions on use of force and injuries sustained during arrest and detention.

Tony has been involved in developing strategies to protect health workers against aggression and violence in the health service. He has completed studies for the Department of Health and other national bodies to identify ways of improving staff and subject safety. He is engaged in developing safe physical interventions and effective training strategies across a number of agencies. Tony served on the guidelines development group of the Joint Royal Colleges Ambulance Liaison Committee. He has published numerous articles in peer-reviewed professional journals.

Phoebe Bragg LLB

Phoebe Bragg is a barrister at 5 King's Bench Walk, London, where she practises criminal law. Phoebe graduated from the University of Oxford, Trinity College, in 2013 with a law degree and was called to the Bar in 2015. She acts for both prosecution and defence across a wide range of criminal cases. Phoebe has worked as disclosure counsel on high-profile inquiries including the Grenfell Tower Inquiry and the Post Office Horizon IT Inquiry. Phoebe is a member of Gray's Inn and an Ann Goddard Scholar. Prior to being called to the Bar she worked in the Criminal Law and Terrorism Division at the Council of Europe on multi-jurisdictional criminal issues as well as on capital defence appellate work across the United States.

Mark Byers MB, BS, MSc, FRCGP, FIMC, FFSEM, MRCEM, DA(U.K.), DRCOG, DMCC, DUMC

Dr Mark Byers is a Consultant in Pre-hospital Care and a Subject Matter Expert on CBRN having been involved with the subject for over 30 years. An author of several texts on the subject he is actively involved in the development of new and novel medical countermeasures and management practices around a CBRN incident

Juliet Cohen MA, MB, BS, DipRACOG, MRCGP, FFFLM

Dr Juliet Cohen is currently the head of doctors at the charity Freedom from Torture and an independent forensic physician examining patients for evidence of torture, trafficking, and domestic violence. She is a member of the Independent Forensic Experts Group and contributed to the new edition of the Istanbul Protocol, the UN Manual on the Investigation and Documentation of Torture, and other Cruel Inhuman and Degrading Treatment or Punishment as a primary drafter of Chapter 5. She has devised and delivered training on documentation of torture and understanding medical evidence, for doctors, judges, and Home Office asylum decision-makers. She has been an expert consultant revising training for the European Asylum Support Office modules on 'interviewing vulnerable persons' and 'evidence assessment'.

Neil Corney MSc

Neil Corney has been a research associate at the Omega Research Foundation since 1996. His work has included research and analysis of the human rights impacts of less-lethal weapons and analysis of the plastic baton round kinetic impact projectile in Northern Ireland. He was a member of the core Academic Working Group which developed the 2020 UN Guidance on Less Lethal Weapons and is currently the chair of the UK's National Taser Stakeholder Advisory Group (NTSAG) and attends the UK National Police Chiefs Council Less Lethal Weapon Working Group.

Stevan R. Emmett MB, ChB, DoHNS, DPM, D.Phil, MRSB, FFPM

Dr Steve Emmett is Chief Medical Officer at Defence Science and Technology Laboratory, Porton Down and is a Consultant & Fellow in Pharmaceutical Physician specialising in medical countermeasure drug development, pharmacology and toxicology. Steve is medical advisor to the advance development programme delivering future therapeutic interventions for historic and emerging chemical and biological threats to the UK. He qualified in medicine from the University of Warwick, has a D.Phil in Clinical Pharmacology from the University of Oxford and has held a number of academic positions and industry posts. Currently he supports and leads teams in medical aspects of research and development across the Dstl portfolio from the laboratory, through clinical development to licensure. Steve is a member of the UK Advisory Group on Military and Emergency Response Medicine and has over 20 years experience in CBRN medicine; incident management, clinical management, teaching and training. He continues to practice clinically in Emergency Medicine in the NHS and is honorary Senior Fellow in Medicine at the University of Bristol.

Sarah Forshaw LLB (Hons), QC

Sarah Forshaw QC is a leading barrister practising in criminal law. She was appointed Queen's Counsel in 2008 (interestingly, only the 193rd woman ever to be awarded silk). She acts for both prosecution and defence.

Sarah is the co-head of Chambers at 5 King's Bench Walk in Temple, London. In 2013, she was appointed leader of the South Eastern Circuit (the largest of the six geographical circuits of the Bar of England and Wales). In 2014, she was nominated for Crime Silk of the Year by Chambers and Partners and is listed as a Band 1 Silk in the current edition. Sarah has also occupied roles as chair of the Central London Bar Mess and director of education and training on the South Eastern Circuit. She was on the Working Group responsible for drafting The Advocate's Gateway toolkit on Memory and Sensory Issues in Witnesses and Defendants with Autism. She has been asked to speak on numerous occasions both within the profession and without, including at the Oxford Criminology Series, All Souls College.

Sarah Forshaw's busy practice is now broadly divided between murder trials and representing professionals facing other (often sexual) allegations, who would not otherwise be able to secure the services of Queen's Counsel under the Legal Aid regulations. In that latter capacity, she also routinely represents individuals facing regulatory proceedings following the conclusion of the criminal trial.

John A.M. Gall BSc (Hons), MB, BS, PhD, FFFLM, FFCFM (RCPA), DMJ (Clin & Path)
Dr John Gall is a consultant forensic physician; an associate professor, Department of Paediatrics, The University of Melbourne; director of Southern Medical Services; principal of Era Health; and is a staff specialist forensic physician with the Victorian Paediatric Forensic Medical Service, located at the Royal Children's Hospital and Monash Children's Hospital, Melbourne. He is the president of the International Association of Clinical Forensic Medicine. John qualified initially as a biochemist, completed his doctorate in the Department of Pathology at the University of Melbourne, and engaged in postdoctoral research in anatomy. He later read medicine at the University of Melbourne and undertook training in anatomical and forensic pathology and clinical forensic medicine. He has practised clinical forensic medicine for almost 30 years, initially as a forensic medical officer with Victoria Police and later as a consultant at the Victorian Institute of Forensic Medicine. John has been extensively involved in under- and postgraduate education both at the University of Melbourne and Monash University. He was an honorary senior lecturer in the Department of Forensic Medicine, Monash University, during which time he taught custodial medicine in the University's Graduate Diploma of Forensic Medicine. He also devised, developed, and administered an international continuing education programme in forensic medicine and, with co-authors, wrote and edited *Forensic Medicine Colour Guide* and the previous two volumes of this series. John has been involved in forensic medical research, and much of these findings have been published. In addition to forensic medicine, he practises in occupational medicine.

Peter Green MBBS, DMJ, FFFLM, FACLM, FCLM
Dr Peter Green is a forensic and legal medicine specialist. He works as a forensic physician and a lead for child safeguarding (designated doctor) in London. He has trained on various aspects of torture prevention in multiple international events and chaired the Independent Medical Advisory Panel to the European NPM project. He co-chaired with Dr Cohen the creation of Quality Standards for Healthcare Professionals Who Work with Victims of Torture. He remains committed to the thesis that the best way to prevent torture is by education.

Felicity Goodyear-Smith MB, ChB, MD, FRNZCGP (Dist), FFFLM
Professor Felicity Goodyear-Smith is a general practitioner and distinguished fellow of the Royal New Zealand College of General Practitioners and a professor in general practice and primary health care at the University of Auckland, New Zealand. She is also a qualified forensic physician and has worked as police doctor and as expert witness for both the prosecution and the defence in the past, with particular expertise in issues around sexual assault. In 2008, she became a member and in 2014 a fellow of the Faculty of Forensic and Legal Medicine of the Royal College of Physicians. Felicity has published over 260 scientific papers in peer-reviewed journals plus five books, 14 book chapters, and a number of other publications. Over 30 of her journal publications and three of her books pertain to forensic issues. The latter range from 'Sexual Assault Examinations – A Guide for Medical Practitioners', published in 1987 (second edition 1990) to 'Murder That Wasn't: The Case of George Gwaze' (2015), which explored the factors leading to George Gwaze being twice falsely charged and twice acquitted of the rape and murder of his 10-year-old niece. In 2015, Felicity also

published a chapter entitled 'Understanding Why and How False Allegations of Abuse Occur' in Ros Burnett's 'Wrongful Allegations of Sexual and Child Abuse'.

Nicholas Hallett MB, ChB, BSc, LLM, MRCPsych
Dr Nicholas Hallett is a consultant forensic psychiatrist at Essex Partnership University NHS Foundation Trust. He works on an acute inpatient medium secure admissions ward. He qualified in medicine in 2011 at the University of Bristol where he also completed a BSc in bioethics during his medical training. He has undertaken a postgraduate qualification at the Northumbria Law School in Newcastle where he was awarded an LLM in mental health law.

He has published a number of articles on medicolegal issues including the role of psychiatric evidence in the defence of diminished responsibility and the relevance of psychiatric evidence in the determination of criminal culpability. He has provided written and oral expert psychiatric evidence in criminal cases on numerous occasions including for cases of homicide. He teaches regularly on the Maudsley Training Programme in London on medicolegal topics including on giving medical evidence at Mental Health Tribunals. He is currently co-editing a new book on clinical topics in forensic psychiatry and criminal law.

Stanislav Lifshitz MD
Dr Stanislav Lifshitz MD is an emergency physician with broad experience in trauma management, military medicine, and pre-hospital care. Combining experience of trauma with a solid grounding in physical interventions, he provides biomechanical support to medical risk assessments of use of force interventions. He has assisted in medical and safety reviews of physical and mechanical interventions across a broad range of organisations that include children's homes, mental health, and custodial services. His knowledge and experience allow him to provide advice on physical intervention in specific vulnerable populations in care.

Matthew McEvoy LLM
Matthew McEvoy has been a research associate at the Omega Research Foundation since 2015, where his work has included the provision of training and technical assistance to human rights monitors and legal professionals to aid their documentation of 'less-lethal' weapons, human rights-based analysis of use of force protocols, and the development of resources concerning the documentation of law enforcement equipment used in the context of public assemblies.

Jane Monckton-Smith BA (Hons), PhD, PGCAP
Jane Monckton Smith is Professor of Public Protection at the University of Gloucestershire with a specialism in homicide, and especially domestic or intimate partner homicides. She has published a number of books on the homicide and forensic investigation. She has an international reputation as an expert in the field and has developed a new theoretical framework for understanding a perpetrator's psychological progression to homicide that is being used widely by multi-agency professionals internationally.

In addition to her academic work she also maintains a diverse portfolio of professional and case work. She works with families bereaved through homicide helping them with criminal justice and other processes; she advises statutory homicide review panels, as well as chairing statutory domestic homicide reviews; she advises police on current and cold investigations, and crisis risk assessments; she trains police and other professionals in assessing threat and risk in cases of domestic violence and stalking, and recognising and identifying suspicious deaths.

Curtis E. Offiah BSc, MB, ChB, FRCS, FRCR
Dr Curtis Offiah is a consultant radiologist with subspecialty expertise in adult and paediatric neuroradiology, head and neck radiology, trauma radiology, and forensic radiology. He has been employed as a full-time consultant position at the Royal London Hospital, part of Barts Health NHS Trust since 2006. The Royal London Hospital is a teaching hospital associated with Queen Mary's University London and Barts and The London School of Medicine and Dentistry and is one of the busiest major trauma centres in the United Kingdom. He has extensive experience in forensic radiology and post-mortem imaging with initial experience gained in North America and Europe. Dr Offiah holds the degrees of Bachelor of Science, Bachelor of Medicine, and Bachelor of Surgery and is a fellow of the Royal College of Radiologists and previously of the Royal College of Surgeons of Edinburgh.

Dr Offiah has research and expert experience in forensic issues relating to radiology, including imaging of blunt and penetrating trauma, radiological wound and injury mechanism profiling, post-mortem radiology, non-accidental injury, elder abuse, and intimate-partner violence. He lectures at national, post-graduate, and undergraduate levels and has lectured at the Royal Society of Medicine, Royal College of Radiologists Annual Scientific Meetings, Faculty of Forensic and Legal Medicine, and the British Association of Forensic Scientists. He holds a senior lecturer appointment at Cameron Forensic Medical Sciences, William Harvey Research Institute of Barts, and The London School of Medicine and Dentistry, Queen Mary University, London. He has published widely including a number of publications including peer-reviewed scientific research papers, reviews, and textbook chapters. He is on the Editorial Board of the National Journal of the Royal College of Radiologists (Clinical Radiology). Dr Offiah undertakes expert forensic radiology case work for numerous police authorities in the United Kingdom and abroad as well as on behalf of the British Military Police, the National Crime Agency, Independent Office of Police Conduct, and a number of HM Senior Coroners in the United Kingdom. He regularly gives expert evidence when required at court on behalf of the Crown Prosecution Service and on behalf of defence as well as at HM Coroner's Court.

J. Jason Payne-James LLM, MSc, FFFLM, FRCS, FRCP, FCSFS, FFCFM(RCPA), RCPathME, DFM, LBIPP Mediator
Professor Jason Payne-James is Specialist in Forensic & Legal Medicine and a Consultant Forensic Physician. He has been a forensic physician for over 30 years. His clinical and research interests are wide-ranging and include documentation and interpretation of injury, evidential sampling, wound and scar interpretation, sexual assault, intimate partner violence, clinical aspects of healthcare in custody, non-fatal strangulation, complaints against healthcare professionals, restraint and less-lethal systems, miscarriages of justice, death and harm in custody and torture and cruel, inhuman and degrading treatment. He is Lead Medical Examiner at the Norfolk & Norwich Hospitals University NHS Trust, Norwich, UK; expert adviser to the

UK National Crime Agency; Chair of the UK Scientific Advisory Committee on the Medical Implications of Less-Lethal weapons. He is Honorary Clinical Professor, Queen Mary University of London and Medical Director of Forensic Healthcare Services Ltd. He is an Executive Board Member of the European Council of Legal Medicine. He created the ForensiGraph® scales and the ForensiDoc® App. He has co-authored & co-edited (amongst others) the Encyclopedia of Forensic & Legal Medicine (1st & 2nd editions); the 13th and 14th Editions of Simpson's Forensic Medicine and the Medical Examiner System in England & Wales: A Practical Guide. His clinical and expert practice is based in the UK but he reviews many cases including those in the criminal justice system, deaths and care in state custody and possible miscarriages of justice, both in the UK and internationally.

Keith J.B. Rix BMedBiol (Hons), MPhil, LLM, MD, MAE, FEWI, FRCPsych, Hon FFFLM

Prof Keith Rix's involvement in the forensic field began in the 1960s when he lived in hostels in London with ex-offenders and assessed prisoners for hostel admission. In 1983, he moved to Leeds as a senior lecturer in psychiatry. There he became a visiting consultant psychiatrist at HM Prison, Leeds, and established the Leeds Magistrates' Court Mental Health Assessment and Diversion Scheme and the city's forensic psychiatry service. He has provided expert evidence for over 35 years, including on a *pro bono* basis in capital cases in the Caribbean and Africa. He is also the editor of *A Handbook for Trainee Psychiatrists* and the co-author, with his wife Elizabeth Lumsden Rix, of *Alcohol Problems*. His RCPsych Publications textbook *Expert Psychiatric Evidence* (2011) has just been published in its second edition by Cambridge University Press as *Rix's Expert Psychiatric Evidence* on an edited and multi-author basis. He has also devised for the Royal College of Psychiatrists the *Multi-Source Assessment Tool for Expert Psychiatric Witnesses* (MAEP) which can be used by all expert witnesses. He is a former chairman of the Fitness to Practise Panel of the Medical Practitioners Tribunal and part-time lecturer in the Department of Law, De Montfort University, Leicester. He is a longstanding member of the Academy of Experts and a founding member and fellow of the Expert Witness Institute. He is now a visiting professor of Medical Jurisprudence, University of Chester, Honorary Consultant Forensic Psychiatrist, Norfolk and Suffolk NHS Foundation Trust and Mental Health and Intellectual Disability Lead, Faculty of Forensic and Legal Medicine of the Royal College of Physicians.

Denise Syndercombe-Court MRSB, CBiol, FIBMS, CSci, DMedT, MCSFS, PhD

A scientist, geneticist, statistician, academic, editor, and author of a prize-winning medical textbook and published author of peer-reviewed scientific publications, Denise Syndercombe Court was trained in systematic reviews and evidence-based approaches of medical and scientific publications. From 1990, she was a senior lecturer in forensic haematology at Barts and The London School of Medicine and Dentistry. In 2012, she moved to King's College London where she is now the professor of forensic genetics. She has more than 30 years of experience in scientific research, forensic evidence examination, and DNA interpretation with a sound knowledge of the civil and criminal justice process, including court presentation as an accredited expert witness. She runs an ISO17025 laboratory dealing with all matters of DNA and is an active researcher in new molecular techniques for human identification. She represents the United Kingdom on the European DNA Profiling Group (EDNAP), is a member of the International Society for Forensic Genetics, secretary general of the British Academy of Forensic Sciences, a member of the Forensic Regulator's DNA Working Group, a member of the Home Office Biometrics and Forensic Ethics Group, and a member of the BSI Committee on Standards in Forensic Science.

Caroline Watson BA Hons (Cantab), MB BChir, MRCGP, DRCOG, QTS
Caroline Watson is a general practitioner and qualified school teacher. She studied medical sciences at Gonville and Caius College, Cambridge University, and qualified from Cambridge University School of Clinical Medicine in 1993. She developed an interest in health education and communication, working with under- and postgraduate medical students, patient groups, community volunteers, and school children. She then trained as a primary school teacher, qualifying in 2010, and combined careers in medicine and teaching. Caroline worked with homeless patients in Cambridge before starting in prison medicine in 2011. She set up a multi-disciplinary prison pain clinic in 2014 and developed an interest in providing multi-faceted support for patients with complex pain, while reducing the risk caused by dependence-forming medicines. She co-authored RCGP Safer Prescribing in Prisons, Second Edition (2019), and was appointed RCGP Clinical Champion for Healthcare in Secure Environments in the same year to produce a toolkit of resources for clinicians working in secure environments. During 2020, she wrote COVID-19 guidance for clinicians working in secure environments, co-authored a COVID-19 pictorial resource for patients, aimed at overcoming language, literacy, and learning disability barriers and produced Covid information for patients published in *Inside Time*, the national prison newspaper. Since then, she has continued writing a monthly health column for *Inside Time*.

She was a co-author of *Tackling Causes and Consequences of Health Inequalities: A Practical Guide (Matheson J et al, 2020)* and has contributed to work on dependence-forming medicines for Public Health England, NHS England, and NICE. Caroline is a clinical advisor for RCGP and for NHSEI Health and Justice team and contributes to the Royal College of Psychiatrists Advisory Group of Quality Network of Prison Mental Health Services. She is on the steering committee of the NHS East of England Palliative and End of Life Network. Caroline has led workshops on leadership, pain management, safer prescribing in prisons, and educational resources. She has spoken at the 6th and 7th RCGP SEG Health and Justice Summits 2019, 1st Virtual Health and Justice Summit 2020, 24th Annual Conference SMMGP RCGP 2020, and NHS Education for Scotland Education and Training Group for Prison Healthcare. She has also contributed to a number of webinars with RCPsych QNPMHS, SMMGP, and NHSEI.

Preface

Forensic medicine continues to be an evolving field in which many issues of controversy arise, certain subjects become the focus of attention, and some subjects arise that had been rarely considered previously. This may be because of new research, new technology, new laws or regulations, and a revision of old concepts and beliefs. There is considerable overlap between the clinical and pathological aspects of forensic medicine and the more general fields of toxicology, fitness to drive, forensic psychiatry and psychology, and forensic biology. This third volume provides a practical update on areas relevant to contemporaneous clinical practice and with a focus for debate in selected topics. We hope the content reflects our wish to ensure that all chapters are either of direct relevance, specific interest or bring new knowledge to readers. Each chapter is written by those with particular expertise or interest in the field and who have been directly involved in the matters about which they write. Every chapter reflects topics that have come to the fore in the past two years and which, in our opinion, are going to be of great relevance to healthcare professionals and other practitioners from a multi-professional and international audience. Some chapters may be jurisdiction specific, but all have been chosen because they have wider applicability.

This volume contains a range of current, new, and controversial subjects, including chapters that provide information on the new Medical Examiner system in England and Wales, riot-control weapons, chemical warfare, non-fatal strangulation, coercive control and the homicide timeline as part of intimate partner abuse, the current use of DNA in crime detection, and the expansion of radiological imaging to assist in the assessment of soft tissue injuries. The needs and problems of the older person in detention are addressed as is the increasingly discussed issue of elder abuse. The controversial areas of false allegations of sexual assault and abusive head trauma in young children are covered and intended to stimulate discussion.

As always, the views expressed in this volume are those of the chapter authors and do not necessarily represent those of the editors or the publishers. We hope that this new volume will once again stimulate discussion and reflection on practice, even if the reader may have different opinions from some of the views expressed here. We are always happy to hear from you.

John AM Gall, Melbourne

J. Jason Payne-James, Southminster

May 2022

1 The new Medical Examiner System in England and Wales: its role in the medicolegal investigation of death

J. Jason Payne-James[1,2,3]
[1] Norfolk & Norwich University Hospital, Norwich, UK
[2] William Harvey Research Institute, Queen Mary University of London, London, UK
[3] Forensic Healthcare Services Ltd., Southminster, UK

Introduction

Perhaps, the first thing to mention when discussing the new Medical Examiner (ME) System in England and Wales is that MEs do not do autopsies and may come from a wide variety of clinical backgrounds. This may cause confusion in many other jurisdictions, with other healthcare professionals and the lay public, where the term 'ME' is often used interchangeably with 'forensic pathologist'.

Similarly, an explanation is needed to ensure that the role of Her Majesty's Coroner (HM Coroner) in England and Wales is not confused with other coronial systems in other jurisdictions. In England and Wales, a coroner holds a judicial post and requires legal experience and qualifications. The office of the coroner was originally established in 1194 as a form of tax collector but over the years has evolved into an independent judicial officer, charged with the investigation of sudden, violent, or unnatural death.

An ME in England and Wales is an independent, senior doctor who reviews (scrutinises) deaths that are not investigated by the England and Wales coronial system. Thus, the ME system in England and Wales has become an essential part of the medicolegal investigation for all deaths which are not overseen by the coronial system (the majority of deaths). The ME system works very closely with coronial services. There were over 500 000 deaths in England and Wales in 2019 (ONS 2020a), and over 200 000 were reported to HM Coroners of which over 80 000 had post-mortem examinations (Ministry of Justice 2020). Those not reported to HM Coroner will in future all be reviewed by the ME system. Overall, the numbers of deaths have been distorted upwards by the coronavirus pandemic for 2020/2021 and are in excess of 600 000. Most coronavirus deaths are considered as natural deaths unless other factors (e.g. failure to supply appropriate personal protective equipment by employers) may be involved in which case coroners may become involved.

Current Practice in Forensic Medicine, Volume 3, First Edition. Edited by John A.M. Gall and J. Jason Payne-James.

This chapter explains the role, function, and aims of the newly introduced ME system in England and Wales. This system is a new process established to improve the medicolegal investigation of death. It has been developed to address perceived gaps in the review of all deaths, to identify patient safety issues, and to prevent previously identified scenarios where the concerns of families and whistleblowers about care of the deceased have been ignored. The ME system in England and Wales is currently on a non-statutory footing, but that is likely to change to a statutory basis by 2022/2023 (Payne-James and Lishman 2022).

Background

There have been many concerns expressed over the years about the medicolegal investigation of death, and a number of UK reports and committees have identified shortcomings in the process suggesting, in particular, that opportunities for improving patient safety have been missed with the potential for concealed homicide.

All doctors in the United Kingdom should be aware of Harold Shipman, an apparently respected general practitioner (family physician) from Hyde in Greater Manchester, UK, who, over a period of 20 or more years, was likely responsible for the murder of around 250 of his patients. The events raised two main questions – what made an apparently respectable doctor turn to murder on such an horrific scale, and most significantly, why did nobody in authority realise what was happening (Home Office/Department of Health 2007)? The Shipman Inquiry was initiated to investigate Shipman's activities and established that he was able to conceal malpractice and kill many of his patients because the systems in place permitted him to avoid questions and suspicion despite him certifying the causes of death of many of his patients.

The Shipman Inquiry was established in January 2001, following Shipman's conviction the previous year for the murder of 15 of his patients (not all cases were included in the criminal proceedings). The Inquiry had a broad remit and was tasked with investigating the extent of Shipman's unlawful activities, enquiring into the activities of the statutory authorities and other organisations involved, and making recommendations on the steps needed to protect patients for the future. In a series of six reports published between 2001 and 2003, the Inquiry made a number of recommendations for the reform of various British systems. It called for coroners to be better trained and underlined that better controls on the use of Schedule 2, 3, and 4 drugs by doctors and pharmacists were needed (Misuse of Drugs Regulations 2001). It also recommended that fundamental changes be implemented in the way that doctors were overseen by the General Medical Council (GMC) (the body with responsibility for regulating registered medical practitioners in the United Kingdom).

The Inquiry also established that there were flaws in the system for reviewing cremations where doctors (medical referees), whose role was to independently certify the cause of death (Ministry of Justice 2012), did not recognise Dr Shipman as anything but a respected colleague and thus perpetuated his dishonest accounts of his patients' deaths. For those patients undergoing burial, Dr Shipman was not required to consult any other medical practitioner and utilised the lack of medical knowledge of registrars of births and deaths for his causes of death to be accepted and registered. The system, as it was, depended on the integrity and honesty of a doctor, and there was no robust and independent oversight. These concerns reiterated and reinforced other reports or inquiries which had also noted that existing arrangements for death certification were confusing and provided inadequate safeguards against possible criminal activity.

The term ME is referred to at para 17.29 of Dame Janet Smith's third report (The Shipman Enquiry 2002) in which reference is made to establishing the role of Medical Coroner (*'17.29 The Society of Registration Officers suggested that the office of medical coroner should be a statutory post, independent from the NHS, with accountability passing up to a Chief Medical Coroner (the Society favoured the term 'Medical Examiner') at the head of a free-standing national agency').* That same report also referred to the Finnish system of death certification in the following terms at 18.122 '*The most impressive aspect of the Finnish system of death certification was the emphasis on the importance of accurately ascertaining the cause of death, even where the death was apparently natural. This is of considerable significance, not only for the deceased's family, but also for society generally; it has significant implications for public health*'. The Shipman Enquiry recommended that a new national coroners' service under a chief coroner should be established at arm's length from national government, replacing the current system of local coroners appointed and funded by local authorities. This service would be responsible for the final certification of death and for deciding whether further investigation was necessary in all deaths, and the new system would contain both medical coroners who would be responsible for establishing the medical cause of death and judicial coroners who would carry out further investigations where necessary (e.g. in the case of suspicious deaths). It was these proposed 'medical coroners' which evolved into the present ME with the role of ME being formally introduced to the England and Wales jurisdiction by the Coroners and Justice Act 2009 (Table 1.1). The 'Luce Review' – 'Death

Table 1.1 Relevant sections in the 2009 Coroners and Justice Act introducing the Medical Examiner into legislation.

Medical examiners

19 (1) [Local authorities] (in England) and Local Health Boards (in Wales) must appoint persons as medical examiners to discharge the functions conferred on medical examiners by or under this chapter.

(2) Each [local authority] or Board must—

(a) appoint enough medical examiners, and make available enough funds and other resources, to enable those functions to be discharged in its area;

(b) monitor the performance of medical examiners appointed by the [local authority] or Board by reference to any standards or levels of performance that those examiners are expected to attain.

(3) A person may be appointed as a medical examiner only if, at the time of the appointment, he or she—

(a) is a registered medical practitioner and has been throughout the previous 5 years and

(b) practises as such or has done within the previous 5 years.

Table 1.2 Recommendations from 2013 Report of the Mid-Staffordshire NHS Foundation Trust Public Inquiry relating to Medical Examiners.

275	Independent medical examiners. It is of considerable importance that independent medical examiners are independent of the organisations whose patients' deaths are being scrutinised.
276	Sufficient numbers of independent medical examiners need to be appointed and resourced to ensure that they can give proper attention to the workload.
277	Death certification. National guidance should set out standard methodologies for approaching the certification of the cause of death to ensure, so far as possible, that similar approaches are universal.
278	It should be a routine part of an independent medical examiner's role to seek out and consider any serious untoward incidents or adverse incident reports relating to the deceased to ensure that all circumstances are taken into account whether or not referred to in the medical records.
279	So far as is practicable, the responsibility for certifying the cause of death should be undertaken and fulfilled by the consultant or another senior and fully qualified clinician in charge of a patient's case or treatment.
280	Appropriate and sensitive contact with bereaved families. Both the bereaved family and the certifying doctor should be asked whether they have any concerns about the death or the circumstances surrounding it, and guidance should be given to hospital staff encouraging them to raise any concerns they may have with the independent medical examiner.
281	It is important that independent medical examiners and any others having to approach families for this purpose have careful training in how to undertake this sensitive task in a manner least likely to cause additional and unnecessary distress.

certification and investigation in England, Wales, and Northern Ireland: the report of a fundamental review' (TSO 2003) came to broadly similar conclusions as the Shipman Inquiry. It was, however, to be more than 20 years after Harold Shipman was convicted and 10 years after the Coroners and Justice Act before a national roll-out of MEs was begun, and then, not in the structure envisaged in the Act. In the interim, other hospital-based scandals were the subject of major enquiries.

Perhaps, the most significant for the (at that time non-existent) ME system was the Report of the Mid-Staffordshire NHS Foundation Trust Public Inquiry (Francis 2013) which identified numerous, serious failing in care between 2005 and 2009. The Report chaired by Robert Francis QC made 290 recommendations, of which a number made specific reference to the need for an Independent Medical Examiner (IME) system. Table 1.2 shows the recommendations about MEs. The Report recognised that '*Significant changes have occurred in the coronial court system since the events under review, including the appointment of a Chief Coroner and the creation of the new post of Independent Medical Examiner (IME)*'.

Other hospital-based scandals have highlighted poor care or deaths that may have been prevented had an effective system of independent scrutiny been in place at the time.

Table 1.3 Intended benefits of the introduction of the Medical Examiner System to England and Wales.

- It will be fair – all deaths will be scrutinised in a robust and proportionate way regardless of whether they are followed by burial or cremation;
- It will be independent – a medical examiner will scrutinise all medical certificates of cause of death (MCCD) prepared by the attending doctor;
- It will be transparent – families will have the cause of death explained to them, including clarification of medical terms, and be able to ask questions or raise concerns;
- It will be robust – there will be a protocol that recognises different levels of risks depending on the circumstances and stated cause of death;
- It will be accurate – the medical examiner will be an experienced doctor, capable of ensuring that the MCCD is completed fully and accurately, providing the NHS, the Office for National Statistics, local authorities, and wide range of other users with better quality cause of death statistics to inform health policy, the planning and evaluation of health services and international comparisons;
- It will be efficient – it will help to make sure that the right cases are reported to coroners; and
- It will improve safety – the new system will allow easier identification of trends, unusual patterns, and local clinical governance issues and make malpractice easier to detect.

Source: Taken from Department of Health (2016). Public Domain(OGL).

The Report of the Morecambe Bay Investigation in 2015 (Kirkup 2015) which examined concerns raised by the occurrence of serious incidents in maternity services including the deaths of mothers and babies concluded: '. . . *a mechanism already in use in other countries has been put forward to scrutinise all deaths in this way that would by its nature pick up maternal and neonatal deaths. This is the appointment of medical examiners, initially proposed by Dame Janet Smith as a recommendation of the Shipman Inquiry, subsequently endorsed by the Luce review, put into enabling legislation in 2009 but not yet implemented. It is our view that implementing these proposals should be reactivated as the best means to provide the necessary scrutiny, not just of maternity-related deaths, but of all deaths*'.

The Department of Health (2016) published a consultation: Introduction of Medical Examiners and Reforms to Death Certification in England and Wales: Policy and Draft Regulations. The intended benefits of the new ME system to the public, health service, and local authorities were as listed in Table 1.3.

After the 2016 Consultation, an inquiry into the Gosport War Memorial Hospital (2018) found that lives of over 450 patients were shortened while in the hospital, despite repeated concerns raised by families who questioned about how their deceased loved ones had been treated and also similar concerns having been raised by nurses about prescribing practices of opiate medicine.

In June 2018, Jeremy Hunt (then Health Secretary) announced that he was rolling out the appointment of the MEs (The Guardian 2018). This implementation appears to have been finally precipitated by the case of Hadiza Bawa-Garba, a trainee paediatrician, convicted of gross negligence manslaughter and struck off following the death of a child in her care, Jack Adcock. The Medical Practitioners Tribunal Service suspended Dr Bawa-Garba for 12 months on 13 June 2017. The GMC successfully appealed, and Dr Bawa-Garba was struck off on 25 January 2018. On 13 August 2018, Dr Bawa-Garba won an appeal against being struck off, restoring the one-year suspension. Many healthcare professionals have raised concerns that Dr Bawa-Garba was being unduly punished for failings in the system, notably the understaffing on the day and inadequate supervision. Linked with this, Sir Norman Williams' report 'Gross negligence manslaughter in healthcare. The report of a rapid policy review' (2018) was instigated to consider the wider patient safety impact resulting from concerns among healthcare professionals that simple errors could result in prosecution for gross negligence manslaughter, even if they occur in the context of broader organisation and system failings. Amongst other recommendations, the Williams report noted '*The Government is introducing a system in England and Wales, where all non-coronial deaths are subject to a medical examiner's scrutiny. The introduction of medical examiners is designed to deliver a more comprehensive system of assurance for all non-coronial deaths. While not specifically concerned with gross negligence manslaughter, the introduction of medical examiners aims to improve the quality and appropriateness of referrals of deaths to coroners and to increase transparency for the bereaved and offer them an opportunity to raise any concerns. The panel supports this aim and the introduction of medical examiners*'.

Thus, from April 2019, a national system of MEs was introduced to acute NHS trusts (and some specialist trusts) in England and local health boards in Wales. These Medical Examiner Services (MESs) were to be provided by ME offices based within (predominantly) the hospital settings. It is fair to say that this action to provide support for bereaved families and to improve patient safety has to be considered a direct response to the repeated and (in some cases) historic recommendations in reports and public inquiries. And although Shipman and Mid Staffordshire and Morecambe Bay were the key drivers there were other examples where concerns raised by families and/or healthcare professional whistleblowers had been repeatedly ignored (Ockenden 2022).

It will be noted that the new ME system was being established in 2019/2020, and barely 9 months into this development health services were suddenly facing unprecedented pressures caused by the Covid-19 (coronavirus) pandemic. In response to the pandemic, the Coronavirus Act 2020 provided easements to improve the flow of excessive deaths. Despite the massively increased workload, and the option of pausing development, many MESs opted to progress throughout, and MESs played an important role in the pandemic response in a variety of ways including supporting frontline clinicians in writing Medical Certificate of Cause of Deaths (MCCDs) or becoming full-time certifiers releasing frontline doctors from an administrative task so that they could prioritise frontline caring duties. In part, this was driven by the consideration that at times of pressure, more mistakes or errors might be made, and this was exactly the time when competent and independent oversight was required. Guidance was issued by NHS England and NHS

Improvement about Coronavirus Act easements which simplified and streamlined many death certification functions and enabled the ME system to be even more relevant (NHS England & NHS Improvement 2020; ONS 2020b). In March 2022 some of these easements were removed (for example allowing any medical practitioner to complete the MCCD), whilst some were retained including the requirement for a deceased patient to have seen a doctor within 28 days of death (previously 14 days) and the permanent abandonment of the Cremation Form 5. (National Health Service 2022).

Structure and function of the Medical Examiner system in England and Wales

The ME role was formally introduced in the Coroners and Justice Act 2009. Its nature has changed and developed since then, prior to implementation – and MESs have been based within National Health Service acute hospitals (by whom MEs and Medical Examiner Officers (MEOs) are employed) reviewing deaths within acute hospital services. The changes to the proposed service (i.e. not including the community initially) did not go unnoticed, and the (then) Chief Coroner stated that he was '*disappointed that the scheme that was consulted on in 2016 which covered all deaths will not currently be implemented*' (*Health Services Journal* 2018) and felt that the ultimate objective should be a structure as envisaged in the 2009 Act. It also appears for pragmatic, practical, and financial reasons that the ME system was moved to the NHS rather than being funded by local authorities.

However, at last, a national service (rather than a small number of local pilot services) was to be introduced across the NHS in England and Wales. The ME system introduced in 2019 aims (National Medical Examiner 2020) to:

- provide bereaved families with greater transparency and opportunities to raise concerns;
- improve the quality/accuracy of medical certification of cause of death;
- ensure referrals to coroners that are appropriate;
- support local learning/improvement by identifying matters in need of clinical governance and related processes;
- provide the public with greater safeguards through improved and consistent scrutiny of all non-coronial deaths and support healthcare providers to improve care through better learning; and
- align with related systems such as the National Learning from Deaths Framework and Universal Mortality Reviews.

MEs supported by MEOs scrutinise (review) all deaths that do not fall under the HM Coroner's jurisdiction across a local area. MEs are trained, independent, senior doctors. Any practising, or recently retired, medical practitioner who has been fully registered for at least five years and has a licence to practise with the GMC can apply to be an ME, but the National Medical Examiner (NME) advises that MEs should be consultant grade doctors or other senior doctors from a range of disciplines or GPs with an equivalent level of experience. The Royal College of Pathologists in the United Kingdom is the lead medical Royal College for MEs and is currently responsible for training MEs and MEOs. Training is currently a combination of e-learning and face-to-face, and successful completion permits ME membership of the Royal College of Pathologists. MEs and MEOs are employed in the NHS system but have an additional, separate professional line of accountability to regional and national ME teams. Independence is overseen by the NME supported by seven regional teams of Regional MEs and Regional MEOs.

The role of the NME is to provide professional and strategic leadership to the Regional ME teams who in turn support a network of MEs at acute hospital trusts. The NME is intended to support safeguards for public, patient safety monitoring and informs the national learning from deaths agenda and will produce an annual report (National Medical Examiner 2020, 2021a). Most trusts will have a lead ME and a number of other MEs who may come from a range of different medical specialties. MEs generally work part time.

Current guidance suggests that in order to provide adequate cover to scrutinise 100% of death, one whole-time equivalent ME and three whole-time MEOs will be require to adequately cover 3000 deaths. These figures have been determined from pilot studies that have been in place since about 2008. As the MES can be considered to be in the 'adolescent' stage of development, it is likely that these figures will be refined as the availability and expertise of the MES become more widely recognised.

The initial phase of introduction of MESs has been to England and Wales acute hospital, and all were asked to set up (starting in April 2019) MEOs focussing on deaths within their own organisation on a non-statutory basis. Initially, adult deaths were the priority, with only some MES teams reviewing neonatal and child deaths. In an ambitious plan, it was originally intended that every non-coronial death in England and Wales would be scrutinised by MES teams by the end of 2022. Unsurprisingly this target has probably slipped to the latter part of 2023. In June 2021, a circular was sent out (NHS England & NHS Improvement 2021) outlining what local health systems must do to enable consistent scrutiny of deaths across all healthcare settings. Some MESs are on track to achieve this, but some have not yet achieved 100% scrutiny of adult hospital deaths, let alone those of the community. For practical reasons, the MES teams in hospital will also act as the hubs for scrutiny of community deaths (e.g. at home, in nursing homes, and private and community hospitals). There are a number of hurdles to the community roll-out, compounded by multiple electronic notes systems, varying IT governance issues, including access to community medical records and communicating with healthcare professionals in the community. Multi-professional working groups developing pilot studies have made progress with this. The NME has provided information for primary care physicians about the progression of the MES (NME 2021b).

In February 2021, the government published a White Paper ('Integration and Innovation: working together to improve health and social care for all') which makes provision for the role of the ME to be put on a statutory footing (Secretary of Health and Social Care. Integration and Innovation: working together to improve health and social care for all. 2021). The timeline for this is not fixed but is likely to be within the next two years.

The overarching ME's roles within each local MES are for every death to consider the following issues during scrutiny:

- What did the person die from? (ensuring accuracy of cause of death on the MCCD)
- Does the case need to be notified to an HM Coroner? (ensuring timely, appropriate referral)
- Determining whether the bereaved have questions about the cause or circumstances of death or concerns about the care before death (ensuring the relevant authority is notified).

Medical Examiners

MEs are trained in the legal and clinical elements of death certification processes. MEs are senior doctors who in the immediate period before the death is registered (within five days) independently scrutinise the causes of death. In all cases not investigated by HM Coroner, the ME needs to address the following issues:

- What did the person die from? (ensuring accuracy of cause of death on the MCCD)
- Does this case need to be reported to a coroner? (ensuring timely, accurate referral)
- Are there any clinical governance concerns? (ensuring the relevant authority is notified).

To achieve this, MEs are responsible for completing the following steps to arrive at their decision:

- speaking to the doctor (the qualified attending practitioner) who was treating the patient on their final illness to discuss the proposed MCCD and enquire about any concerns of care;
- reviewing the medical records and any supporting diagnostic information;
- agreeing the proposed cause of death and the overall accuracy of the MCCD; and
- discussing the cause of death with the next of kin/informant and establishing if they have any concerns with care that could have impacted/led to death.

Further roles that the MES provide include:

- acting as a medical advice resource for HM Coroner;
- acting as a medical advice resource for registration services;
- acting as a medical advice resource for organ donation teams;
- initiating escalations for deaths where learning might be gained (e.g. structured judgement reviews and serious untoward incident investigations);
- ensuring that patterns or concerns about care by clinicians and others are raised appropriately; and
- feeding back praise or compliments to individuals and teams.

As the national ME system develops and there is wider awareness, the role expands into areas that are extensions of the primary roles above. It is also important to recognise that the ME system is intended to flag and identify concerns and not to investigate. Unsurprisingly, there is no clear line between making adequate enquiry to determine whether something needs escalation and investigation itself. As with HM Coroners, and their decision-making, there is substantial variability in how different MEs might approach a case, and that might in turn be influenced by the stance of the local coroner. Interpretation of the notification requirements for deaths to coroners (Ministry of Justice 2019) is an art rather than a science, and a local example of a chart to assist clinical teams in knowing when to notify is provided in Figure 1.1. Most coroners, however, will now require a clinical team to have discussed any cases with the MES before notification is commenced.

Areas that develop and rapidly increase workload once awareness that a local MES is up and running are the medical advice resource for HM Coroner; medical advice resource for registration services; and medical advice resource for organ donation teams. It is essential that each MES devotes adequate time to education and training, as despite a considerable time having passed since MES were introduced there is still a lack of widespread understanding of the nature and role of the ME system. It is very important for those healthcare professionals in contact with MES to realise that MEs and MEOs often identify praise and compliments about teams, individuals, and hospitals, and disseminating that praise to the individuals and teams is another very important part of the role of the ME. Most MES are likely to find that praise about care of the deceased considerably outweighs criticism.

The MES will need to liaise locally with Mortality Governance Systems teams, and this may provide an opportunity to develop electronic versions of documents, pending the introduction of a national ME software system.

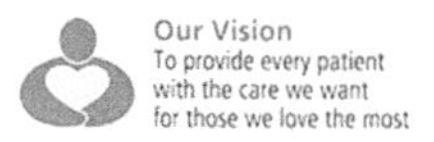

Norfolk and Norwich University Hospitals NHS
NHS Foundation Trust

The NNUH Medical Examiner Service

The Medical Examiner (ME) role was introduced in the Coroners & Justice Act 2009 and provides independent:

- Safeguards for the public by ensuring proper scrutiny of all non-coronial deaths.
- Appropriate referral of deaths to the Coroner.
- Services for the bereaved to give an opportunity for them to raise any concerns to a Doctor not involved in the care of the deceased.
- Advice to improve the quality of death certification on the medical certificate of cause of death (MCCD).

Notification of deaths to HM Coroner - basic principles

- A death must be notified to the relevant Senior Coroner where there is reasonable cause to suspect that the death was due to (that is, more than minimally, negligibly or trivially), caused or contributed to, by circumstances specified in the list to the right.
- A death under the circumstances set out as follows should always be notified, regardless of how much time has passed since the death*.
- HM Senior Coroner for Norfolk requests that <u>all</u> potential notifications, when any of the circumstances listed to the right apply, should be reviewed by the Medical Examiner Service (contact details below) <u>prior</u> to the notification being sent.

*The Notification of Deaths Regulations 2019 are modified when specific provisions in the Coronavirus Act 2020 are in place.

When to notify deaths to HM Coroner

- The death was due to poisoning, including by an otherwise benign substance.
- The death was due to exposure to, or contact with, a toxic substance.
- The death was due to the use of a medicinal product, the use of a controlled drug, or psychoactive substance.
- The death was due to violence, trauma or injury.
- The death was due to self-harm.
- The death was due to neglect, including self-neglect.
- The death was due to a person undergoing any treatment or procedure of a medical or similar nature.
- The death was due to an injury or disease attributable to any employment held by the person during the person's lifetime.
- The person's death was unnatural, but does not fall within any of the above circumstances.
- The cause of death is unknown.
- The registered medical practitioner suspects that the person died while in custody or otherwise in state detention.

The Medical Examiner Service can be contacted on DECT ███ or Extension ███. A Medical Examiner can be contacted out of hours on ██████. Please leave a message and contact details if no response. The written notification to HM Senior Coroner should state whether a Medical Examiner has been consulted (and which one).

Figure 1.1 Basic principles of notification to HM Coroner and utilised at the Norfolk and Norwich University Hospitals NHS Foundation Trust. *Source:* J. Payne-James.

Medical Examiner Officers

MEOs are the golden thread or glue that provide the continuity between (mostly part-time) MEs and all other parties involved in the care of the deceased and bereaved. Their role can be compared with that provided by coroners' officers for coroners, and via delegated authority they can carry out several of the functions of the ME. MEOs are generally full-time employees (unlike the predominantly part-time MEs) and manage the MES in their hospital.

MEOs cannot perform the scrutiny of the medical records of deceased patients but may undertake a screening of the notes to support and assist the MEs and make contact with qualified attending practitioners (QAPs) and bereaved families. The extent of their involvement (and thus what tasks can be undertaken via delegated authority) may depend on their previous clinic and other experience. MEOs come from a wide range of backgrounds, including nursing, bereavement, and mortuary. The broader the range of backgrounds of the MES team, the better that the MES can service the needs of other stakeholders. A MEO may or may not have a clinical background. Those appointed to MEO roles (whether with a clinical background or not) should encompass the following (non-exhaustive) skills and competences:

- previously have experience in a patient- or customer-facing role and of working in either current death certification systems or a clinical or NHS setting;
- have an understanding of medical records and disease pathology;
- be able to provide advice on terminology and causes of death and to explain these and the ME's thoughts and rationale to coroner's officers, doctors, and those with no medical or healthcare background;
- have strong interpersonal skills and be comfortable working with people following a bereavement; and
- be able to build and maintain effective relationships with other stakeholders such as faith groups, funeral directors, and legal services.

The MEO has a very responsible role and will be interacting with a range of professionals including doctors, allied health professionals, registration services, coronial services, crematoria, medical referees, faith organisations, organ donation teams, and funeral directors, in addition to families. The MEO needs to understand the respective roles of all these groups to try and ensure expeditious and timely scrutiny of deaths and to keep families informed with sensitivity and compassions. Many of the roles of these groups overlap, and it is essential that the sensitivities of all those contributing to the care of the deceased and the bereaved are respected.

MEOs undergo core training and face-to-face training in the same way as MEs. This training has been delivered by the Royal College of Pathologists, and training has continued (remotely and successfully) throughout the COVID-19 pandemic.

How does a Medical Examiner Service work?

Every MES will have different local processes and protocols to best service the needs of their host organisation, whilst running an effective scrutiny process. MEs and MES are responsible for the scrutiny of every death within a hospital and will be responsible for scrutiny of every out-of-hospital death in the future. In order to achieve the aspirational target of 100% of deaths being scrutinised within 24 hours of a death, adequate personnel and resources are required. The ME system is, to some extent, being developed from the ground up, using principles established by the NME office.

As the MES is undertaking contact with clinicians and others, such as families, the documentary evidence of any contact is essential to enable the review of decision-making and assist in further investigations. Two main documents are common to most ME offices, the ME-1A and ME-1B forms (Figures 1.2 and 1.3). These may be locally adapted. Form ME-1A (not always used) may be adapted for local use but

National Exemplar Form

This form may be used and evaluated by pilot areas working with the Department of Health to improve the process of death certification.

If urgent: Response required by: ..../..../.....

Reference No.: ..../..../.....
(To be completed by medical examiner's office.)

Reason

Administrative Information

Form ME-1 (Part A)

To be provided to a medical examiner or coroner following a death.

The information provided in this form is confidential.

The administrative information to be provided to a medical examiner or, where necessary, a coroner is prescribed by Regulations made under the Coroners and Justice Act 2009. It can be documented in the deceased person's clinical records or given on this form.

A1. Name of deceased person and the date and time of death

Name: *(Forename)* *(Family name)*	Date and time of death: *(Date)* *(Time)*

A2. Key information about the deceased person

M/F D.O.B: NHS No.: Residential address:	Name of organisation responsible for care prior to death: Last occupation *(and any relevant work history)*:
Place and address where death occurred *(If address is same as residential, state 'As above'.)* ☐ Home ☐ Hospital ☐ Hospice ☐ Nursing or care home ☐ Other	

A3. Next of kin, partner, relative or representative of the deceased person *(Include more than one if appropriate.)*

Name *(Provide names of best contacts, see A5.)*	Relationship *(Note if expected to register death.)*	Phone No. *(Include mobile No. if possible.)*	Present at death	Informed *(if not present)*

Form ME-1 (Part A) Administrative Information – Draft v2.4 (02-Oct-18) Page 1 of 2

Draft National Exemplar Form

This form may be used and evaluated by pilot areas working with the Department of Health to improve the process of death certification.

Reference No: ..../..../.....
(To be completed by medical examiner's office.)

A4. Names and contact details for medical practitioners, clinicians etc. *(Complete where applicable.)*
(Use 'N/A' for not applicable and 'N/K' for not known. If the person's details have been provided in a previous row, state 'As above'.)

Role	Name	Organisation/location	Personal phone/ bleep No. *(Not the on-call number.)*
Doctor(s) able to write a MCCD* *(Must be qualified to certify.)*			
Usual GP *(or alternative GP at practice)*			
Person responsible for nursing or care before death			
Person who verified fact of death			
Hospital consultant responsible for care *(where applicable)*			

* *Medical Certificate of Cause of Death*

A5. Information provided by or regarding next of kin
(E.g. concerns expressed about the circumstances or cause of death, who to contact/not contact, people who may be vulnerable, etc.)

☐ A formal complaint has been (or is expected to be) received about care or treatment *(include relevant information above)*

A6. Other relevant administrative information

Form ME-1 (Part A) Administrative Information – Draft v2.4 (02-Oct-18) Page 2 of 2

Figure 1.2 The National Medical Examiner exemplar ME-1A form (may not always be used or adapted for local use). *Source:* NHS England.

National Exemplar Form

This form may be used and evaluated by pilot areas working with the Department of Health to improve the process of death certification.

Notified on: ..../..../..... at .....

Reference No.: ..../..../.....
(To be completed by medical examiner's office.)

Required by: ..../..../.....

Medical Examiner's Advice and Scrutiny

Form ME-1 (Part B)

The information provided in this form is confidential.

Information in Sections B2, B3, B4, B10 must be recorded by a medical examiner. Other information may be recorded by a medical examiner's officer (MEO) or another person acting on behalf of a medical examiner.

B1. Name of deceased person and the date and time of death

Name: *(Forename)* *(Family name)*	Date and time of death: *(Date)* *(Time)*

B2. Scrutiny of clinical records and other documented information

Information scrutinised: ☐ Full clinical record ☐ Summary clinical record ☐ Coroner documentation ☐ **Other**

Notes made by medical examiner during scrutiny:

Form ME-1 (Part B) Medical Examiner's Advice and Scrutiny – Draft v2.4 (02-Oct-18)

Draft National Exemplar Form

This form may be used and evaluated by pilot areas working with the Department of Health to improve the process of death certification.

Reference No.: ..../..../.....
(To be completed by medical examiner's office.)

B3. Outcome of scrutiny by medical examiner

Death: ☐ Unexpected ☐ Sudden but not unexpected ☐ Expected ☐ Individualised end of life care plan

Case to be referred to HMC ☐ Yes ☐ No

Reason

Potential learning identified ☐ Yes ☐ No

Refer to ☐ Speciality M&M ☐ Clinical Governance ☐ Medical team ☐ Nursing ☐ Other please specify

Reason for review

Structured Judgment Review case ☐ Yes ☐ No

Reason:

☐ Deaths where the bereaved or staff raise significant concerns about the care

☐ Deaths of those with learning disabilities or severe mental illness

☐ Deaths in a specialty, diagnosis or treatment group where an 'alarm' has been raised (for example, an elevated mortality rate, concerns from audit, CQC concerns)

☐ Deaths where the patient was not expected to die –for example, in elective procedures

☐ Deaths where learning will inform the provider's quality improvement work.

☐ Maternal or neonatal deaths

B4. Cause of death established during scrutiny by the medical examiner

Approximate interval between onset and death

1a

1b

1c

2

Form ME-1 (Part B) Medical Examiner's Advice and Scrutiny – Draft v2.4 (02-Oct-18)

Figure 1.3 (a) The National Medical Examiner exemplar ME-1B form (may be adapted for local use – see Figure 1.3b). (b) Locally adapted ME-1B form (currently in use by the Norfolk & Norwich University Hospitals NHS Trust MES). *Source:* (a, b) NHS England.

Draft National Exemplar Form
This form may be used and evaluated by pilot areas working with the Department of Health to improve the process of death certification.

Reference No.:
(To be completed by medical examiner's office.)

B5. Discussion with qualified attending practitioner (QAP) - *if required*

(If this discussion takes place before certification and the doctor has not provided in writing a preliminary view of the cause of death – or reason why no such view has been formed – then this information must be obtained and noted below at the outset of the discussion.)

QAP talked with: Name   Role   Date: Time:

Notes: *(If no preliminary view can be formed before requesting advice, make a note of the reason.)*

continuation sheet

☐ Cause of death provided before scrutiny or noted above is accepted without change
☐ Cause established by the medical examiner and documented is accepted by doctor
☐ Doctor and medical examiner have agreed the following alternative cause of death
☐ Death needs to be discussed with a coroner for reasons noted in B2

Approximate interval between onset and death

1a
1b
1c
2

B6. Discussion with coroner/coroner's office *(if required)*

Notes:

☐ Coroner does not need to investigate the death and has agreed to issue a 100A
☐ Coroner has agreed to conduct an investigation

Form ME-1 (Part B) Medical Examiner's Advice and Scrutiny – Draft v2.4 (02-Oct-18)

Draft National Exemplar Form
This form may be used and evaluated by pilot areas working with the Department of Health to improve the process of death certification.

Reference No.:
(To be completed by medical examiner's office.)

B7. Discussion of cause of death with informant/next of kin or other appropriate person

Cause of death discussed with: *(name)*   *(relationship)*

Cause of death discussed by: *(name)*   on *(date)*   at *(time)*

Notes

☐ continuation sheet

☐ Cause of death accepted without any concerns being raised *(Continue at B8)*
☐ Concerns raised and addressed without requiring discussion with a coroner *(Continue at B8)*
☐ Concerns raised that require the death to be discussed with a coroner *(Record details in B6)*

B8. Medical examiner's details and signature

I confirm that I have carried out an independent and proportionate scrutiny of this death in a way that complies with the relevant standards and procedures.

Name of medical examiner *(print)*:   Office:

Signature: Date:

(Where the information on this form is provided electronically, the signature may also be electronic.)

Form ME-1 (Part B) Medical Examiner's Advice and Scrutiny – Draft v2.4 (02-Oct-18)

Figure 1.3 *(Continued)*

predominantly contains basic administrative data. The ME-1A may not always be used in hospitals but may be more useful for remote working and for assisting with community deaths. Form ME-1B, which may also be adapted locally, provides information about the scrutiny undertaken about the MES including the date and times of the deceased's death; the ME review of notes; the sources of information used; any discussions with the QAP, HM Coroner, and the family; and any record of escalation. It records the outcomes of these discussions, including any modification to the MCCD. It is an important record that summarises the entire scrutiny process and may be discoverable in any litigation.

Once the MES has been informed of a death, the MEO (under delegated authority) may, if suitably competent, undertake a pre-review of the deceased's medical records, contact the QAP to determine what is proposed for the MCCD, and identify any concerns. These will be then discussed with the MEs who may make their own further enquiries having fully scrutinised the deceased's last medical admission. Sometimes (for example, if there is lack of clarity about previous employment and the patient has died with a possible employment-related condition such as pulmonary fibrosis) further review of historical medical notes and discussion with the clinical team will be required. The scrutiny should always be proportionate. It is not expected to be a detailed forensic analysis.

Once the MCCD wording is agreed with the QAP (suggestions on the wording or formulation of the MCCD are only advisory but rarely ignored), then the bereaved family can be contacted for their comments. There are three specific elements which are important when speaking with the bereaved, which are offering condolences, asking whether there were any concerns about their relative's care, and explaining the wording on the MCCD.

The wording on the MCCD is also subject to review by the local registration team. When an informant (normally a relative) registers a death, the registrar will review the MCCD. The Royal College of Pathologists now publishes the 'Cause of Death List' (Royal College of Pathologists 2020), a document which is intended not to reflect every possible cause of death but addresses those conditions which often result in discussion between the MCCD writing doctor, registrar, and HM Coroners.

Should a registrar identify that a cause of death written on the MCCD is not an 'acceptable' cause (for example, renal failure, with no underlying condition identified), then they would be required to notify that death to HM Coroner, and registration would be paused pending review. With a MES in place, such instances should be rare, as the MCCD will have been reviewed and discussed by the ME who will provide guidance as to appropriate completion.

In some register office areas, registrars will accept a cause of death that is not on the 'acceptable' list, if the MCCD is countersigned by the ME, confirming that it has been subject to scrutiny and there are no concerns.

The 'Cause of Death List' document also provides very useful advice on completion of MCCDs and for those informants registering deaths. The Notification of Deaths Regulations 2019, the Cause of Death List, and a further publication 'Guidance for doctors completing Medical Certificates of Cause of Death in England and Wales Office of National Statistics' (2020b) are documents with which MEs and MEOs must be familiar to undertake their functions and to appropriately advise others. These are sometimes modified (the latter being modified for the emergency period of the coronavirus pandemic), and so the MES needs to monitor any changes in guidance closely.

If concerns about care are raised either by the bereaved or QAP, it is for the MES to identify, and escalate, or on some occasions direct the family to a body, person, or organisation most appropriate for their concerns. All of this should be documented in detail on the ME-1B. Under delegated authority, the MEOs can carry out many of the functions of the ME with the exception of the full review of the medical records and signing off the whole process (scrutiny) as complete. The outcome measures of importance at present are the number of MCCDs where wording is changed, identifying concerns of next of kin, and ensuring appropriate notification to HM Coroner. Figure 1.4 shows the workload of a MES working at a large acute hospital in 2020/2021 during the coronavirus pandemic. In this period, there were 2856 deaths in the trust (excluding stillbirths). The MES reviewed and scrutinised a total of 2687 (94.08%) of those deaths.

Within this period, 458 deaths (17.04%) were notified to HM Coroner; 55% of these continued to inquest or post-mortem; 2394 (89%) bereaved families were spoken to during this period; 25 cases reviewed were patients with learning disabilities or severe mental illness; 81 cases were recommended for a Structured Judgement Review (Royal College of Physicians 2016) based on concerns from staff, bereaved families, or were unexpected deaths (e.g. elective surgeries); and 22 deaths were escalated for further review as a potential 'patient safety incident' (Payne-James et al. 2022).

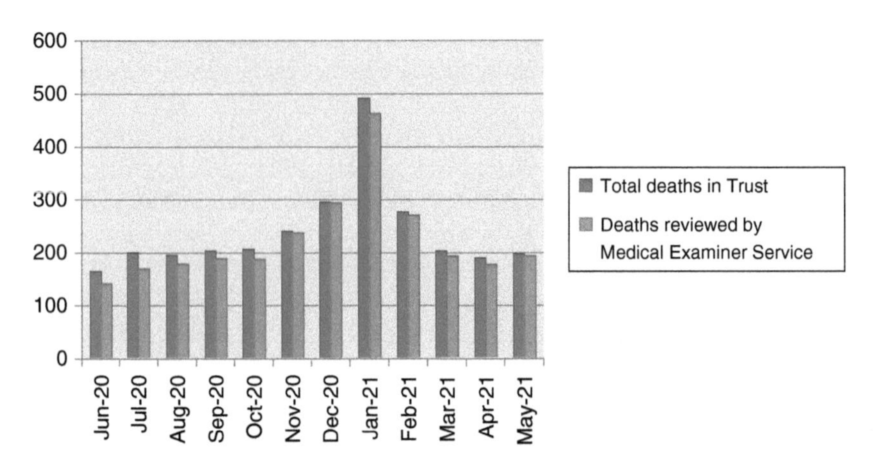

Figure 1.4 Deaths scrutinised on a monthly basis in relation to total deaths within the Norfolk and Norwich University Hospitals NHS Trust from June 2020 to May 2021.

Relationships with other teams supporting the deceased and bereaved

The NME has been developing a Good Practice Series focussing on specific aspects of the MES role. The first – 'How MEs can support people of Black, Asian, and minority ethnic heritage and their relatives' – provides recommendations including 'ME should actively monitor trends and patterns for further action/health planning response, including whether outcomes for patients and relatives of those from Black, Asian, and minority ethnic communities differ from those of other communities' and that 'MEs should be sensitive to the cultural and religious expectations and needs of all those who have suffered loss. . .'.

Good Practice Series Number 2 provides guidance on 'How medical examiners can facilitate urgent release of a body'. This advises on identifying which communities may have particular needs or wishes regarding timely release of a body and recommends working closely with other departments such as bereavement or the mortuary to undertake this. This may be used and adapted for those where death is anticipated and may be assisted by discussing the proposed cause of death in advance. Other topics covered by the Good Practice Series include Learning Disability & Autism; Organ and tissue donation; Post-mortem examinations; and Child Deaths.

Conclusion

Despite a decade or more of delays and some change in form, a new, independent ME system has been introduced in England and Wales and is likely to be the most significant advance in the medicolegal investigation and review of death within living memory. Scandals and poor care continue to be identified from hospitals prior to the introduction of MEs. Paradoxically, although the drivers for its introduction were, in part, community and neonatal/child deaths, these were not the first patient groups to be subject to ME scrutiny. As recently as 2022, the Ockenden Report investigated after bereaved families had raised concerns where babies and mothers died or potentially suffered significant harm whilst receiving maternity care at The Shrewsbury and Telford Hospital NHS Trust. For pragmatic reasons, adult, in-hospital deaths were the first cohort of deaths where the system could be introduced, and since 2019, and despite a pandemic, the ME system has advances and developed. It is likely that by 2023 all non-coronial deaths in England and Wales (about 400 000 per annum) will be subject to ME scrutiny. At this stage, it is unclear whether the system will achieve one of its stated aims, namely '*[allowing] easier identification of trends, unusual patterns and local clinical governance issues and make malpractice easier to detect*', but as the system progresses, there is optimism that this will be achieved and will be a model for other jurisdictions to follow. Whether the ME system can identify and prevent future scandals about patient safety and care is the big unknown, but the initial development is very promising.

References

Coroners & Justice Act 2009. https://www.legislation.gov.uk/ukpga/2009/25/data.pdf (accessed 21 June 2021).

Department of Health (2016). Introduction of Medical Examiners and Reforms to Death Certification in England & Wales: Policy and Draft Regulations. https://assets.publishing.service.gov.uk/government/uploads/system/uploads/attachment_data/file/517184/DCR_Consultion_Document.pdf (accessed 22 June 2021).

Francis R. (2013). Report of the Mid-Staffordshire NHS Foundation Trust Public Inquiry. https://www.gov.uk/government/publications/report-of-the-mid-staffordshire-nhs-foundation-trust-public-inquiry (accessed 21 June 2021).

Gosport War Memorial Hospital (2018). The Report of the Gosport Independent Panel. HMSO. https://www.gosportpanel.independent.gov.uk/panel-report/ (accessed 21 June 2021).

Health Services Journal (2018). Chief coroner warns over watered down medical examiner role. https://www.hsj.co.uk/policy-and-regulation/chief-coroner-warns-over-watered-down-medical-examiner-role/7024072.article (accessed 23 June 2021).

Home Office/Department of Health (2007). Learning from tragedy, keeping patients safe. Overview of the Government's action programme in response to the recommendations of the Shipman Inquiry. HMSO. https://assets.publishing.service.gov.uk/government/uploads/system/uploads/attachment_data/file/228886/7014.pdf (accessed 21 June 2021).

Kirkup, B. (2015). *The Report of the Morecambe Bay Investigation*. The Stationery Office https://assets.publishing.service.gov.uk/government/uploads/system/uploads/attachment_data/file/408480/47487_MBI_Accessible_v0.1.pdf (accessed 22 June 2021).

Ministry of Justice (2012). Medical practitioners: guidance on completing cremation forms. HMSO.

Ministry of Justice (2019). Notification of Deaths Regulations 2019 guidance. https://www.gov.uk/government/publications/notification-of-deaths-regulations-2019-guidance (accessed 22 June 2021).

Ministry of Justice (2020). Coroner Statistics Annual 2019, England and Wales. https://assets.publishing.service.gov.uk/government/uploads/system/uploads/attachment_data/file/888314/Coroners_Statistics_Annual_2019_.pdf (accessed 22 June 2019).

Misuse of Drugs Regulations (2001). https://www.legislation.gov.uk/uksi/2001/3998/schedule/1/made (accessed 5 October 2021)..

National Medical Examiner (2020). Implementing the medical examiner system: National Medical Examiner's good practice guidelines. https://www.england.nhs.uk/wp-content/uploads/2020/08/National_Medical_Examiner_-_good_practice_guidelines.pdf (accessed 22 June 2021).

National Medical Examiner (2021a). Information for primary care on extending medical examiner scrutiny to non-coronial deaths in the community. https://www.england.nhs.uk/establishing-medical-examiner-system-nhs/non-coronial-deaths-in-the-community/ (accessed 22 June 2021).

National Medical Examiner (2021b). National Medical Examiner's Report 2020. https://www.england.nhs.uk/wp-content/uploads/2021/04/B0413_NME-Report-2020-FINAL.pdf (accessed 21 June 2021).

National Health Service. Coronavirus Act expiry. Death certification and registration easements from 25 March 2022. March 2022

NHS England & NHS Improvement (2020). Coronavirus Act – excess death provisions: information and guidance for medical practitioners. https://www.england.nhs.uk/coronavirus/wp-content/uploads/sites/52/2020/03/COVID-19-Act-excess-death-provisions-info-and-guidance-31-03-20.pdf (accessed 22 June 2021).

NHS England & NHS Improvement (2021). System letter: extending medical examiner scrutiny to non-acute settings. https://www.england.nhs.uk/wp-content/uploads/2021/06/B0477-extending-medical-examiner-scrutiny-to-non-acute-settings.pdf (accessed 22 June 2021).

Ockenden, D. (2022). Ockenden Report: Findings, Conclusions & Essential Actions from the Independent Review of Maternity Services at the Shrewsbury & Telford NHS Hospital Trust.

Office of National Statistics (2020a). Deaths registered in England & Wales. https://www.ons.gov.uk/peoplepopulationandcommunity/birthsdeathsandmarriages/deaths/datasets/deathsregisteredinenglandandwalesseriesdrreferencetables (accessed 22 June 2021).

Office of National Statistics (2020b). Guidance for doctors completing Medical Certificates of Cause of Death in England and Wales. https://assets.publishing.service.gov.uk/government/uploads/system/uploads/attachment_data/file/877302/guidance-for-doctors-completing-medical-certificates-of-cause-of-death-covid-19.pdf (accessed 22 June 2021).

Payne-James, J.J. and Lishman, S.C. (2022). *The Medical Examiner System in England & Wales: A Practical Guide*. Taylor & Francis.

Payne-James JJ, Parapanos L, Carpenter K, Lopez B. The workload of a medical examiner service at an acute National Health Service hospital during the COVID-19 pandemic: The Norfolk & Norwich University Hospital experience. Medicine, Science & the Law. Published March 15, 2022 https://doi.org/10.1177/00258024221087005

Royal College of Pathologists (2020). Cause of death list. https://www.rcpath.org/uploads/assets/c16ae453-6c63-47ff-8c45fd2c56521ab9/G199-Cause-of-death-list.pdf (accessed 22 June 2021).

Royal College of Physicians (2016). Using the structured judgement review method: a guide for reviewers (England). https://www.rcplondon.ac.uk/sites/default/files/media/Documents/NMCRR%20guide%20England_0.pdf (accessed 11 October 2021).

Secretary of Health and Social Care (2021). Integration and Innovation: working together to improve health and social care for all. https://www.gov.uk/government/publications/working-together-to-improve-health-and-social-care-for-all/integration-and-innovation-working-together-to-improve-health-and-social-care-for-all-html-version (accessed 22 June 2021).

The Guardian (2018). NHS patient deaths to be investigated by medical examiner. Health secretary makes announcement after BMA feared doctors were being criminalised for errors. https://www.theguardian.com/society/2018/jun/11/nhs-patient-deaths-investigated-medical-examiners-jeremy-hunt (accessed 31 December 2021).

The Shipman Enquiry (2002). Third report. Death certification and investigation of deaths by coroners. HMSO.

TSO (2003). *Death Certification and Investigation in England, Wales and Northern Ireland: The Report of a Fundamental Review*. TSO.

Williams, N. (2018). Gross negligence manslaughter in healthcare. The report of a rapid policy review. https://assets.publishing.service.gov.uk/government/uploads/system/uploads/attachment_data/file/717946/Williams_Report.pdf (accessed 22 June 2021).

2 Who makes false allegations and why? The nature, motives, and mental health status of those who wrongly allege sexual assault

Felicity Goodyear-Smith
University of Auckland, Auckland, New Zealand

The true prevalence of sexual assault is unknown. There is evidence that child sexual abuse, sexual assault, and rape are all both underreported (true cases are not reported to the authorities) and overreported (many suspected cases turn out to be based on wrongful suspicions or erroneous reports) (Herman 2016). The issue is confounded by the problem of definition. For example, what is defined as rape has changed and generally broadened over time (Bierie and Davis-Siegel 2015) and varies by the nature of the act, the circumstances in which it occurred, and perceived consent (Cook et al. 2011). There may be jurisdictional inconsistencies. The age of consent, and hence what may be considered statutory rape, varies between 12 (for example the Philippines) and 18 years of age (South Korea) in different countries (The Week Staff 2021). Unwanted sexual experiences can include lewd exposures, being kissed, fondled or touched sexually, and sexual penetration (Harned 2004), and while researchers may define an incident as rape, in many cases the woman experiencing it does not (Newins et al. 2018). Conversely, false allegations may be defined as only those proven to be deliberately fictitious, with unsubstantiated and recanted allegations excluded from the count (O'Donohue et al. 2018). Wrongful claims that are unknowingly untrue or non-malicious are not usually considered false (Ferguson and Malouff 2016). Regardless, both true and false allegations of sexual assault are common, and both cause considerable harm.

The nature of false allegations

Some false allegations are made knowingly and deliberately, whereby accusers wrongfully claim that events took place that they know did not. Others can be truly inadvertent. There may a misunderstanding or misinterpretation of an event or mistaken identity regarding a perpetrator. Recall may not provide an accurate record of

Current Practice in Forensic Medicine, Volume 3, First Edition. Edited by John A.M. Gall and J. Jason Payne-James.

what has occurred and is subject to distortion through post-event misinformation. In many cases, people may sincerely believe something happened that has not. Furthermore, an allegation may be partially true, but some elements may be distorted or embellished, either deliberately or from an honest mistake (Goodyear-Smith 2016a).

Given the variation in the types of false allegations, and the different ways these may come about, the nature, motive, and mental health status of those who make false allegations of sexual assault will differ, depending on the category within which the case falls.

Deliberate fabrication

People tell lies either because they believe they have something to benefit by doing this, or because they are incapable of discerning the truth (Mares and Turvey 2018). A person intentionally making an untrue allegation of sexual assault is usually motived by personal gain (Kanin 1994). Kanin lists possible motives for false allegations as revenge, to produce an alibi, or to get sympathy. In 2017, DeZutter et al. postulated that motivation can be subdivided into eight different categories: material gain, alibi, revenge, sympathy, attention, a disturbed mental state, relabelling, or regret (DeZutter et al. 2018).

Such cases may be reported by the media, but in contrast to the considerable attention often afforded to sexual abuse cases, acquittals are given little publicity (Samuels 2003). Reports of false allegations are generally confined to a sentence or two, often with anonymity afforded the accuser but not the accused. In the case examples given, where the accused or accuser are identified, their names have been reported in the media and hence are in the public domain.

Material gain

Financial reward may be a driver for false allegations. In the 1990s, the New Zealand Accident Compensation Corporation (ACC) provided $10 000 lump-sum awards per alleged past sexual abuse incident, with no corroborative evidence nor concomitant report to police required (Goodyear-Smith 2016b). Lump sums were paid even when the case went to trial and the accused was acquitted (Sunday News 1991). This fuelled an exponential growth of sexual abuse claims, which waned once the regulations were changed, coming into effect in 1995. In the year (July to June) 1987–1988, there were 221 sexual abuse claims, with $1.9 million paid out. By 1992–1993, there were 13 000 claims costing ACC $43.55 million. This figure had dropped to $4.65 million in 1995–1996, and by 1997–1998, claims were down to 4419. Police trawling operations and litigation lawyers working for contingency fees also may fuel false allegations. In the United Kingdom, former residents of care homes made false sexual abuse allegations when posters appeared on their prison walls offering free lawyers and substantial compensation packages (Webster 2005; Smith 2016).

There are other ways to gain financial rewards. In 2012, a Scottish nurse Natalie Mortimer reported to police that she had been repeatedly raped by her grandfather,

Gordon Ritchie, as a child (Thompson 2014). He was arrested and held in custody awaiting trial. However, Mortimer's mother began to doubt the allegations, and eventually Mortimer admitted she had lied for financial gain. She understood that she and her half-sister were to share their grandfather's inheritance, but she would get all the money if he was sent to prison. Her grandfather was exonerated, and she was charged with making a false accusation.

Celebrities may be targets. In 2014, police raided the home of Sir Cliff Richard in Berkshire, UK, following a child sex assault allegation. Richard denied the allegation and was never charged. However, the case was covered extensively by the BBC (the UK national broadcaster) regarding claims made by four unnamed men of sexual offences perpetrated on them by Richard between 1958 and 1983. After two years of investigation, police conceded that they had found no evidence to lay charges. Subsequently, it was revealed that one of the accusers, a man in his 40s, had made a blackmail attempt before going to the police, threatening to spread 'false stories' unless he received a sum of money (Young 2016). Richard successfully sued the BBC for helicopter filming and publicising the police raid on his home (BBC 2019; Sir Cliff Richard: BBC pays £2m in final settlement after privacy case – BBC News).

In England, 51-year-old Carl Beech was jailed for 18 years in 2019 after his allegations of a paedophile ring of Westminster 'Very Important Persons' (VIPs) were found to be false (Evans 2019). Those named included former Prime Minister Edward Heath and former members of parliament, the home secretary, chief of defence, director of the secret service, and the Director-General of M15. These were historical claims said to have taken place in the 1980s and 1990s, accepted uncritically by the Metropolitan Police Service who believed his stories were 'credible and true'. Under the code-named Operation Midland, the police mounted a large-scale investigation that lasted over two years and cost over £2 million as well as an additional million investigating Beech's false claims and compensating the accused Lord Bramall and his wife, whose home they had entered without a warrant. When instructed to objectively examine Beech's allegations, the police found that the 'key elements of the story were totally unfounded, hopelessly compromised, and irredeemably contradicted by other testimony'. His motivation appeared to be the £22 000 he received in compensation as well as the attention he received. Retired High Court Judge Sir Richard Henriques carried out an independent review of the handling of the case by the Metropolitan Police Service. His report was scathing, and his 25 concluding recommendations included that the instruction to 'believe a victim's account' should cease, and that 'investigators should be informed that false complaints are made from time to time and should not be regarded as a remote possibility. They may be malicious, mistaken, designed to support others, financially motivated, or inexplicable' (Henriques 2019).

Revenge

False claims may be made as retribution against a boyfriend, ex-spouse, employer, teacher, or someone else who has caused upset. A 23-year-old woman was convicted of false rape complaint in which she had named a suspect to the police (NZPA 1996d). Later, she admitted that the complaint was false, but she wanted to get revenge on the accused who had upset her. She had previously made false assault charges against two other people for the same motives. In another instance, a 21-year-old student reported to the police that her boyfriend had raped her. Later, she admitted she had made up the story because she was angry with him (Court reporter 1998).

Nineteen-year-old Taia Melrose-Cooper similarly reported to police that she had been raped at knifepoint by two men, one of whom she identified. After surveillance cameras proved her story false, she admitted she fabricated the story to get back at the alleged perpetrator because she did not like him (Newbold 2000). Tracey Kettle (34) was found guilty of false complaint against a man with whom she had consensual sex after meeting him at a nightclub. She accused him of rape two weeks later when the relationship broke up, saying she wanted to get at him because he was 'scum' (NZPA 1996c). In another example, an English woman slashed her body, hair, and clothing to convince police that she had been attacked and raped by four men while walking home. A large-scale manhunt costing more than £150 000 ensued, until police discovered that she had simulated the attack after the accused had refused to give her a ride home (Pilling 2010).

Teachers are often the victims of false complaints. A survey conducted by the Association of Teachers and Lecturers found one in five school staff had had a false allegation made against them by a pupil (The Guardian 2015). Seven 11-year-old girls in a middle school in Washington DC filed charges that their gym teacher, Ronald Heller, had fondled them in the locker room. In this case, all the girls recanted after a month, admitting that they had lied to get the teacher removed from the school (Ricchissin 2000).

Unrequited love interest

An allegation may be made in response to an unrequited romantic interest.

One example involves two brothers, Paul and Jason Dale, and a third man, Callum McLeod, all in their 20s, convicted and sentenced to eight years for raping, sodomising, and violating a young woman with two carrots, a broomstick, and a candle (Taylor 1999). The woman subsequently signed an affidavit witnessed by a Justice of the Peace in which she confessed that all the allegations were lies. She admitted that the acts she had described in her statements did not happen, and that any sexual activity that had occurred was consensual. She gave her motive as feeling animosity towards Jason Dale because she wanted a permanent relationship with him to which he would not agree. Despite her confession, an appeal and subsequently a request for pardon were denied, based on the police having 'serious doubts' about her retraction. The men remained in jail and served their full sentences. This was despite of the fact the woman subsequently accused two further men of drugging and raping her, allegations proven false at trial (NZPA 2000).

Archana Gupta, a 27-year-old woman, alleged that her ex-lover Michael Lewis had kidnapped, then raped and sodomised her at knifepoint in a car park. Police ignored his protest that he had been in another city that week. His parents mortgaged their home to fund his defence. After four months in custody, tests showed that his blood did not match the evidence taken from her body and clothes, and she admitted that she had lied (Masters 1999).

Alibi

A false allegation might be made to account for an unexplained absence or to explain unfaithfulness or unexpected pregnancy. A woman slipped a note to a service station attendant saying she had been raped and needed help. She later admitted to having made the story up because her partner was coming out of jail and she wanted to hide her affair with another man – the alleged rapist (NZPA 1996b). Heather Atkins, a 26-year-old Columbia woman, claimed that two unknown men had violently raped her and given her a black eye, whereas the injury actually had been inflicted by a woman who came home to find her boyfriend in the shower with Atkins (Kuenzie 2012).

In Spain, a 27-year-old woman seeking an abortion invented a story of rape to prevent her husband knowing that she was pregnant to her lover (Govan 2009), and in the United States Corl Keith (22) reported to police that she had been raped in the snow by a man in a ski mask. Eventually, she admitted that she had made up the story to hide an illicit affair and to get treatment for a possible pregnancy (Ganim 2010).

Sympathy and attention

Sexual assault victims may get considerable sympathy and media attention, and this has been shown to be the motivation in many cases (Hewson 2016). In 2007, a 41-year-old pharmacist Nicola May claimed she had been raped. After police had spent 670 hours investigating the case, she admitted she had made up the story, self-inflicted her injuries and cut her own phone line. May laid the complaint to get her ex-lover's attention after he had jilted her (NZPA 2007). In another case, an 11-year-old girl claimed that she had been sexually attacked by a man who had jumped on her out of the bushes. Police subsequently reported that the complaint was false, made to get attention (NZPA 1997). Twenty-year-old Natasha Statesman-West was sentenced to community service after an 'attention-seeking' false rape allegation (Court reporter 2004), and student Michelle Grafton (19) alleged that she was forced into a car by a man she knew and taken to a house where she was bound to a bed and repeatedly raped. Eventually, her allegations were shown to be completely fabricated. On pleading guilty, she said she lied because she liked the attention (NZPA 2010).

Occasionally, a rape report in the media will generate a copycat false complaint, with someone else claiming to be another victim of the same perpetrator. In one such case, a 45-year-old woman reported being tied up and sexually assaulted outside her home by a Māori or Pacific Island man in his 20s. Subsequently, a 25-year-old woman from the same street reported to the police that she had been grabbed by an attacker just outside her front door. He had slashed her clothes off with a knife and tried to rape her. She had fought him off but suffered cuts to her face and hands. From her description, the police believed it was the same attacker, and residents in the area were warned to secure their homes and protect themselves against this serial rapist. The woman was interviewed in the media about her experience. A month later she told police that she had come home to find a note made of cut-out newspaper letters in her kitchen from the offender, threatening to come back and 'do her properly' (NZPA 1998b). Police kept guard outside her home that night to protect her, but the following day they charged her with making a false complaint. She remained unnamed. The original sex attack on the 45-year-old woman was unsolved (NZPA 1998a).

The stories may be elaborate. They may include claims of home invasion, such as a British woman who falsely claimed that a black man forced his way into her home and attacked her (Newsroom 2016). Similarly, a 12-year-old girl claimed that a 44-year-old taxi-driver Ateara Taiaroa had driven her to his home, parked in his garage, and had sex with her on a bed there. He was acquitted at trial when it was shown that it was impossible to drive the car into the garage because it was full of junk, and that it had never contained a bed (Christchurch Press 1997).

Some resort to self-inflicted injuries to support their case. A 15-year-old teenager complained to the police that a man had entered her home, hit her with a piece of timber, and sexually assaulted her. She gave a detailed description, and police issued an identikit picture, complete with a dog-collar tattoo around his neck. The detective found the girl very convincing, and 13 police were assigned to the case. However, flaws in her story mounted. Two weeks and 235 investigating hours later, it was established that she had made up the story, and there was no offender (NZPA 1999). Another 12-year-old claimed that a man had grabbed her through a bedroom window; she had struggled and escaped to a neighbour's place, but scene examination did not corroborate her story (NZPA 1996a), and in yet another case, a 13-year-old sporting injuries falsely claimed pack-rape by three men, until evidence showed that she had caused the harms herself (Newbold 2016).

An accuser may attract considerable sympathy and attention from authorities, the media, and peers, particularly when an accusation is made against an authority figure or a celebrity. Accusers may be congratulated for their courage, supported, and encouraged by police and social services and gain recognition and status as members of a victim advocacy group. This may encourage the fabrication of more and more stories of abuse. For people disenfranchised from society, such as previous residents of boys' homes who have ended up on the wrong side of the law, this emotional attention can bring great psychological satisfaction (Webster 2005).

Political manipulation

Historical allegations may be made as character assassination to prevent someone being appointed to a high office. While some claims may be true, others may be fabrications designed to capture media attention and interfere with election or appointment to a position. No charges may end up being made, but reputations can be ruined.

It is often not possible to determine whether such claims are true or false, given the passage of time and lack of possible corroboration. However, where the motivation is to derail an election or an appointment it may be effective, even in cases where evidence subsequently points to a false allegation. In 2017, a list was circulated alleging various sexual improprieties of 36 British Member of Parliament (MPs) of all political persuasions (Payne 2017). Most were furiously denied, and there was no way of knowing how much of it was true. Similarly, multiple allegations of sexual misconduct have been made against more than 80 US politicians, both republican and democrat, and judges. Examples of sexual allegations made at critical times in a political career include Clarence Thomas when nominated as Supreme Court judge (Committee on the Judiciary 1991); Brett Kavanaugh when he was shortlisted for nomination to the US Supreme Court; Republican Donald Trump during his election campaign (Bradner et al. 2018); Chief Justice of the Supreme Court Roy Moore, when making a bid for Alabama Senate (Watson and Tillett 2019); and most recently, Democrat presidential candidate Jo Biden (Villa and Alter 2020).

Mental disorder

Some people who make false allegations may be suffering from mental health issues. They may be pathological liars, sometimes telling complex fantasies for no discernible reason. Law student Nick Wills (22) was accused by a 17-year-old fellow student of raping her. He had an alibi that was not investigated by police. He lost his job, was subject to harassment on campus, and had to face nine years in prison if convicted. Extensive inquiries by his family, including hiring a private investigator, established a watertight alibi. It was also discovered that the accuser had a rich fantasy life and told friends many untrue stories, such as her father owning his own plane. She admitted she had made the story up and pleaded guilty to making a false complaint. She received name suppression (Newbold 2016). There is no evidence that she was assessed for, nor diagnosed with, a mental disorder. However, while pathological lying is not a mental diagnosis *per se*, it is listed in the Diagnostic and Statistical Manual of Mental Disorders as a symptom of several personality disorders.

Another example is Debra Wood (20), who accused a 30-year-of man of sexual assault. Several months later, it was established that she had ripped off her own bra and used a key to scratch her breasts, chest, and neck before laying the false complaint. She was said to be suffering from a personality disorder, and she claimed to be depressed at the time of the incident (Brown 2010).

Inadvertent allegations

Not all false allegations are intentional. Many cases of wrongful allegations arise when the accusers sincerely but falsely believe they have been sexually assaulted by someone, whereas either they are naming the wrong person, or the event as recounted did not occur.

Mistaken identity

Mistaken identity may occur when police are eager to solve a case and present the victims with pictures or line-ups of suspects. Since 1992, the Innocence Project has exonerated 367 people in the United States, mostly through DNA testing. The majority of cases involved the victim identifying the wrong man from a book of mug shots and/or at a line-up (Innocence Project 2021). Eye-witness testimony is often fallible, and a victim who believes that the perpetrator must be one of those presented by the police may make a relative (he looks the closest) rather than an absolute (he is definitely the one) judgement. Confidence levels that the identification is accurate may get increasingly inflated by the time the case comes to trial (Loftus 2019).

In 1992, an 11-year-old Auckland girl was abducted and raped. Although blindfolded, she identified her assailant as her neighbour, David Dougherty. He was charged and convicted, although he strongly protested his innocence, and his girlfriend testified as his alibi for that night. In 1997, DNA re-testing of semen on the girl's underpants demonstrated that it could not have come from Dougherty. Following an appeal and retrial, Dougherty was acquitted and released from prison. The Minister of Justice turned down his bid for compensation on the grounds that Dougherty has not proven his innocence. His reasons were that the complainant had never retracted her allegation that Dougherty was the person who had abducted and sexually violated her, and that the jury acquittal was because the charges had not been proven, rather than Dougherty being 'truly innocent' (Raea and Fergusson 1997). Five years later, in 2003, the DNA was matched to that of Nicholas Reekie, a convicted rapist of three other women. Dougherty received a public apology and paid compensation for wrongful imprisonment. For the victim though, in her mind Dougherty was still the offender. She said that she never meant to cause trouble for anyone. *'If I was wrong, then I feel really bad for him'*, she said, but *'for me, he's still the scary man. It's hard to get that out of your mind'* (Cleave and Gower 2003).

Misunderstanding or misinterpretation of real events

Sometimes sexual activity took place, but one party subsequently alleges that this occurred without consent. In broad terms (and dependent on some jurisdictional variation) individuals can only consent when deemed capable of doing so – for example, over the age of consent, awake, not under the influence of drugs or alcohol, and with no mental impairment. Consent cannot be granted where there is threat or use of force (House of Commons of Canada 1992; Ministry of Justice 2005). It has recently been argued that consent is not 'a discrete event that happens once during a sexual encounter, rather a process that occurs throughout the sexual activity' (Beres 2014). This can be problematic. In many sexual encounters, consent may be implicitly portrayed by both verbal and non-verbal behaviours, and if there is no evidence of resistance, consent is unlikely to be repeatedly sought and given throughout the sexual activity. A man may perceive he has consent, but retrospectively a woman may decide that he did not. For many cases of sexual assault, the sexual activity is agreed, but the dispute relates to whether or not the activity was undertaken with consent.

American professional basketball player Kobe Bryant was accused of raping a 19-year-old hotel worker in his room and faced a life in prison. He admitted to the sexual encounter but was adamant that it was consensual. The woman agreed to drop criminal charges, provided he issued an apology to be read in court. Kobe Bryant gave the following statement:

> 'First, I want to apologize directly to the young woman involved in this incident. I want to apologize to her for my behavior that night and for the consequences she has suffered in the past year. Although this year has been incredibly difficult for me personally, I can only imagine the pain she has had to endure. . .. I also want to apologize to her parents and family members, and to my family and friends and supporters, and to the citizens of Eagle, Colorado. I also want to make it clear that I do not question the motives of this young woman. No money has been paid to this woman. She has agreed that this statement will not be used against me in the civil case. . .. Although I truly believe this encounter between us was consensual, I recognize now that she did not and does not view this incident the same way I did. After months of reviewing discovery, listening to her attorney, and even her testimony in person, I now understand how she feels that she did not consent to this encounter. I issue this statement today fully aware that while one part of this ends today, another remains. I understand that the civil case against me will go forward. That part of this case will be decided by and between the parties directly involved in the incident and will no longer be a financial or emotional drain on the citizens of the state of Colorado' (CCN 2004).

While the criminal case was dismissed, the woman pursued a civil suit which was settled for an undisclosed sum (Henson 2004).

Of particular concern is the wave of acquaintance rape allegations that have hit the United States and Australian colleges and universities in recent years. In response to female complainants that their allegations of sexual assault on campus were ignored, campaigns were mounted to address this, which some say overcorrected the problem (New Zealand Government 1989). These cases generally occur in the context of excessive alcohol use and dates or 'hook-ups', and usually the issue is one of consent. Most are dealt with by the college rather than the criminal system, with the standard using 'balance of probabilities' rather than 'beyond reasonable doubt'. Young women are encouraged to lay charges when they are concerned about a previous sexual encounter and told that their college has 'the power to suspend or expel assailants without waiting for the verdict of a lengthy criminal trial. Swift punishment of offenders enables survivors to more comfortably resume their studies'. However, these cases are often based only on 'she said', 'he said' testimony.

There have been numerous cases where young men have been suspended or expelled from college after sexual encounters that they perceived as consensual, but the young women involved subsequently decided were not, both parties were so intoxicated, neither could remember clearly what had taken place, or sometimes a third party decided a woman had been abused despite her denial (Yoffe 2017). Two very drunk students at Occidental College in the United States had sex which neither could completely remember, and which both regretted (Banks 2014). Encouraged by an activist member of staff the young woman subsequently filed a sexual misconduct complaint. The District Attorney decided not to lay charges as they were both 'willing participants exercising bad judgement'. The College decided that while the young woman 'engaged in conduct and made statements that would indicate she consented to sexual intercourse', she did not have the capacity to consent because she was too drunk. Although they acknowledged that the young man was 'too drunk to recognise that she was too drunk to consent', he was considered responsible for his actions and expelled.

Sincerely believed-in false allegations

Claimants may make allegations in which they sincerely believe but which are not true (Goodyear-Smith 2016a). Memories may not be, or may not remain, accurate, and new post-event information can easily become incorporated to create a false memory (Loftus 1997). This may be from something they have imagined, dreamt, heard, seen on film, or read about and then become to believe that this actually happened to them (Laney and Loftus 2013).

During the so-called memory wars of the 1980s and 1990s (Crews 1995), there was widespread misconception that children routinely 'repress' brutal abuse, the memories of which can only be retrieved through special techniques (Loftus 2004). Counsellors and psychotherapists used highly suggestive methods such as guided imagery, age regression, dream interpretation, and frank hypnosis to excavate buried trauma memories of childhood abuse. The debate peaked in the 1990s, with thousands of predominantly women recovering memories of alleged satanic ritual abuse as children in the course of therapy (Ross 1995).

By the beginning of the 1990s, suddenly thousands of elderly parents, carers, doctors, teachers, and others were claiming to be falsely accused of sexual crimes (Boakes 1999). Described as an epidemic, the cases started in the United States and spread to the United Kingdom, Canada, Australia, and New Zealand. Some of these allegations were bizarre and highly improbable or impossible, such as abuse taking place within satanic cults with ritualised murders, cannibalism, and abortions, or remembering multiple penetrative sexual intercourse and sodomy events occurring in infancy (Spanos et al. 1994). In the United States, over 12 000 reported cases of satanic ritual abuse were investigated by law enforcement agencies, with no substantiated evidence found in any case (Lanning 1991). Cases involving allegations of ritual or satanic abuse in England and Wales were shown to be unfounded in a similar study (La Fontaine 1994).

Subsequent extensive and robust research has demonstrated the susceptibility of memory recovery to create sincerely believed-in pseudomemories (Laney and Loftus 2013). Memory is malleable, and it is possible to 'manufacture entirely false events and pasts that never occurred' (Loftus 2004). These memories are often detailed, experienced with emotion, and held with confidence, and without independent corroboration there is no reliable way to distinguish between a true memory and a false one (French and Ost 2016).

This phenomenon was largely limited to English-speaking Western countries (United States, Canada, United Kingdom, Australia, and New Zealand) where literature such as the *Courage to Heal* (Bass and Davis 1988) and training programmes on memory repression and recovery were widely disseminated. Accusers in such cases were generally white and well educated from middle-class families, who appeared to be vulnerable people with troubled lives. The typical pattern involved their memories of sexual abuse surfacing when they attended psychotherapy for unrelated issues such as relationship breakdown or depression (Goodyear-Smith et al. 1998).

Some women later realised that their memories were untrue and retracted. In 1995, Vynnette Hamanne successfully sued her psychiatrist, Dianne Humenansky, for implanting false memories during psychotherapy (Loftus 1996). She had developed 'multiple personality disorder' through therapy, come to believe that she was victim of bizarre childhood sexual abuse involving satanic rituals, and that she had seen her grandmother stirring a cauldron of dead babies. Professional bodies, including the UK Royal College of Psychiatrists (Brandon et al. 1997), the British Psychological Society (Working Party 1994), the American Medical Association (Coble 1994), the American Psychiatric Association (American Psychiatric Association 1993), the National Association of Social Workers (NASW National Council on the Practice of Clinical Social Work 1996), the European Therapy Studies Institute (European Therapy Studies Institute 1996), the Canadian Psychiatric Association (Blackshaw et al. 1996), and the Australian Psychological Society (Australian Psychological Society 1994), among others, all issued statements warning on the use of memory recovery techniques, citing a lack of evidence to support the accuracy of memories recovered in this way and the risk that false memories may be implanted.

While explicit 'memory recovery' is no longer practised, it was demonstrated during this era that memories can easily be created or distorted by misinformation. This highlights the need for charges to be laid only where corroborative evidence is available. Where two parties report opposing memories, in the absence of external corroboration there is no way to determine whether or not an event actually happened. It is unsafe to convict based solely on believing the accuser, regardless of how confident or credible they may appear. If we have no way of determining the veracity or otherwise of an allegation, natural justice demands that we err on the side of the innocent until proven guilty.

Cases arising from cognitive biases of investigators

When an investigator approaches a case with the belief that abuse has occurred, confirmatory bias may mean that only evidence supporting the allegation is looked for, and any evidence contradicting the claim is ignored or misinterpreted (O'Brien 2009). There may be an initial allegation from the supposed victim, but in other cases the victim may not have disclosed abuse at the time, as in the cases of 'recovered' false memories of incest and rape generated by misguided psychotherapy described above. It can become increasingly difficult to see that the initial belief was mistaken, and counterevidence can cause cognitive dissonance (Tavris and Aronson 2007).

In New Zealand, an anal finding in a 10-year-old girl with HIV/AIDS dying of natural causes was misinterpreted by doctors as indicating sodomy (Goodyear-Smith et al. 2014). This led to the child's uncle wrongly charged with her rape and murder, for which he was twice acquitted (Goodyear-Smith 2015). Even when evidence proved that there was no tear to the child's anus, five years later, the medical and nursing staff, including the charge paediatrician and the head of the intensive care unit, still had a shared idée fixe that they had seen a huge gaping, meaty wound. The child was moribund, and what they saw was prolapse of her rectal lining, abnormally red and thinned from the ravages of HIV, because her anal sphincter had relaxed. What they thought they saw was a huge tear of her anus extending back several centimetres. This wrongly interpreted, horrifying image became an enduring memory.

There is overwhelming evidence that prepubertal children can contract gonorrhoea both sexually and non-sexually (Goodyear-Smith 2007). However, generally authorities assume that all such cases mean that sexual abuse of the child is confirmed. This cognitive bias can lead to children removed from their families, and parents facing criminal charges in cases where it can be demonstrated that fomite transmission (e.g. by a towel or washcloth) is likely. The American Academy of Paediatrics assumes that with very rare exceptions, all such infections in children, be they 'genital, rectal, oral, or ophthalmologic', are diagnostic of sexual abuse (Committee on Child Abuse and Neglect 2020). This is despite, for example large epidemics of gonococcal conjunctivitis recorded in African and Australian Aboriginal communities during hot and humid conditions with a high density of flies (Miller et al. 1999; Matters 1981; Brennan et al. 1989). Lack of concurrent gonorrhoeal genital outbreaks suggests that horizontal transmission by fomite is much more likely than widespread infection of children's eyes from sexual abuse (Merianos et al. 1995).

Concomitant with the wave of 'recovered memories', a moral panic ensued regarding ritual abuse occurring in day-care centres. This started in the United States but spread to the United Kingdom, New Zealand, and Australia. Widespread suggestive interviewing of small children by parents, social workers, and police investigators led to children inadvertently fabricating stories, often involving bizarre and terrible abuse. Workers in many crèches were wrongly charged with thousands of counts of ritual sexual abuse, and many were convicted, until it was recognised that the allegations had arisen through overzealous interviewing, and most convictions were subsequently overturned (Nathan and Snedeker 2001). One highly publicised case is the McMartin preschool in California, where the Buckley family running the centre was charged with hundreds of acts of sexual abuse to children but later acquitted on all counts (de Young 2004). This case was followed by dozens of others, including in Bakersfield (California), Jordan (Minnesota), Edenton, (North Carolina), Martensville (Canada), and Wenatchee (Washington) (Jenkins 2004). Bob Kelly was convicted of 99 counts of sexual abuse of children at the Little Rascals Day Care in North Carolina in1992. In 1995, all charges were dismissed on appeal (Ceci and Bruck 1995). At the Wee Care Nursery school in New Jersey, Kelly Michaels was indicted on 299 offences relating to 33 children, including many bizarre acts such as making children eat her faeces and sexually assaulting them with knives, spoons, forks, and Lego blocks. She was sentenced to 47 years in prison in 1988, but after appeals all charges were dropped in 1994 (Ceci and Bruck 1995). In Wenatchee, Washington, 43 Sunday school teachers, a pastor, and parents were arrested on 29 726 charges of child sex abuse involving 60 children after an extensive investigation by police and social workers. Eventually, all were acquitted (Egan 1995). Many other cases were documented in the 1980s and early 1990s, with most of those accused ultimately freed (de Young 2007).

Such cases are not exclusive to childcare centres, and there are numerous circumstances whereby belief that children have been abused leads to interviewers unintentionally coaching children to create false stories. One example was the removal of nine children from their families in the Orkney Islands, where social workers conducted lengthy interviews to get the satanic ritual allegations they were expecting. Eventually the stories were shown to be unfounded, a judicial inquiry by Lord Clyde determined that handling of the case was fatally flawed, and the children were returned home (Clyde 1992).

Child custody disputes

The 1980s and 1990s also saw a dramatic increase in reported cases of incest and sexual abuse in the context of parental custody disputes (Green 1991a). While of course couples may separate where genuine cases of incest have come to light, false allegations may also occur during disputes about custody of children when parents split up (Meadow 1993). A sexual abuse allegation is a powerful weapon which may facilitate a woman and her children having no further contact with a hated ex-partner (Goodyear-Smith 2016a). When a couple separates, one may make a false allegation to discredit an ex-partner in the eyes of family, friends, or the courts, particularly when seeking sole child custody (Grattagliano et al. 2014). False claims may be made by one parent, usually mother, to influence the family court and prevent the child having access to the other, usually father (Ehrenberg and Elterman 1995).

False allegations in such contexts are more likely to be raised by a parent rather than a child. Confirmation bias is a significant risk here. Allegations arising in the context of acrimonious disputes may have been contaminated by suggestive interviewing (Mikkelsen et al. 1992; Webster 2005). Informal questioning by a parent, social worker, or teacher or rumours and overheard conversations, prior to any formal interviewing, may taint a child's testimony. Once such misinformation has been introduced, erroneous reporting can appear as credible as a true disclosure (Ceci et al. 2007).

Both the experience of parental divorce and sexual abuse can lead to non-specific psychological and behavioural sequelae such as depression, anxiety, and conduct disorders in children. Care needs to be taken to ensure that signs of emotional distress in the child are not assumed to be diagnostic of sexual abuse. When there is interparental conflict after separation, an angry and anxious parent may also misinterpret normal childcare practises of bathing, toileting, or dressing occurring during visitation, or affectionate contact between a parent and a child such as kissing and hugging, as abuse (Green 1991b).

After separation from his wife, a man was charged with previous sodomy of his two sons, then aged six and eight (Goodyear-Smith 1997). The elder boy had received counselling for disruptive behaviour. His therapist decided this must be due to sexual abuse by his father. The boys underwent a number of interviews, and after initial denials, both eventually claimed that their father had anally abused them in the shower. Despite changing and contradictory testimony, and the impossibility of a large man doing this to a small boy when both were standing in a shower box, the man was convicted and sentenced to six years imprisonment. Nine months later, the elder boy started asking 'Does God know if you are telling lies?'. He insisted that his father had never done any of the things alleged, but he had finally 'disclosed' after the interviewer had kept asking the same questions and would not accept 'no' for an answer. The conviction was quashed, and the father was awarded joint custody of his sons.

Conclusion

In the laudable attempt to improve sexual assault conviction rates and bring perpetrators to justice, jurisdictions around the world have made sexual offences a *crimen exceptum*, where corroboration is no longer required for juries to convict

(Burnett 2016). A doctrine operating from the premise that false allegations rarely happen, and that all sexual offence complainants must be believed (Faller 1984; Weaver 2016), has led to prosecution policy assuming presumptive guilt of the defendant. The pendulum has swung from one institutional bias to another.

A raft of reforms to make it easier for complainants to testify include removal or extension of statute of limitations for sexual crimes, whereby charges may be laid decades after the alleged event. Charges may be representative in historical charges, whereby 'the complainant cannot reasonably be expected to particularise dates or other details of the offence' (New Zealand Legislation 2011). The passage of time and lack of a definitive date can make seeking evidence to counter the claim an impossible task for a defendant. Conviction will rely solely on believing the complainant.

Those bringing charges may be given guaranteed anonymity, whereas the accused is publicly identified from the outset (Samuels 2003). Giving testimony behind screens or via video link may violate the ancient judicial guarantee for persons accused of a crime the right to look their accusers in the eye (Herrmann and Speer 1994). Complainants are referred to as victims, and suspects as perpetrators at the time charges are laid, pre-empting the role of a trial to determine guilt or innocence (Weaver 2016).

While we do not want to return to times where sexually assaulted women and children were seldom awarded justice, the removal of safeguards to protect innocent defendants has increased the risk of wrongful allegations and convictions. Belief should not replace forensic examination of evidence. This chapter has demonstrated that complainants may be telling the truth, but they may also be lying or mistaken. There may be no way to tell the difference. Complainants should be treated with compassion and respect but should neither be believed nor disbelieved. These same courtesies should be extended to the accused. The police and other investigators should approach each allegation from a neutral stance of 'Here is an allegation – what evidence is there to confirm or refute?'

Convictions based solely on believing the complainant not the defendant in the absence of any corroborative evidence are counter to the premise of guilty beyond reasonable doubt. In the witch hunts of the seventeenth century, Increase Mather argued that 'better that ten witches go free, than the blood of a single innocent be shed', and decried convictions based solely on the 'spectral evidence' of the accusers (Mather 1693). Presumption of innocence is a time-honoured principle of natural justice (Cascarelli 1996). Given that there are many instances and circumstances where people are falsely accused, even if it means that some guilty men go unpunished, a fair and just system must act to protect the innocent.

References

Christchurch Press (1997). Man not guilty of intercourse with 12 year old. *The Press* (4 July).

Committee on the Judiciary (1991). First session on the nomination of Clarence Thomas to be Associate Justice of the Supreme Court of the United States. *Hearings Before the Committee on the Judiciary United States Senate One Hundred Second Congress*, Washington, DC.

American Psychiatric Association (1993). *Statement on Memories of Sexual Abuse.* Washington, DC.

Australian Psychological Society (1994). *Guidelines Relating to Recovered Memories.* Melbourne: APS.

Banks, S. (2014). Campuses must distinguish between assault and youthful bad judgment. *Los Angeles Times* (9 June).

Bass, E. and Davis, L. (1988). *The Courage to Heal.* New York: HarperCollins Publishers.

BBC (2019). BBC pays out £2m in legal costs to Sir Cliff Richard. *The Guardian* (4 September). https://www.theguardian.com/music/2019/sep/04/cliff-richard-bbc-pays-millions-legal-costs.

Beres, M.A. (2014). Rethinking the concept of consent for anti-sexual violence activism and education. *Feminism & Psychology* 24: 373–389.

Bierie, D.M. and Davis-Siegel, J.C. (2015). Measurement matters: comparing old and new definitions of rape in federal statistical reporting. *Sexual Abuse: Journal of Research & Treatment* 27: 443–459.

Blackshaw, S., Chandarana, P., Garneau, Y. et al. (1996). Adult recovered memories of childhood sexual abuse: position statement. *Canadian Journal of Psychiatry* 41: 305–306.

Boakes, J. (1999). False complaints of sexual assault: recovered memories of childhood sexual abuse. *Medicine, Science, and the Law* 39: 112–120.

Bradner, E., Rju, M., Mattingly, P., and Bash, D. (2018). Trump orders FBI probe into Kavanaugh; senate vote delayed. *CNN Politics* (29 September)

Brandon, S., Boakes, J., Glaser, D. et al. (1997). Reported recovered memories of child sexual abuse: recommendations for good practice and implications for training, continuing professional development and research. *Psychiatric Bulletin* 21: 663–665.

Brennan, R., Patel, M., and Hope, A. (1989). Gonococcal conjunctivitis in Central Australia. *Medical Journal of Australia* 150: 48–49.

Brown, G. (2010). Relief as fictitious victim sentenced. *The Press* (2 July).

Burnett, R. (2016). Reducing the incidence and harms of wrongful allegations of abuse. In: *Wrongful Allegations of Sexual and Child Abuse* (ed. R. Burnett). Oxford, UK: Oxford University Press.

Cascarelli, J. (1996). Presumption of innocence and natural law: Machiavelli and Aquinas. *American Journal of Jurisprudence* 41: 229–269.

CCN (2004). Bryant: 'I want to apologize' to the young woman. CNN.com (2 September). https://edition.cnn.com/2004/LAW/09/01/bryant.statement/index.html

Ceci, S. and Bruck, M. (1995). *Jeopardy in the Courtroom: A Scientific Analysis of Children's Testimony.* Washington, DC: American Psychological Association.

Ceci, S.J., Kulkofsky, S., Klemfuss, J. et al. (2007). Unwarranted assumptions about children's testimonial accuracy. *Annual Review of Clinical Psychology 3* (2007): 311–328.

Cleave, L. and Gower, P. (2003). 10 years of guilt over for rape victim. *New Zealand Herald* (31 May).

Clyde, J. (1992). *The Report of the Inquiry into the Removal of Children from Orkney in February 1991.* Edinburgh: HMSO.

Coble, Y. (1994). Report of the Council of Scientific Affairs: memories of childhood abuse. *CSA report 5-A-94.* American Medical Association.

Committee on Child Abuse and Neglect (2020). *Gonorrhea in Prepubertal Children.* American Academy of Pediatrics.

Cook, S.L., Gidycz, C.A., Koss, M.P., and Murphy, M. (2011). Emerging issues in the measurement of rape victimization. *Violence Against Women* 17: 201–218.

Court Reporter (1998). False rape complaint. *Coaster* (23 September), p. 5.

Court Reporter (2004). False rape allegation 'attention seeking'. *The Press.*

Crews, F. (1995). *The Memory Wars: Freud's Legacy in Dispute.* London, England: Granta Books.

De Young, M. (2004). *The Day Care Ritual Abuse Moral Panic.* McFarland.

De Young, M. (2007). Two decades after McMartin: a follow-up of 22 convicted day care employees. *Journal of Sociology & Social Welfare* 34: 9–33.

Dezutter, A., Horselenberg, R., and Van Koppen, P.J. (2018). Motives for filing a false allegation of rape. *Archives of Sexual Behavior* 47: 457–464.
Egan, T. (1995). Pastor and wife are acquitted on all charges in sex-abuse case. *New York Times* (12 December).
Ehrenberg, M. and Elterman, M. (1995). Evaluating allegations of sexual abuse in the context of divorce, child custody, and access disputes. In: *True and False Allegations of Child Sexual Abuse: Assessment & Case Management* (ed. T. Nay). New York: Routledge.
European Therapy Studies Institute (1996). Recovered memoires of trauma and sexual abuse: ETSI guidelines for therapists. *Medicine, Science and the Law* 36: 106–109.
Evans, M. (2019). Man known as 'Nick' filmed crying as he falsely tells police of murdered schoolmate, Westminster VIP trial hears. *The Telegraph* (14 May).
Faller, K.C. (1984). Is the child victim of sexual abuse telling the truth? *Child Abuse and Neglect* 8: 473–481.
Ferguson, C.E. and Malouff, J.M. (2016). Assessing police classifications of sexual assault reports: a meta-analysis of false reporting rates. *Archives of Sexual Behavior* 45: 1185–1193.
French, C. and Ost, J. (2016). Beliefs about memory, childhood abuse, and hypnosis among clinicians, legal professionals, and the general public. In: *Wrongful Allegations of Sexual and Child Abuse* (ed. R. Burnett). Oxford, UK: Oxford University Press.
Ganim, S. (2010). Woman charged in false report. *Centre Daily Times* (12 March).
Goodyear-Smith, F. (1997). Roger & Heather Smythe – case history. *COSA Newsletter* 4: 7–10.
Goodyear-Smith, F.A., Laidlaw, T.M., and Large, R.G. (1998). Parents and other relatives accused of sexual abuse on the basis of recovered memories: a New Zealand family survey. *New Zealand Medical Journal* 111: 225–228.
Goodyear-Smith, F. (2007). What is the evidence for non-sexual transmission of gonorrhoea in children after the neonatal period? A systematic review. *Journal of Forensic and Legal Medicine* 14: 489–502.
Goodyear-Smith, F., Sharland, M., and Nadel, S. (2014). Think Hickam's dictum not Occam's razor in paediatric HIV. *BMJ case reports.*
Goodyear-Smith, F. (2015). *Murder That Wasn't: The Case of George Gwaze.* Dunedin: Otago University Press.
Goodyear-Smith, F. (2016a). Why and how false allegations occur. In: *Wrongful Allegations of Sexual and Child Abuse* (ed. R. Burnett). Oxford, UK: Oxford University Press.
Goodyear-Smith, F. (2016b). Why and how false allegations of sexual abuse occur. In: *Wrongful Allegations of Sexual and Child Abuse* (ed. R. Burnett). Oxford: Oxford University Press.
Govan, F. (2009). Woman faces jail for lying about rape to get abortion in Spain. *The Telegraph* (19 August).
Grattagliano, I., Corbi, G., Catanesi, R. et al. (2014). False accusations of sexual abuse as a means of revenge in couple disputes. *Clinica Terapeutica* 165: e119–e124.
Green, A. (1991a). Factors contributing to false allegations of child sexual abuse in custody disputes. *Child & Youth Services* 15: 177–189.
Green, A.H. (1991b). Factors contributing to false allegations of child sexual abuse in custody disputes. *Child & Youth Services* 15: 177–189.
Harned, M.S. (2004). Does it matter what you call it? The relationship between labeling unwanted sexual experiences and distress. *Journal of Consulting & Clinical Psychology* 72: 1090–1099.
Henriques, R. (2019). *The Independent Review of the Metropolitan Police Service's handling of Non-recent Sexual Offence Investigations Alleged Against Persons of Public Prominence.* London: Metropolitan Police Service.
Henson, S. (2004). Bryant's accuser files civil suit. *Los Angeles Times* (11 August).

Herman, S. (2016). Reducing harm from false allegations of child sexual abuse: the importance of corroboration. In: *Wrongful Allegations of Sexual and Child Abuse* (ed. R. Burnett). Oxford: Oxford University Press.

Herrmann, F. and Speer, B. (1994). Facing the accuser: ancient and medieval precursors of the confrontation clause. *Virginia Journal of International Law* 34: 481–552.

Hewson, B. (2016). The compensations of being a victim. In: *Wrongful Allegations of Sexual and Child Abuse* (ed. R. Burnett). Oxford, UK: Oxford University Press.

House of Commons of Canada (1992). An act to amend the criminal code (sexual assault). Sections 273 to 276. In: *Bill C-49* (ed. 34TH PARLIAMENT, R. S). Ottawa, Canada: Canadian Government Publishing Centre.

Innocence Project (2021). DNA exonerations in the United States [Online]. New York. https://innocenceproject.org/dna-exonerations-in-the-united-states/ (accessed 7 September 2021).

Jenkins, P. (2004). *Moral Panic: Changing Concepts of the Child Molester in Modern America.* New Haven, CT: Yale University Press.

Kanin, E.J. (1994). False rape allegations. *Archives of Sexual Behavior* 23: 81–92.

Kuenzie, J. (2012). Police: woman faked story about rape, assault in Five Points [Online]. Columbia, SC. http://www.wistv.com/story/19905381/police-woman-faked-story-about-rape-assault-in-five-points (accessed 7 September 2021).

La Fontaine, J. (1994). *The Extent and Nature of Organised and Ritual Abuse.* London: HMSO.

Laney, C. and Loftus, E.F. (2013). Recent advances in false memory research. *South African Journal of Psychology* 43: 137–146.

Lanning, K.V. (1991). Ritual abuse: a law enforcement view or perspective. *Child Abuse & Neglect* 15: 171–173.

Loftus, E. (1996). Memory distortion and false memory creation. *Bulletin of the American Academy of Psychiatry and the Law* 24: 281–295.

Loftus, E. (1997). Creating false memories. *Scientific American* 27: 70–75.

Loftus, E. (2004). Dispatch from the (un)civil memory wars. *The Lancet* 364: 20–21.

Loftus, E.F. (2019). Eyewitness testimony. *Applied Cognitive Psychology* 33: 498–503.

Mares, A. and Turvey, B. (2018). Chapter 2 - The psychology of lying. In: *False Allegations* (ed. B. Turvey, J. Savino and A. Mares). San Diego, CA: Academic Press.

Masters, B. (1999). Woman was charged after tests cleared former boyfriend of attacking her. *Washington Post* (1 March).

Mather, I. (1693). *Cases of Conscience Concerning Evil Spirits.* Boston, MA: Benjamin Harris.

Matters, R. (1981). Non-sexually transmitted gonococcal conjunctivitis in Central Australia. *Communicable Diseases Intelligence* 13: 3.

Meadow, R. (1993). False allegations of abuse and Munchausen syndrome by proxy. *Archives of Disease in Childhood* 68: 444–447.

Merianos, A., Condon, R.J., Tapsall, J.W. et al. (1995). Epidemic gonococcal conjunctivitis in Central Australia. *Medical Journal of Australia* 162: 178–181.

Mikkelsen, E.J., Gutheil, T.G., and Emens, M. (1992). False sexual-abuse allegations by children and adolescents: contextual factors and clinical subtypes. *American Journal of Psychotherapy* 46: 556–570.

Miller, C., Miller, H.L., Kenney, L., and Tasheff, J. (1999). Issues in balancing teenage clients' confidentiality and reporting statutory rape among Kansas Title X clinic staff. *Public Health Nursing* 16: 329–336.

Ministry of Justice (2005). Crimes Act 1961 No 43. Section 128 replaced by Section 7 of the Crimes Amendment Act 2005 No 41 [Online]. Wellington. http://www.legislation.govt.nz/act/public/2006/0069/latest/DLM393463.html (accessed 7 September 2021).

NASW National Council on the Practice of Clinical Social Work (1996). *Evaluation and Treatment of Adults with the Possibility of Recovered Memories of Childhood Sexual Abuse*. Washington, DC: National Association of Social Workers.

Nathan, D. and Snedeker, M. (2001). *Satan's Silence: Ritual Abuse and the Making of a Modern American Witch Hunt*. Basic Books.

New Zealand Government (1989). Children, Young Persons, and Their Families Act 1989. *Public Act 1989 No 24*. Wellington.

New Zealand Legislation (2011). Criminal Procedure Act. In: *20 Charge May Be Representative* (ed. N. Goverment). Wellington: New Zealand Government.

Newbold, G. (2000). *Crime in New Zealand*. Palmerston North: Dunmore Press.

Newbold, G. (2016). *Crime, Law and Justice in New Zealand*. New York: Routledge.

Newins, A.R., Wilson, L.C., and White, S.W. (2018). Rape myth acceptance and rape acknowledgment: the mediating role of sexual refusal assertiveness. *Psychiatry Research* 263: 15–21.

Newsroom (2016). Police discontinue rape investigation. *Milton Keys Citizen* (18 January).

NZPA (1996a). Girl claims false attack. *Taranaki Daily News* (26 August).

NZPA (1996b). Many false rape complaints. *Western Leader* (6 April).

NZPA (1996c). Woman charged with false rape complaint. *Waikato Times* (13 April).

NZPA (1996d). Woman makes false rape claim. *New Zealand Herald* (6 March).

NZPA (1997). False claim. *North Shore Times Advertiser* (12 August).

NZPA (1998a). Residents shocked at false claim charges. *NZ Herald* (14 May).

NZPA (1998b). Sex attack suspect leaves calling card at victim's door. *NZ Herald* (13 May).

NZPA (1999). The man who never was. *Waikato Times* (20 February).

NZPA (2000). Woman sought in false claims case. *NZ Herald* (13 September).

NZPA (2007). False rape claim woman spared jail due to mental health. *NZ Herald* (8 June).

NZPA (2010). Woman admits making up rape claim. *New Zealand Herald* (4 May).

O'Brien, B. (2009). Prime suspect: an examination of factors that aggravate and counteract confirmation bias in criminal investigations. *Psychology, Public Policy, and Law* 15: 315–334.

O'Donohue, W., Cummings, C., and Willis, B. (2018). The frequency of false allegations of child sexual abuse: a critical review. *Journal of Child Sexual Abuse* 27: 459–475.

Payne, S. (2017). Will sexual harassment in Westminster be as big as MPs expenses? *Financial Times* (31 October).

Pilling, K. (2010). 'Wicked' woman jailed over false rape claim. *Independent* (15 July).

Raea, S. and Fergusson, A. (1997). Knowledge set Dougherty free. *New Zealand Herald* (19 April), p. A15.

Ricchissin, T. (2000). Legacy of a lie. *Baltimore Sun* (17 December).

Ross, C. (1995). *Satanic Ritual Abuse: Principles of Treatment*. Toronto, Canada: University of Toronto Press.

Samuels, A. (2003). Anonymity for the rape accused? *Journal of Criminal Law* 67: 492–494.

Smith, M. (2016). Telling stories? Adults' retrospective narratives of abuse in residential homes. In: *Wrongful Allegations of Sexual and Child Abuse* (ed. R. Burnett). Oxford: Oxford University Press.

Spanos, N.P., Burgess, C.A., and Burgess, M.F. (1994). Past-life identities, UFO abductions, and satanic ritual abuse: the social construction of memories. *International Journal of Clinical & Experimental Hypnosis* 42: 433–446.

Sunday News (1991). We'll still pay, ACC says after rape acquittal. *Sunday News* (24 November).

Tavris, C. and Aronson, E. (2007). *Mistakes Were Made (But Not By Me)*. Harcourt: Orlando, FL.

Taylor, P. (1999). Victim admits sordid sex allegations a lie. *Truth*.

The Guardian (2015). One in five school staff victims of false claims, survey shows. *The Guardian* (30 March).
The Week Staff (2021). The ages of consent around the world. *The Week* (16 March).
Thompson, C. (2014). Nurse jailed after she lied to police that her grandfather had raped her when she was a child to try to get her inheritance early. *Daily Mail* (8 January).
Villa, L. and Alter, C. 2020. What we know about Tara Reade's allegation that Joe Biden sexually assaulted her. *Time* (8 May).
Watson, E. and Tillett, E. (2019). Take 2: embattled Roy Moore launches second campaign for Alabama Senate seat. *CBS News* (20 June).
Weaver, M. (2016). Met chief: police should not believe all sexual offence complainants. *The Guardian.*
Webster, R. (2005). *The Secret of Bryn Estyn.* Oxford: Orwell Press.
Working Party (1994). *Recovered Memories: Report of the Working Party of the British Psychological Society.* Leister: British Psychological Society.
Yoffe, E. (2017). The uncomfortable truth about campus rape policy. *The Atlantic* (7 September).
Young, P. (2016). Cliff Richard sex abuse accuser 'previously arrested over blackmail plot'. *The Independent* (19 June).

3 Disclosure of evidence in sexual assault cases

Sarah Forshaw QC[1] *and Phoebe Bragg*[2]
[1] *King's College, London, UK*
[2] *Trinity College, University of Oxford, Oxford, UK*

Introduction

In England and Wales, the disclosure exercise has typically been the least publicised part of a criminal trial. It takes place, for the most part, behind closed doors and before a trial starts. It relates to the obligation on the prosecution in a criminal trial to disclose material which may undermine the prosecution case or provide a measure of support for the defence at trial. The disclosure process ought to be straightforward. In practice, it almost invariably is not. The Judicial Disclosure Protocol (2013) and the Criminal Cases Review Commission (2021) have warned that failure to disclose material to the defence remains the biggest single cause of miscarriages of justice.

The basic premise behind the prosecution's disclosure obligation is rooted in fairness and common sense. When it is suspected that an individual may have committed a serious criminal offence, the police have a duty to investigate. It is not for the accused to establish his innocence by pursuing his own lines of enquiry. Nor will the accused have the resources, the capability, or even necessarily the know-how to secure evidence that might help to exonerate them. Accordingly, the police have a duty, with the benefit of expertise, manpower, and early access, to *pursue all reasonable lines of enquiry* when they investigate an allegation, *whether those lines of enquiry point towards or away from guilt of the suspect*. When that investigation uncovers material tending to point towards guilt, it will be relied upon by the prosecution and served as evidence in the case. Material that does not assist the prosecution is not served; it is called 'unused material'. If any of that 'unused material' might assist the defence or undermine the prosecution, it must be disclosed to the defence so they may deploy it at trial.

It is the failure of the prosecution to fulfil their disclosure obligations that has brought the issue to the fore in recent years in a series of highly publicised criminal cases. By way of example, in December 2017 the trial of a young student, Liam Allan, who had been wrongly charged with six counts of rape and six counts of sexual assault, came to an abrupt halt when text messages were finally found on the complainant's telephone that exonerated him (BBC News 2018). Police had insisted there was nothing of interest on the phone download. But for the persistence of trial

Current Practice in Forensic Medicine, Volume 3, First Edition. Edited by John A.M. Gall and J. Jason Payne-James.

counsel, Mr. Allan may have spent many years in prison. A mixture of police incompetence and Crown Prosecution Service (CPS) apathy very nearly led to a terrible miscarriage of justice. Liam Allan's case is by no means unique.

In 2018, the CPS ordered a review of all rape and serious sexual assault (RASSO) cases where the accused had been charged but had not yet reached trial in the period between January and mid-February. It was found that, over that period of just five weeks, 47 prosecutions had been stopped following the discovery that material that ought to have been disclosed to the defence had either not been identified, not been obtained, or had arisen after charge and not been dealt with in proper or timely fashion (Davies and Dodd 2018).

The fact that the cases that triggered this systemic review were sexual allegations was no coincidence. Rape and serious sexual offences involve sensitive information, often a relationship history, sometimes pre- and post-allegation digital contact, and, in recent report cases, they almost invariably involve forensic examination. Disclosure failings have understandably come to the attention of the criminal courts who must grapple with striking the proper balance between, on the one hand, placing an unachievable burden on the CPS when it comes to disclosure and, on the other, ensuring that a realistic and proportionate approach does not result in any miscarriage of justice.

There have been recent developments regarding the attitude to disclosure in respect of the ever-increasing amount of material falling to be disclosed in a digital era. That relates principally to texts, emails, calls, and social media accounts. It is not proposed to focus on that aspect of disclosure here. Rather, focus will concentrate on the aspects of disclosure most pertinent to the medical professional.

Definition and interpretation

Disclosure is defined within section 3 Criminal Procedure and Investigations Act 1996 (CPIA) and refers to provision by the prosecution to the defence of '*material which might reasonably be considered capable of undermining the case for the prosecution against the accused or of assisting the case for the accused, and which has not previously been disclosed*'. This includes an obligation on the prosecution to take reasonable steps to secure material in the hands of third parties (material held by other Government departments, health and education authorities, financial institutions, and other third parties).

The importance of disclosure has repeatedly been emphasised by the Court of Appeal, no more vigorously than in R v H and C:

> 'Fairness ordinarily requires that any material held by the prosecution which weakens its case or strengthens that of the defendant, if not relied on as part of its formal case against the defendant, should be disclosed to the defence. Bitter experience has shown that miscarriages of justice may occur where such material is withheld from disclosure. The golden rule is that full disclosure of such material should be made. ([2004] 2 AC 134, at 147)'.

In 2013, the chair of the Criminal Cases Review Commission, Richard Foster, pointed to disclosure failures as the 'continuing biggest single cause of miscarriages of justice'. Five years later, he once again echoed this sentiment. There have been countless reviews and reports designed to tackle disclosure issues in criminal proceedings, and on 28 July 2018, a Justice Select Committee Report on 'Disclosure of evidence in criminal cases' concluded that

> 'problems with the practice of disclosure have persisted for far too long, in clear sight of people working within the system. Disclosure of unused material sits at the centre of every criminal justice case that goes through the courts and as such it is not an issue which can be isolated, ring fenced, or quickly resolved' (Justice Select Committee 2017).

The Court of Appeal made the disclosure process clear in R v Olu stating that it is

> 'the task of the prosecutor to identify the issues in the case and for the disclosure officer to act under the prosecutor's guidance; the disclosure regime will not work in practice unless the disclosure officer is directed by the prosecutor as to what is likely to be most relevant and important so that the officer approaches the matter through the exercise of judgement not simply a schedule-completing exercise'.

The fact of the matter is that a police officer's task when undertaking disclosure in complex cases such as those involving sexual assault allegations can be voluminous. It would be easy to overlook a detail buried in many pages of original handwritten unused material that just might be explosive in the hands of the defence. Moreover, often the disclosure officer is also the officer in charge of the case and the officer who has most contact with the complainant. While the disclosure officer has a duty of objectivity, that part of his role is undeniably counterintuitive. The police and CPS are encouraged to call those who make an allegation 'the victim' from the outset of an investigation. That is despite the recommendations of Sir Richard Henriques, a widely respected retired judge, who was asked to conduct an independent review of the police's handling of non-recent sexual allegations following the Carl Beech ('Nick') debacle (Henriques 2019). Sir Richard recommended that those tasked with investigating should adopt the correct terminology and not label every accuser a 'victim' from the outset – for that reverses the burden of proof. In his report he said 'There is plain evidence, in the cases that I have reviewed, that an instruction to believe complainants has over-ridden a duty to investigate cases objectively and effectively. An instruction to remain objective and impartial whilst interviewing a complainant will not detract from the obligation to support complainants through the criminal justice process nor deprive any complainant of rights under the Victims Charter. . .. Any process that imposes an artificial state of mind upon an investigator is, necessarily, a flawed process. An investigator, in any reputable system of justice, must be impartial. The imposed "obligation to believe" removes that impartiality'. Sir Richard's logic was unarguable. Yet, his recommendation was never implemented.

Although it would, in our experience, be a rare case in which the police deliberately and in bad faith conceal material for fear that it may undermine the prosecution, it is never easy to look at a case objectively from an opposite viewpoint and to ensure that all material capable of assisting the defence is identified. If such a task is to be done properly and not simply be a 'schedule-completing exercise', it must be met with resources, training, and support to assist disclosure officers in conducting the exercise properly. This is something that has been neglected and led the Court of Appeal to publicly emphasise the need for disclosure officers to receive proper training in R v Malook.

Disclosure and the medical professional

Expert witnesses, whether instructed by the prosecution or the defence, have additional disclosure obligations. They have an overriding duty to assist the court and are under an obligation to disclose relevant information about themselves and their work. Expert witnesses instructed by the prosecution must reveal the existence of any unused material to the prosecutor, and the prosecutor will then decide whether it needs to be disclosed to the accused. This obligation takes precedence over any internal codes of practice or other standards set by the expert's own professional organisation.

In the case of R v Ward, the court placed a further important duty upon the expert, as follows:

> '. . . an expert witness who has carried out or knows of experiments or tests which tend to cast doubt upon the opinion he is expressing is under a clear obligation to bring records of those tests to the attention of the solicitor instructing him . . .'

An expert who fails to make clear where authors of papers s/he relies on for one purpose have expressed views of their own which are inconsistent with his/hers – even where there is no dishonest motive – may expect to be roundly criticised. The experiment, test material, or undermining opinion should be supplied to the disclosure officer and prosecutor.

In medicolegal terms, the topic of disclosure is a particularly thorny issue. On the one hand, communications between the doctor and patient and medical records do not attract privilege and therefore cannot be relied upon to withhold disclosure of material. On the other, the Court of Appeal has said

> 'Victims do not waive the confidentiality of their medical records or their right to privacy under Article 8 of the ECHR by the mere fact of making a complaint against the accused. Judges should be alert to balance the rights of victims against the real and proven needs of the defence. . . . General and unspecified requests to trawl through such records should be refused'.

It is for the CPS to seek the consent of the complainant for disclosure of his or her medical records, and it is their responsibility to make the appropriate application if that consent is not forthcoming. Accordingly, this article concentrates on record keeping and the disclosure-savvy responsibilities of the expert rather than ethical considerations of confidentiality as between the patient and the medical professional. The medical professional is not the arbiter of what may or may not be withheld for the purposes of criminal proceedings. Rather, it is the duty of the medical professional whose involvement is relied upon by the Crown to keep careful, legible notes and to provide them, together with any conclusions – and any matter that may cast doubt upon them – to the prosecuting authorities in transparent and independent, non-partisan fashion.

In the context of sexual assault investigations there are a number of common threads linked to disclosure that the prosecution and ultimately the courts have to grapple with. The nature of sexual assault allegations means that a complainant is likely to come into contact with a number of people in the period following a complaint and during an investigation, including forensic medical professionals. Their involvement will undoubtedly create a wealth of information that, more often than not, falls to be disclosed to the defence prior to trial. Issues have arisen where extensive investigations have not properly been documented and record keeping has been lax. The resultant disclosure has therefore been poor, forfeiting any opportunity the defence might otherwise have of challenging other aspects of the case on the basis of that material.

By way of example, forensic medical examiners who fail accurately to record the account given to them by the complainant when that complainant is first examined following an allegation of rape are inevitably laying themselves open to criticism from all sides. If the account was consistent with the complainant's later police statement or video interview, it might be relied upon by the prosecution at trial as 'complaint evidence'. If it was inconsistent and unrecorded, it would deprive the defence of valuable disclosable material and may lead to a miscarriage of justice. In either case, the keeping of careful notes is vital to the proper working of the criminal justice system.

The issue of record keeping has been addressed to an extent by the forensic medical profession themselves by the adoption of various guidelines and procedures. When looking retrospectively at an investigation, defence teams can be cognizant of such procedures and guidance in place in order to identify the type of material that may have been generated during such examinations. If it is missing, it may not have been properly itemised for disclosure, or the proper procedures may simply not have been adhered to.

In the case of R v DS, TS, a trial involving allegations of rape and sexual assault which was adjourned three times over a significant period of time due to non-disclosure, it was defence counsel instructed just prior to the first trial who was able to identify swathes of material that should have been listed on the unused schedule but were not. As a result, disclosure requests were made, and thereafter, additional material was repeatedly drip-fed to the schedule of unused material, ultimately leading to legal argument as to whether or not the case should be stayed as an abuse of process. This is analysed later in the article.

In the context of forensic and legal medicine, experts instructed by the prosecution, such as forensic clinicians in sexual assault cases, now have a duty 'to retain, to record, and to reveal'. The CPS themselves have guidance for experts more generally listing what records should, as a minimum, contain. This list includes strict record keeping of the receipt and movement of any materials received, provision of detailed legible notes made by experts or their assistants, which includes notes of verbal and other communications. Most importantly, the detailed taking of notes of any witness accounts or explanations given to any medical expert.

Guidance on note taking and record keeping in sexual assault cases is also contained within the 'Forensic Science Regulator Guidance' (2020), the relevant parts of which are shown in Table 3.1.

The Faculty of Forensic and Legal Medicine (FFLM) has also produced a number of pro forma templates in sexual assault cases including a 'Pro forma for adult

Table 3.1 Relevant paragraphs on guidance on note taking and record keeping in sexual assault cases from the 'Forensic Science Regulator Guidance' (2020).

a. 10.1.1 Each contact with the complainant by any professional shall be recorded in the set of case notes pertaining to that complainant. All notes shall be clear, accurate, and legible and include details of all activities that have taken place that are directly relevant to contact with the complainant at the facility.

b. 10.1.2 Notes shall be recorded contemporaneously, but where this is not possible, notes shall be made as soon as possible after the activity has taken place.

c. 10.1.3 All manual notes shall be made in permanent ink, signed, and dated including time if appropriate by the professional recording the notes. The name, role, and professional registration/identification number of the professional shall be included and legible. If the recorder is not the practitioner undertaking the tasks, then the practitioner's details are recorded, and the practitioner reviews, signs, and dates the notes as a true and accurate record.

d. 10.1.4 For electronic notes, the name, role, and professional registration/identification number and date, including time if appropriate, of the professional undertaking the tasks are recorded. These details are reviewed by the practitioner for accuracy.

e. 10.1.5 Where any additions or amendments are made to the notes by any person, the amendment shall be clear and signed and dated. If the amendment is made by someone other than original professional, the name, role, and professional registration/identification number of that individual shall be recorded in the notes.

f. 10.1.6 Where a correction to the notes is required, a single line shall be run through the correction so that the original note can still be read.

g. 10.1.7 Where abbreviations are included in notes, they shall be unambiguous and easily understood, for example LVS for low vaginal swab.

h. 10.1.8 It is important that any decision made by the professional is recorded along with the reason for making the decision. Where there is an expected course of action, the reason for making the decision not to follow the expected course shall be detailed in the record.

i. 10.1.11 All notes (including permanent records such as colposcope images) shall be retained by the facility in a secure location that complies with data protection requirements. The notes shall be available and accessible when they are required for the purpose of second opinion, peer review, the investigation, and/or any criminal justice proceedings.

j. 10.1.12 Where notes are required to be removed from the facility, the reason for removal shall be documented, and a record kept by the facility of the professional removing and returning the notes. It is preferable for copies or secure electronic access (with audit tracking) to records to be used so that the potential to lose records is eliminated.

forensic sexual assault examinations' (FFLM 2021). This includes details of the all-important first account to be taken from the complainant.

Notwithstanding the procedures that are put in place, disclosure failures continue. Where there is material the defence suspect is in existence, they have a duty under the Criminal Procedure Rules (CrimPR) to alert the court to problems of non-disclosure. The result is an inevitable tension between the potential disclosure failures of the prosecution and the unhappy position of those defending who suggest that there are items missing without knowing what they may reveal. Is it a 'fishing expedition' or a sensible request for material that the defence should have been provided with? It is the courts that have been left to attempt to strike the proper balance.

The role of the court in disclosure

The court's role in resolving disclosure issues has been formalised following the introduction of the CrimPR, CrimPR3.2(1), which states that the court must further the overriding objective by 'actively managing the case'. This aim underpinned the recommendations of the Auld Report (Lord Justice Auld 2001; especially at ch 10, paras 229–231), and The Leveson Review further highlighted the importance of the court's case management (para 38) stating that case management '*cannot be seen as a tick box exercise* . . . To that end, all parties must be required to comply with the CrimPR and to work to identify the issues so as to ensure that court time is deployed to maximum effectiveness and efficiency'.

The role of the court with regard to disclosure was highlighted in two decisions of the Court of Appeal in quick succession in 2015: R v Boardman and R v DS, TS. The former was an appeal by the Crown against a judge's decision to exclude the entirety of the phone evidence in a trial where there had been grave disclosure failings. This resulted in the prosecution being unable to prove their case, and they appealed against what was essentially a terminatory ruling. The latter related to an appeal by the Crown against a decision of a judge to stay the case (relating to sexual offences) as an abuse of process because of the consistent and wholesale disclosure failings by the police and CPS.

Although the Court of Appeal ultimately came to different decisions in each case, both were highly critical of the disclosure process. In R v Boardman, the defendant was charged with stalking by sending abusive messages and making telephone calls to a number of women. His defence statement (a legally required document setting out the nature of his defence) raised that he denied ever using the phone said to have been involved in the commission of the offences. The Crown had a master CD ROM of the call data from the phone which was not served on the defence until a few days before the trial was due to start despite repeated requests. The defence sought an adjournment on the grounds that a cell site expert needed to consider the call data from the disputed phone against that of a second phone accepted as his, and that this could not be done in time for trial. The trial judge refused the adjournment but excluded the call and cell site data, thereby effectively ending the prosecution.

The prosecution appealed the ruling of the trial judge. The Court of Appeal upheld the original decision, and the appeal failed. Two notable comments were made in the appellate court's judgment:

> 'the CPS were not entitled to expect that no sanction would follow (a failure to meet a requirement) unless the case had been brought back to the court for a further order: the resources of the court cannot be expected necessarily to extend to what might be described as the provision of a "yellow card"' [38]; *and* 'to create a trap for the prosecution generally, or the CPS in particular, by the over-zealous pursuit of inconsequential material which does not go to the issue, . . .in the hope that the CPS will fall down and that an application can be made which has the effect of bringing the prosecution to an end. . .is itself an abuse of process of the court and judges will be assiduous to identify it and impose sanctions on those who seek to manipulate the system'. [42]

The Court of Appeal therefore, in the same breath, both condemned the prosecution's failings and then issued a warning to defendants not to take advantage of them.

In R v DS and TS, where there had been multiple adjournments of the trial and a trial being aborted on the eighth day due to non-disclosure, the trial judge ultimately stayed the proceedings as an abuse of process. The case concerned allegations of rape and sexual assault. There had been multiple disclosure failings. However, the defence had also failed to comply with directions of the court regarding provision of a defence statement or identifying disclosure requests until late in the day. The crucial moment, however, came when the OIC attended on the first day of trial with three new boxes of materials that neither prosecution counsel nor the CPS had ever seen. The trial continued with the disclosure exercise continuing in the interim, including senior police officers being asked to attend court to carry out a full and independent disclosure exercise. On the eighth day of the trial, the judge discharged the jury on the basis that a fair trial demanded disclosure had to be completed before the defence could properly cross-examine.

The defence subsequently applied to stay the case an abuse of process due to the disclosure failures. That application was successful in front of the trial judge who concluded that

> 'Notwithstanding the seriousness of the charges, I take the view that this abuse is so exceptional the court ought to mark its wholesale condemnation of the prosecution by allowing a stay and refusing the prosecution the right to pursue the case'. [27]

The prosecution appealed this decision to the Court of Appeal, who overturned it, holding that this was not a case where the judge had concluded that the trial would be unfair if it went to retrial (the disclosure exercise now having properly been completed). While a stay of proceedings acted as a sanction to the prosecution and was vital in protecting the integrity of the criminal justice system, it had not been appropriate here. The Court held that, unlike in Boardman, the late-disclosed documents were of limited materiality and were only relevant to credibility. The failures of the defence to identify and raise disclosure issues in advance of trial were also noted as a critical factor in determining where failings lay.

Over five years on, and the balance as to the responsibility for identifying disclosure remains one which the courts continue to grapple with, particularly with technological advances and the reality that most people have huge amounts of digitally stored material that may be relevant to an investigation and fall to be disclosed. In the recent case of R v CB and another, the Court listed two otherwise unrelated cases in order to consider various issues relating to the retention, inspection, copying, disclosure, and deletion of electronic records. This judgement came after a flurry of separate judgements in recent years relating to disclosure which were quoted by the Court of Appeal including:

R v McPartland and another at [74]

'It was suggested on behalf of the defence in the course of argument that it is now entirely usual practice in cases involving allegations of sexual assault, that the mobile phone of a

complainant should be examined. This is not and should not be thought to be correct. What is a reasonable line of enquiry depends on the facts of each case'.

R v E at [75]

'It was submitted by Mr Wright that, in reaching his decision to stay the prosecution, the judge did so on the basis that there was always a duty on investigators to seize and interrogate the phone of any complainant who makes an allegation of a sexual offence. [. . .] For our part, however, we do not accept that the police were or are under such a duty. If the judge had made his finding on that basis then it may well have been that he did so based on an error of law which impacted on his own assessment of the position'.

Four issues of principle were set out as follows:

First: Identify the circumstances when it is *necessary* for investigators to seek details of a witness's digital communications

Second: When it is necessary, how should the review of the witness's electronic communications be conducted?

Third: What reassurances should be provided to the complainant as to the ambit of the review and the circumstances of any disclosure of material that is relevant to the case?

Fourth: What is the consequence if the complainant refuses to permit access to potentially relevant device? Similarly, what are the consequences if the complainant deletes the relevant material?

The Court noted that while these issues were raised in the context of digital material, they were equally applicable in the wider context of disclosure and will clearly be drawn upon by trial judges when arguments are raised as to what should and what should not be disclosed.

The Court of Appeal judgements in the context of forensic and legal medicine

Looking at the four principles raised in R v CB and another in the context of forensic and legal medicine, it is clear that the 'necessity' test for disclosure of forensic material is likely to be clearer than in the sphere of digital technology. However, forensic clinicians should draw on their expertise and knowledge of proper procedures and processes to identify what detail has not been provided that should have. The necessity element must be identified from the outset.

Let us take, by way of example, the forensic examination of a complainant in a sexual assault case. Was she subject to blood sampling or hair sampling? If the complainant was a child and there was no suggestion of drugs or alcohol playing a part in the alleged offence, then perhaps not. But what if the complainant alleges that her drink was spiked with drugs before the alleged assault? Necessity will undoubtedly apply. If no drugs were found in the bloodstream, then that material must be disclosed to the defence, together with any possible explanation (timing being of the essence). If the blood sample was negative, the drug may be found in a hair sample. If a sample was taken and not examined, it should be documented as in existence and made available to the defence, should they wish to conduct their own tests (if not too late). A simple request by the defence that they be provided with any hair sample would likely be questioned for necessity without the relevant context under the new guidance. If both samples were tested and both were negative, then that too must be disclosed to the defence.

The issue of consent, however, is equally applicable to forensic medicine. Just as with digital material, disclosure of medical records relating to a complainant will engage the Article 8 right to privacy, and the appropriate consent must be given to gain access to records and to enable disclosure if appropriate.

The complainant has the right to decline consent if they so wish, but this may have a consequence on the outcome of the case. The Court of Appeal in R v CB and another specifically addressed the impact of a refusal (in this instance of disclosing digital material) on the criminal procedure at [100]:

> 'The court should not be drawn into guessing at the content and significance of the material that may have become unavailable. Instead, the court must assess the impact of the absence of the particular missing evidence and whether the trial process can sufficiently compensate for its absence. An application can be made for a witness summons for the mobile telephone or other device to be produced. If the witness deletes material, although each case will need to be assessed on its own facts, we stress the potential utility of cross-examination and carefully crafted judicial directions. If the proceedings are not stayed and the trial proceeds, the uncooperative stance by the witness, investigated by appropriate questioning, will be an important factor that the jury will be directed to take into account when deciding, first, whether to accept the evidence of the witness and, second, whether they are sure of the defendant's guilt'.

Applications for witness summons in the context of medical documentation are not applicable in the same way as with digital material. Prosecutors must satisfy themselves that consent to disclosure of the medical records/counselling notes has been obtained by the police from the person to whom the notes refer. It is only if the complainant or witness does not consent to the release of their medical records, and there are reasonable grounds to believe that the disclosable material is contained within the medical records, prosecutors will need to consider whether it is appropriate to use the witness summons procedure to gain access to these records. The complainant or witness has a right to make representations to the court as to why the records should not be disclosed. CPS Guidance states that '*In these circumstances, the prosecutor would not usually be able to represent the interests of the victim or witness at the hearing being unaware of the content of the records, and their relevance (or otherwise) to the proceedings*'.

Where forensic material is missing or has been lost, this will not automatically lead to the prosecution being stayed as an abuse of process. The Court in R v CB cited with approval at [99] the case of R v PR in which it was said:

> 'The question of whether the defendant could receive a fair trial when relevant material had been accidentally destroyed would depend on the particular circumstances of the case, the focus being on the nature and extent of the prejudice to the defendant. In many instances, a

careful judicial direction would ensure the integrity of the proceedings by focusing the jury's attention on the critical issues that they needed to have in mind. Imposing a stay where records were missing was not a step that would be taken lightly; it would only occur when the trial process, including the judge's directions, was unable adequately to deal with the prejudice caused to the defence by the absence of the materials that had been lost, R. (on the application of Ebrahim) v Feltham Magistrates' Court [2001] EWHC Admin 130, [2001] 1 W.L.R. 1293, [2001] 2 WLUK 597, DPP v Fell [2013] EWHC 562 (Admin), [2013] 1 WLUK 605, R. v D [2013] EWCA Crim 1592, [2013] 9 WLUK 146 and R. v Allan (Christopher Mero) [2017] EWCA Crim 2396, [2017] 10 WLUK 822 applied. In the instant case, notwithstanding the records that had been destroyed, P had been in possession of a substantial amount of material that could have been used to test the reliability and credibility of the complainant. There was extraneous evidence of real substance to assist the jury in assessing whether her account was to be accepted. The judge's directions to the jury had needed to include the need for them to be aware that the lost material might have put P at a serious disadvantage, in that documents and other materials which he would have wished to deploy had been destroyed. The judge had given an impeccable direction to the jury to take that prejudice into account when considering whether the prosecution had been able to prove that P was guilty. P had received a fair trial (see paras 65-67, 71-74 of judgement)'.

When receiving instructions from the defence in criminal matters, the power of an expert to identify missing material that is likely to be available and may assist, but is not apparent on the face of papers, is indispensable to targeted disclosure requests. That is particularly so because the conduct of the defence, in drawing failings in disclosure to light, is likely to be considered material. As is clear from the pattern of judgements from the Court of Appeal, the courts have become less amenable to entertaining generalised requests for material that 'possibly' exists. The expertise of a medical professional may be required to explain the significance of lost forensic material or material that was simply never obtained and what that could have shown or not shown.

Conclusion

Those who defend in the criminal courts have become all too familiar with the 11th hour dump of pertinent unused material on the first day of trial in a sexual assault case. More often than not, that material is finally disclosed only because counsel for the prosecution has become more closely involved in the disclosure exercise and recognises the potential use to which the material might be put by the defence. Sometimes the phrase 'there is nothing more to disclose' rings distinctly hollow. Non-disclosure rarely comes to light after the event, and even then, instances when the Court of Appeal concludes that it casts doubt upon a conviction are few and far between. Accordingly, it is vital to get disclosure right first time.

To get it right, we suggest, the disclosure exercise is crying out for five things. First, the proper allocation of resources so that, with the ever-growing volume of digital unused material, there is adequate finance and man power to sift and properly analyse material for the purposes of disclosure. Second, proper training of disclosure officers and CPS representatives so that they know, instinctively, what it is that might look at first sight to be insignificant (because it does not square with their case) may in fact become a powerful tool in the hands of the defence. Third, a revisiting of the fact that the police and CPS, whose duty it is to identify disclosable material, are encouraged to call all complainants 'victims' (as opposed to complainants) from the outset and to believe that the allegation is true; 'Believe the Victim'. That vocabulary, and the mindset it carries with it operates against the presumption of innocence and the careful identification of material held by the prosecution that might undermine the Crown's case. Fourth, the early identification of independent prosecution trial counsel (who has a duty to act as a 'minister of justice' as part of their code of ethics) and a means of ensuring their close involvement in the disclosure process from the outset. Fifth, an increased readiness by the courts to make it clear that, when disclosure failings arise, there will be real consequences for any prosecution.

To that list, we add that, in cases involving medical professionals, their expertise is of critical importance in the disclosure process. Trial counsel need to – and do – rely on them.

References

BBC News and Shaw, D. (2018). Met Police apologise for Liam Allan rape case errors. https://www.bbc.com/news/uk-england-42873618 (accessed 2020).

(2018). CPS criminal legal guidance disclosure manual. www.cps.gov.uk/legal-guidance/disclosure-manual

(2019). CPS guidance for experts on disclosure, unused material and case management. www.cps.gov.uk/legal-guidance/disclosure-experts-evidence-case-management-and-unused-material-may-2010-guidance

CPS guidance on rape and sexual offences: disclosure and third party material. https://www.cps.gov.uk/legal-guidance/rape-and-sexual-offences-chapter-15-disclosure-and-third-party-material

(2018). CPS publishes outcome sexual offences review. https://www.cps.gov.uk/cps/news/cps-publishes-outcome-sexual-offences-review

(2019). Criminal Cases Review Commission disclosure review update. https://ccrc.gov.uk/press-releases/ccrc-disclosure-review-update/ (accessed 2020).

Criminal Procedure and Investigations Act 1996. https://www.legislation.gov.uk/ukpga/1996/25/contents

Davies, C. and Dodd V. (2018). CPS chief apologises over disclosure failings in rape cases. https://www.theguardian.com/law/2018/jun/05/scores-of-uk-sexual-offence-cases-stopped-over-evidence-failings (accessed 2020).

Disclosure of evidence in criminal cases. https://publications.parliament.uk/pa/cm201719/cmselect/cmjust/859/85908.htm#_idTextAnchor075

Disclosure – A protocol for the control & management of unused material in the Crown Court" paragraph 52.

Evidence withheld in 47 rape and sexual assault cases, says CPS. https://www.bbc.co.uk/news/uk-44366997

Faculty of Forensic & Legal Medicine of the Royal College of Physicians (2021). Pro forma for adult forensic sexual assault examination. https://fflm.ac.uk/wp-content/uploads/2021/05/Proforma-for-Adult-Forensic-Examination-Dr-S-Paul-and-Dr-ME-Vooijs-May-2021.pdf (accessed 2021).

Failure to disclose vital evidence in criminal cases growing, says watchdog. https://www.theguardian.com/law/2018/oct/11/failure-disclose-evidence-miscarriage-justice-warning-criminal-cases-review-commission

Forensic Science Regulator (2020). Guidance for the assessment, collection and recording of forensic science related evidence in sexual assault examinations. https://assets.publishing.service.gov.uk/government/uploads/system/uploads/attachment_data/file/886647/212_Guidance_Sexual_Assault_Examination.pdf

Henriques, R. (2019). Metropolitan Police Service's handling of non-recent sexual offence investigations alleged against persons of public prominence independent review. https://www.met.police.uk/henriques (accessed December 2013).

(2013). Judicial protocol on the disclosure of unused material in criminal cases. https://zakon.co.uk/admin/resources/downloads/judicial-protocol-on-thedisclosure-of-unused-material-in-criminal-cases-2013-1.pdf

Judiciary of England and Wales (2014). Magistrates' court disclosure review. https://www.judiciary.uk/wp-content/uploads/2014/05/Magistrates'-Court-Disclosure-Review.pdf.

Justice Select Committee (2017). https://publications.parliament.uk/pa/cm201719/cmselect/cmjust/859/859.pdf.

Lord Justice Auld (2001). *Review of the Criminal Courts of England and Wales*. London: TSO. https://www.criminal-courts-review.org.uk/ccr-00.htm.

Report of Sir Richard Henriques. https://www.met.police.uk/SysSiteAssets/foi-media/metropolitan-police/other_information/corporate/mps-publication-chapters-1---3-sir-richard-henriques-report.pdf

R v Boardman [2015] 1 Cr. App. R. 33.

R v CB and another [2020] EWCA Crim 790.

R v DS, TS [2015] EWCA Crim 662; [2015] WLR(D) 281.

R v E [2018] EWCA Crim 2426.

R v Gohil [2018] EWCA Crim 140; [2018] WLR 3697.

R v Liam Allen [2018]. R v Malook [2011] EWCA Crim 254.

R v McPartland and another [2019] EWCA Crim 1782; [2020] 1 Cr. App. R. (S.) 51.

R v Oliver Mears [2018]. R v Olu [2010] EWCA Crim 2975.

R v PR [2019] EWCA Crim 1225.

R v Samson Makele [2018]. Squier v General Medical Council [2016] EWHC 2739R v H and C [2004] 2 Cr App R 10.

The criminal procedure rules expert evidence. http://www.justice.gov.uk/courts/procedure-rules/criminal/docs/2015/crim-proc-rules-2015-part-19.pdf

The Rt Hon. Lord Justice Gross (2011). *Review of Disclosure in Criminal Proceedings*. https://www.judiciary.uk/wp-content/uploads/JCO/Documents/Reports/disclosure-review-september-2011.pdf.

The Rt. Hon. Lord Justice Gross The Rt. Hon. Lord Justice Treacy. (2012). Further review of disclosure in criminal proceedings: sanctions for disclosure failure. https://www.judiciary.uk/wp-content/uploads/JCO/Documents/Reports/disclosure_criminal_courts.pdf.

The Rt Hon Sir Brian Leveson President of the Queen's Bench Division (2015). *Review of Efficiency in Criminal Proceedings*. https://www.judiciary.uk/wp-content/uploads/2015/01/review-of-efficiency-in-criminal-proceedings-20151.pdf.

4 Current perspectives on the type and use of weapons used to police public assemblies around the world

Matthew McEvoy and Neil Corney
Research Associates, Omega Research Foundation, Manchester, UK

Introduction

The use of coercive means to police public protests remains both commonplace and controversial. Dramatic scenes of bloodied civilians and clouds of tear gas provoke outrage among many, and the growth of citizen journalism means that such scenes frequently reach a wide audience, often in real time. In 2019 and 2020 alone, mass protest movements were met with state-sanctioned police violence and excessive use of force in many countries, including Belarus, Bolivia, Chile, Colombia, France, Hong Kong, India, Lebanon, Nigeria, Peru, Sudan, and Thailand, among others. Black Lives Matter protests shook the United States (US) and spread around the world, frequently facing the kind of police violence and repression that spurred the protests in the first place.

Civil unrest is likely to continue in the near future. During 2020, there were large protests against government responses to the COVID-19 pandemic in at least 26 countries, with all regions affected (Carothers and Press 2020). The Omega Research Foundation has mapped incidents of excessive force by law enforcement under Covid-19 restrictions (Omega Research Foundation 2021). The economic fallout from the pandemic is predicted to exacerbate the existing political instability and poverty (Institute for Economics & Peace 2020). Shrinking economies and restricted access to credit will lead to cuts to social services and increases in taxes, impacting on the standard of living in all affected countries. In addition, the level of inequality between the rich and poor is likely to increase even further. Some countries are particularly vulnerable, and the UK Government's recent announcement that it will reduce its overseas aid budget provides an early warning of the hardship that is in store for conflict-affected and other aid-dependent countries. This combination of factors is likely to lead to increased protests and violent responses from repressive state authorities in all regions (Institute for Economics & Peace 2020). This, in turn, will see a range of weapons, including less-lethal weapons (LLWs), used with ever greater frequency and in larger quantities.

Current Practice in Forensic Medicine, Volume 3, First Edition. Edited by John A.M. Gall and J. Jason Payne-James.

Set against this backdrop, it is crucial that medical professionals are familiar with those weapons most frequently used, the manner in which they are being used, and the associated health risks. This chapter focuses on 'LLWs'. These are weapons which, in the course of expected or reasonably foreseen use, carry a lower risk of causing death or serious injury than firearms (United Nations 2020, p. 45). Whilst LLWs are being used with greater frequency to police assemblies, it is important to note that in a significant number of countries, firearms loaded with conventional bullets or other ammunition likely to cause life-threatening injuries are still widely used in the context of public gatherings. For example, in October 2020, Nigerian security forces reportedly used live ammunition during protests against police brutality, killing at least four people (Human Rights Watch 2020a). Similarly, lethal weapons are frequently used in conjunction with LLWs to devastating effect, as exemplified by the Myanmar security forces' deadly reaction to anti-coup protests in the first quarter of 2021.

In Myanmar and many other places, unaccountable and excessive use of force by security forces, including the use of LLWs, is resulting in large numbers of deaths and serious injuries. Such injuries include thousands of instances of life-changing disfigurement and disability every year worldwide. The LLWs industry continues to evolve apace, and new technologies are being developed and marketed for law enforcement use. Notwithstanding this, the types of weapons most frequently used to police public gatherings, and whose use continues to cause serious injuries and death, have generally been in use for decades, and evidence suggests that this use is increasing. This chapter explores the recent trends in the use of striking weapons, kinetic impact projectiles, chemical irritants, and stun grenades.

This chapter does not discuss the legal constraints on the use of force in great detail. Nevertheless, it is important that forensic professionals are aware that there are international human rights instruments and standards by which law enforcement agencies must abide. The UN *Code of Conduct for Law Enforcement Official* (United Nations 1980) and the *Basic Principles on the Use of Force and Firearms by Law Enforcement Officials* (United Nations 1990) set out key rules and principles for public security officials. The 2020 UN *Human Rights Guidance on Less Lethal Weapons in Law Enforcement* (hereinafter, 'UN Guidance') presents detailed guidance on the use of some of the weapons discussed in this chapter.

Forensic professionals should also be aware that the link between excessive force and torture and other cruel, inhuman, or degrading treatment or punishment (hereinafter, 'torture and other ill-treatment') is well established, even when the act is perpetrated outside of custodial settings. The European Court of Human Rights has consistently held that 'recourse to physical force which has not been made strictly necessary by a person's own conduct is in principle an infringement of the (the prohibition of torture and other ill-treatment)', including in the context of a public assembly (Case of İzci v. Turkey 2013). When participants in a public assembly resort to violence, they may forfeit their right to the freedom of peaceful assembly, but they retain their rights to be free from torture and other ill-treatment, and to life, and an effective remedy for all human rights violations (Kiai and Heyns 2016, paras. 8–9). LLWs are frequently used by law enforcement officials to violate these rights, whether it be unnecessarily to discriminate against or punish those not engaged in violence or disproportionately against those who are.

It is not necessarily the responsibility of forensic or healthcare professionals to determine whether a particular incident amounts to torture. Appropriately recorded and documented forensic evidence can play a key role in helping legal professionals to determine whether the use of force was lawful and if the elements of the crime of torture are present. Effective documentation of the use of weapons, including through implementation of the Istanbul (United Nations 2004) and Minnesota (United Nations 2017a) Protocols, is key to achieving accountability for individual cases of abusive use of force. The revised Istanbul Protocol should be published in 2022. On a more systemic level, where medical evidence corroborates multiple allegations of problematic practices or weapons causing a disproportionate amount of harm, this can help to effect change in policy and practice by law enforcement officials.

Before discussing the types of LLWs most frequently used to police public assemblies, it is important to consider four interlinked issues which frequently result in inappropriate or unsafe weapons, equipment, and tactics being deployed by law enforcement officials.

General issues

Firstly, whilst the use of force by law enforcement officials is governed by international law and policing norms and standards, the weapons and equipment they utilise are not. There are no international technical, manufacturing, or performance standards for LLWs. This absence has resulted in a very wide range of products being marketed directly to law enforcement, some of which are inherently unsuitable and should be prohibited (for example, spiked batons and shields or weighted restraints) and others with widely varying performance characteristics, resulting in unacceptable risk of injury (for example, multiple impact projectiles and high concentration chemical irritant sprays). Whilst no international standards exist, some states have their own national standards, for example, the United Kingdom has a technical standard for personal chemical irritant spray (United Kingdom Home Office - Centre for Applied Science and Technology 2014) and the US National Institute of Justice maintains a performance standard for handcuffs (US National Institute of Justice 2019). The absence of international standards, administered by globally recognised standards authorities, continues to result in law enforcement agencies often having to rely on manufacturer data alone.

Secondly, selection and testing of LLWs prior to deployment can reduce the risk of subsequent harms. However, such processes are often inadequate or absent in states with fewer policing resources, resulting in inappropriate or dangerous weapons being deployed. The UN Guidance clearly establishes the responsibility of states to take a precautionary approach and conduct testing, ideally by an independent body, of LLWs and law enforcement equipment prior to its procurement and the equipping of law enforcement officials (United Nations 2020, see section titled 'Legal Review, Testing and Procurement'). Testing should evaluate the effects of all reasonably likely or expected uses of the weapons, in scenarios of the expected use, and particular consideration should be given to assess the potential effects on persons who may be especially at risk, for example children, the elderly, pregnant persons, and the medically unwell. Ongoing monitoring of injuries caused by LLWs is crucial in providing data to feed back into the testing and selection process as well as to inform changes in policy, procedure, and guidelines for use.

Thirdly, the trade in law enforcement equipment is poorly regulated by states, resulting in inappropriate goods being transferred to abusive regimes, where they may be used to commit human rights abuses. Many states do not control the trade (import and export) of law enforcement equipment, beyond those classified as 'conventional arms', such as small arms, ammunition, and certain chemical irritants (see, for example The Wassenaar Arrangement 2020). This often excludes goods such as pepper and tear gas sprays, kinetic impact projectile munitions, electric shock weapons, water cannon, batons, and restraints as well as other weapons used in policing public assemblies. There have, however, been important advances made in this area including at the European Union, and most recently, internationally (see Omega Research Foundation and Amnesty International 2020), such trade control standards are a vital component in controlling the trade (and use) of LLWs.

A fourth key cause underlying violent policing of protests is the militarisation of public security that has occurred in response to perceived 'new threats', particularly terrorism and drug trafficking (Centro de Estudios Legales y Sociales 2018). Militarisation of public security encompasses both military personnel carrying out law enforcement duties and the use of military tactics and equipment by law enforcement officials. The inherent differences between the military and law enforcement paradigms mean that this blurring of roles, tactics, and equipment can be problematic. Where soldiers are primarily trained for combat and to kill enemy combatants, law enforcement officials must instead fulfill their duty 'by serving the[ir own] community and by protecting all persons against illegal acts, consistent with the high degree of responsibility required by their profession' (United Nations 1980).

The crossover of military equipment to law enforcement is not a new, or small-scale, phenomenon. Perhaps, the most high-profile and sustained example of this transfer is the US Department of Defense's *1033 Program*, which facilitates the acquisition of military equipment by law enforcement agencies throughout the United States. Under this program, since 1997, over 8000 law enforcement agencies have acquired more than $7.2 billion of military equipment, including armoured vehicles, assault rifles, ammunition, and night-vision goggles, and equipment which has no legitimate domestic law enforcement function, for example bayonets (Barrett 2020). Figure 4.1. is illustrative of the type of transformation many police forces have undergone, with traditional uniforms being replaced by military-style fatigues, helmets, and combat boots, and assault rifles being carried routinely rather than exceptionally.

Not all military equipment are inherently unsuitable for law enforcement. For example, specialist equipment such as stun grenades, high-powered firearms, and armoured vehicles may be appropriate in exceptional circumstances, such as certain antiterrorist operations. It does become problematic, however, when such equipment is deployed in other settings, such as public protests, where heavy displays of force may be disproportionate, can elevate tensions (United Nations 2020, para. 6.3.1), and can lead to unnecessary injuries or loss of life.

Having considered some of the reasons participants in public assemblies are frequently exposed to unnecessary or disproportionate force, it is important to explore how such force is meted out.

Figure 4.1 Law enforcement officials attired in military-style uniforms with armoured vehicles. *Source:* Jamelle Bouie, "swat team, fully assembled". 13 August 2014. Ferguson, Missouri.

Less-lethal weapons

Kinetic impact/striking weapons

Hand-held kinetic impact weapons, also known as striking weapons, include a range of specially manufactured batons and truncheons. Sticks and clubs are also widely used by law enforcement, and the use of short and long whips (sjamboks) in some countries has also been documented.

Apart from the hands or feet, batons are the oldest type of weapon in use by law enforcement. They are ubiquitous worldwide due to their low cost. The visual message from an officer wielding a baton is universally understood; it is a sign of authority and indicates that pain will result if they strike.

Specially manufactured weapons can be made of rubber, wood, plastic, or metal. They vary in width and length, from 20 cm for close use up to 2 m for greater stand-off distance. Batons can be straight, extendable/expandable (collapsing to a short length for ease of carrying or concealment), or side handled (commonly referred to as a tonfa). Longer batons are able to deliver greater impact energy, increasing the risk of serious injury. Use of overarm strikes or from an elevated position, such as from a police horse, or standing on a police vehicle, increases the risk of fatal head injuries caused by baton strikes.

Other hand-held kinetic impact weapons may be made of natural materials, such as the bamboo cane 'lathi' in India or the rattan cane used in various Asian jurisdictions.

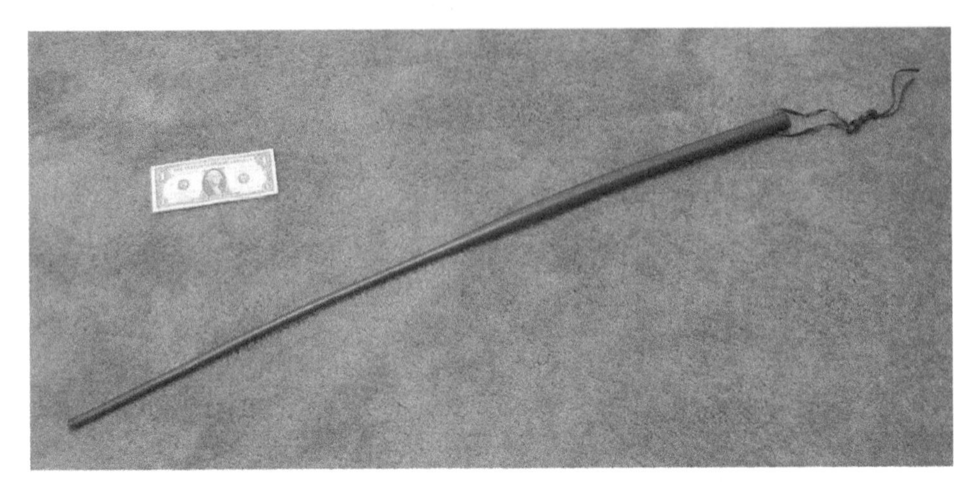

Figure 4.2 Plastic sjambok. *Source:* OwenX. "A 90 cm (3 ft) plastic sjambok used by South African Police". 30 December 2005. Ontario, Canada. CC BY-SA 3.0.

Whips have also been used in certain states, such as by the Central Cossack Battalion in Moscow who attacked and whipped peaceful protestors in May 2018 (Gershkovich 2020). The Council of Europe's Committee for the Prevention of Torture reported that journalists were also attacked with whips at the same event (Council of Europe 2018). Long plastic or animal hide whips (sjamboks), as shown in Figure 4.2, have been used during the policing of Covid-19 restrictions in South Africa, where reporters filmed police assaulting civilians with sjamboks (Reddy and Allison 2020), and in Uganda, where police were photographed using whips to 'clear' market vendors (Olewe 2020).

Batons and striking weapons can be used offensively or defensively. They are used by law enforcement officials to strike a subject to cause physical pain or to threaten physical pain in order to force them to comply or to deter them from an action. Such weapons should only ever be used on the larger muscle groups on the limbs; even then, striking weapons can cause severe bruising, lacerations, and broken bones, especially if the joints are targeted. They can also be used defensively by law enforcement officials as a blocking instrument, for instance to protect themselves from blows from assailants. The use of a baton, and the force inflicted, is greatly dependent on the individual user, their height, strength, technique (e.g. side or overarm strikes), and their training and oversight.

Batons are also used as a tool of restraint, including to hold limbs. Dangerous and potentially lethal 'neck holds' or 'carotid holds' using batons have been used by law enforcement officials to detain and subdue individuals. At least one EU-based company is known to have trained law enforcement officials in countries including Togo, Georgia, and India in such neck restraints using batons (Omega Research Foundation and Amnesty International 2017, p. 19). The UN Guidance states that neck holds should 'not be employed, as they present an especially high risk of death or serious injury as a result of compression of large blood vessels or the airway' (United Nations 2020, para. 7.1.5).

The United Nations *Resource book on the use of force and firearms in law enforcement* (hereafter UN Resource Book) specifies that:

> 'Given the range of functions that a baton can have, training in baton use is an essential part of a law enforcement agency's effort to develop capacity to differentiate the use of force depending on the law enforcement objective and the threat encountered. A baton is not expected to be lethal in the case of proper use'. (United Nations 2017b)

Any more than a single strike from a baton may amount to excessive use of force, which is always unlawful. Officers should reassess the need for any further strikes after each strike to make sure that additional force is strictly necessary and proportional. Nonetheless, media images and footage of protests often show law enforcement officials repeatedly beating people, often individuals who are on the ground, or chasing people whilst repeatedly striking them including on the head. Such actions risk serious injury or even death. Strikes should also be controlled and carefully aimed in order not to impact sensitive body parts, including joints, and the kidney area. A single strike to the head can result in concussion, traumatic brain injury, skull fractures, or death.

Striking weapons, including police batons, have been frequently used for torture and other ill-treatment, including excessive use of force, both in places of detention and outside of custodial settings. The following recent examples illustrate the widespread and contemporary nature of such abusive practices.

In Nigeria, protests began across the country on 8 October 2020, with calls for the authorities to disband and abolish the special anti-robbery squad (SARS), a police unit that had been widely reported to have been involved in abusive use of force. Human rights monitors and Non-Governmental Organisations (NGOs) reported widespread excessive use of force against peaceful protestors, including severe beatings inflicted by officers using both police batons and sticks, often with multiple officers beating an individual on the ground (Human Rights Watch 2020a).

In Belarus, Amnesty International reported numerous accounts of torture and other ill-treatment following post-election protests in August 2020, and many of these cases involved the misuse of batons or truncheons by security forces. Katsyaryna Novikava told Amnesty International that she had been detained for 34 hours at the Centre for Isolation of Offenders, where she saw that the entire yard of the facility was filled with arrested men who had been forced to lie down in the dirt. Inside the centre, dozens of men were told to strip naked and get down on all fours, while officers kicked and beat them with truncheons (Amnesty International 2020a). The OSCE also reported the widespread and systematic use of beatings for punishment of journalists, protestors, and civilians with police batons (Benedek 2020).

In the United States, during the 2020 nationwide Black Lives Matter and antipolice violence protests, human rights monitors reported the widespread abusive use of batons to beat protestors. Such violence included incidents where police, standing above a crowd on police vehicles, used batons to beat people's heads. This violence often occurred when no threat existed or when protestors were detailed in a police cordon or 'kettle' and unable to escape (Human Rights Watch 2020b).

Chemical irritants

A typical image in media reports of the policing of assemblies or protests shows streets filled with clouds of white smoke, or police using aerosol sprays against individuals or crowds. These are two examples of chemical irritants, one of the most widely used LLWs in law enforcement worldwide, with almost every country using some form of chemical irritant weapon.

Chemical irritants are highly potent substances that produce sensory irritation and pain in the eyes and upper respiratory tract. They can also cause inflammation of the mucous membranes, including in the respiratory tract. The chemicals most commonly used are the irritant agents CN or CS – often called tear gas – and the inflammatory agents OC/Pepper or PAVA – often called pepper spray. A number of other irritant agents, such as CR, are marketed and held by different States but are not commonly encountered (see Organisation for the Prohibition of Chemical Weapons 2019, Annex 4).

They are delivered via a wide range of methods and means. These include handheld aerosol sprays, shoulder-worn and backpack sprayers, handheld or vehicle-mounted smoke generators or foggers, hand-thrown grenades, weapon-launched projectiles and grenades, and via water cannon and, more recently, via unmanned air or ground vehicles. Handheld aerosol sprays range in size from 25 to 500 ml, sometimes more, while shoulder-worn and backpack-style sprayers and smoke generators generally have a much larger capacity and can cover a wider area often in a very short time. Hand-thrown and weapon-launched projectiles/grenades and water cannons can be used from greater ranges and can be used to contaminate a wide area.

A particular issue with the majority of chemical irritant sprayers/foggers is that they have no dose control or cutoff trigger mechanism to control the amount dispersed. Under international laws of law enforcement, only minimum amount of force should be used (and only when strictly necessary and proportionate to the threat), and force must cease when the threat from the individual ceases. In operational practice, law enforcement officials should use the minimum 'effective dose', i.e. one very short burst of spray and then reassess the threat. However, often sprays are continually discharged at individuals or groups, potentially dispersing large quantities of chemical irritant. Design changes to sprayer equipment, to introduce 'dosing' triggers that release only small quantities at a time, might help overcome this issue.

Chemical irritants are designed to deter or disable individuals by producing temporary but intense irritation and pain of the eyes, upper respiratory tract, and sometimes skin. A range of factors can determine their effects including the type of chemical agent and the means of delivery used, the location and environmental conditions in which they are used (heat and humidity), and the concentration and quantity of the irritant. The amount of active irritant agent in products available to law enforcement officials varies widely, and manufacturers will often offer a range of percentages for any given product or offer custom or bespoke fills for customers. This results in those using, affected or treating the affected not knowing what amount of chemical irritant has been delivered, or can lead to speculation and confusion as to the type of irritant being used, and may also result in unusual medical effects being encountered. For example, PepperBall, a commonly used compressed gas launcher system, which delivers chemical irritant via 0.68 caliber plastic encapsulated projectiles, advertises a range of projectiles with a wide range of percentages of irritants: 'VXR LIVE-X, a more concentrated formula, containing approximately 10× the PAVA of the VXR LIVE projectile' (PepperBall 2021).

Small handheld sprays are often deployed by law enforcement officials, intended to be used to temporarily deter or disorientate individuals at very close range (0–3 m). Chemical irritants are indiscriminate in nature, and the UN Special Rapporteur on the Rights to Freedom of Peaceful Assembly and of Association has warned that they fail to differentiate 'between demonstrators and non-demonstrators, healthy people and people with health conditions' (Kiai 2012, para. 35). Even the use of small, handheld sprays risks affecting innocent bystanders in a public gathering. Meanwhile, the United Nations has stressed that means of delivery for wider areas should only be used 'for dispersing groups that present an immediate and direct threat and when conventional methods of policing have been tried and have failed, or are unlikely to succeed' (United Nations 2017b, p. 87).

Almost all reports and studies on chemical irritants note the need for more research as well as the lack of statistically significant prospective clinical studies on the morbidity and mortality of relevant weapons and devices. Exposure to chemical irritants produces a wide range of effects including profuse tearing of the eyes, coughing, chest tightness, restricted breathing, vomiting, chemical burns, blistering of the skin/contact dermatitis, and, in extreme cases, death – either through asphyxiation or chemical poisoning, particularly if large quantities are used, or there is no means of escaping the effects, for example in enclosed spaces (see, United Nations Committee against Torture 2017, paras. 24–25). Those affected frequently feel anxiety and panic. The European Court of Human Rights has recognised that large doses of pepper spray 'may cause necrosis of tissue in the respiratory or digestive tract, pulmonary oedema or internal haemorrhaging' (Case of Oya Ataman v. Turkey 2007, paras. 17–18). If launched projectiles containing chemical irritants hit a person directly, they can cause penetration wounds, concussions, other head injuries, or death (Case of Abdullah Yaşa and Others v. Turkey 2013; Omega Research Foundation and OSCE Office of Democratic Institutions and Human Rights 2021).

International human rights bodies have created strict standards concerning the use of chemical irritants. The UN Guidance states that chemical irritants, including those delivered via handheld sprays, should only be used when there is an imminent threat of injury, and warns against repeated or prolonged exposure to irritants (United Nations 2020, paras. 7.2.3 and 7.3.5). The UN Resource Book also recommends against chemical irritants being used against the same people several times in a short time period, or in confined spaces (United Nations 2017b, p. 88).

Certain groups that are particularly at risk from the effects of chemical irritants, and for whom it may be life threatening, include older people, children, pregnant women, and people with respiratory problems. According to the American Academy of Pediatrics, 'Children are uniquely vulnerable to physiological effects of chemical agents. A child's smaller size, more frequent number of breaths per minute and limited cardiovascular stress response compared to adults magnifies the harm of agents such as tear gas'. (Kraft 2018).

The use of chemical irritants during the COVID-19 pandemic has increased the risk of adverse medical effects due to Covid-19's effect on breathing and the lungs as well as the risk of infection through induced coughing or sneezing, particularly for those in detention (Omega Research Foundation 2020). In 2020, the American Thoracic Society called for a moratorium on the use of tear gas and other chemical agents deployed by law enforcement against protestors participating in demonstrations, citing 'the lack of crucial research, the escalation of tear gas use by law enforcement, and the likelihood of compromising lung health and promoting the spread of COVID-19' (American Thoracic Society 2020). In the United States, nearly 1300 medical professionals stated their opposition to the use of chemical irritants, which 'could increase the risk of COVID-19 by making the respiratory tract more susceptible to infection, exacerbating existing inflammation, and inducing coughing' (Greiner et al. 2020).

An area of increasing importance, but where no clinical studies have yet been published, is the growing awareness of the effects of chemical irritants on different human bodies and reproductive health. Reports have suggested that there may be a relationship between use of tear gas and miscarriage (Physicians for Human Rights 2012). Following the widespread use of large quantities of chemical irritants during Black Lives Matter and other protests in the United States in summer 2020, media reports emerged of physiological effects for people who menstruate (Slisco 2020; Stunson 2020; Nowell 2020). As a result of such reports and the lack of research, the organisation Planned Parenthood is undertaking a study into 'Tear Gas and Reproductive Health' (Planned Parenthood North Central States 2021).

National policing and medical and practitioner bodies should offer advice to medical and other professionals who may have to treat or come into contact with those affected by chemical irritants. International policing standards require law enforcement officials to 'ensure that assistance and medical aid are rendered to any injured or affected persons at the earliest possible moment' (United Nations 1990, 5[c]). It is therefore vital that full details of any chemical irritants used are known and published (Faculty of Forensic & Legal Medicine 2021).

In a major study documenting the misuse of tear gas worldwide, published in June 2020 (updated in February 2021), Amnesty International with assistance from Omega Research Foundation verified over 500 videos of more than 100 events in some 31 countries and territories where tear gas has been misused (Amnesty International 2020b). Incidents included cases of security forces firing projectiles and grenades into cars, inside a school bus, and in hospitals, residential buildings, metro stations, and shopping centres as well as directly at individuals, and against others in confined spaces, as well as excessive quantities being fired.

The use of grenades that explode to disperse the chemical irritant has resulted in serious injuries. For example, in France, the GLI-F4 exploding tear gas grenade containing trinitrotoluene (TNT) was eventually withdrawn in early 2020 following numerous serious injuries, where individuals had their hands blown off or were injured by shrapnel from the exploding grenade (Breeden 2020). The GLI-F4 was replaced with the GM2L, which used a pyrotechnic charge rather than TNT to explode and disperse irritant and was stated to be a safer alternative. However, its use has continued to result in serious injuries, including, in December 2020, a participant at an assembly in Paris who had his fingers blown off when he touched the grenade as it exploded (SudOuest.fr 2020).

Direct firing of grenades of projectiles at individuals can result in serious injuries or death. Such launched projectiles or grenades should never be fired directly at persons, they are designed to be launched into open areas to land on the ground and disperse chemical irritant. However, their use against individuals continues to be reported, including, for example the death of Haykel Rachdi in Tunisia, who died of his injuries after reportedly 'being struck on the head by a police tear gas canister' (Cordall 2021). Particularly horrific injuries and deaths were reported in Iraq in 2019, when 'antiriot police' and other security forces in Baghdad fired grenades, referred to by local people as 'smokers', directly at individuals. The 40-mm military grade, heavy, and Iranian- and Serbian-manufactured grenades caused severe injuries and many deaths when the metal projectiles impacted and penetrated the skull (Hoz et al. 2020; Amnesty International 2019).

The use of large quantities of chemical irritant has increasingly been reported, either by the firing of tear gas projectiles and grenades or dispersal using sprayers and other equipment.

During escalating protests in 2019 in Hong Kong, police used chemical irritants in large quantities, firing 800 tear gas projectiles in a single day, and in some cases in confined spaces. For example, on 11 August 2019, police fired multiple rounds of tear gas inside the Kwai Fong Mass Transit Railway Station, a confined space with limited exits (Omega Research Foundation and Amnesty International 2020). Tunisian police used large amounts of launched tear gas cartridges during unrest in the city of Tataouine between 23 and 26 June 2020. Volunteers reportedly collected and counted almost 18 000 cartridges from the city after the operation had finished. People were affected in hospitals, places of worship, and their homes, with at least 180 people reportedly admitted to hospital for emergency treatment after inhaling tear gas, and 26 others were injured when struck with tear gas projectiles (Amnesty International 2020c).

In 2020, Federal Agents from the Department of Homeland Security were deployed to protect federal property in Portland, Oregon, under 'Operation Diligent Valor'. Protests occurred over many weeks, and federal agents repeatedly deployed massive quantities of chemical irritants and smoke including via thermal foggers, as shown in Figure 4.3 – hand-held devices producing large volumes of tear smoke from water- or oil-based solutions as well as from 'smoke pots' – open pots of burning incandescent substance (believed to be HC – hexachloroethane) that produced dense clouds of grey/white smoke. Protesters reported a wide range of respiratory and/or reproductive health impacts after attending protests and being exposed to the chemical irritants, including unusual effects on menstrual cycles (Northwest Center for Alternatives to Pesticides, Williamette Riverkeeper, Cascadia Wildlands et al. v. U.S. Department of Homeland Security, Chad Wolf, in his capacity as Acting Secretary, U.S. Department of Homeland Security 2020; Samayoa 2020).

Kinetic impact projectiles

Kinetic impact projectiles are commonly referred to as rubber or plastic bullets, but they can also be made from other materials including wood. 'Less-lethal' ammunition can contain single or multiple projectiles including, for example pellets, balls, blocks, cylinders, and fabric bags filled with pellets (bean bags). They are designed to cause blunt, non-penetrative trauma, but they can perforate the skin, penetrating the body and vital organs, particularly when fired from close range or at sensitive areas of the body, particularly the eyes. Such projectiles can also cause other serious injuries, including broken bones, concussion and other head injuries, and internal organ injury.

Figure 4.3 Federal law enforcement agents use a thermal fogger during a Black Lives Matter protest. *Source:* ©Doug Brown, ACLU of Oregon. 29 July 2020. Portland, Oregon, USA.

Due to their indiscriminate effects and the danger they present, the Council of Europe Commissioner for Human Rights has expressed concern at the unsuitability of certain weapons that are increasingly used for the policing of protests. In this context, she stated that the number of serious injuries caused by the use of kinetic impact projectiles was 'particularly striking' (Mijatović 2019a).

There is some divergence between international human rights standards on the lawful threshold for using kinetic impact projectiles. The UN Guidance states that they should only be used to strike a violent individual posing 'an imminent threat of injury' (United Nations 2020, para. 7.5.2). Both the ODIHR *Human Rights Handbook on Policing Assemblies* and the UN Resource Book impose a stricter standard, however, stating that kinetic impact projectiles should only be used against individuals posing an immediate threat of *serious* injury or death (Organisation for Security and Cooperation in Europe 2016 p. 81; United Nations 2017b, pp. 94–95).

These international human rights instruments also share important commonalities. All state that kinetic impact projectiles should only be used to target individuals posing a threat, not groups, and that the use of ammunition containing multiple projectiles, such as those shown in Figure 4.4, does not comply with the international human rights standards due to their inherent lack of accuracy. Both the UN Guidance and the ODIHR Handbook stipulate that kinetic impact projectiles should only be aimed at the lower abdomen or legs (United Nations 2020, para. 7.5.2; Organisation for Security and Cooperation in Europe 2016, p. 81), and this is complemented by the Resource Book which states that they should not be fired at sensitive parts of the body, 'in particular the head, neck, chest and groin' (United Nations 2017b, p. 95).

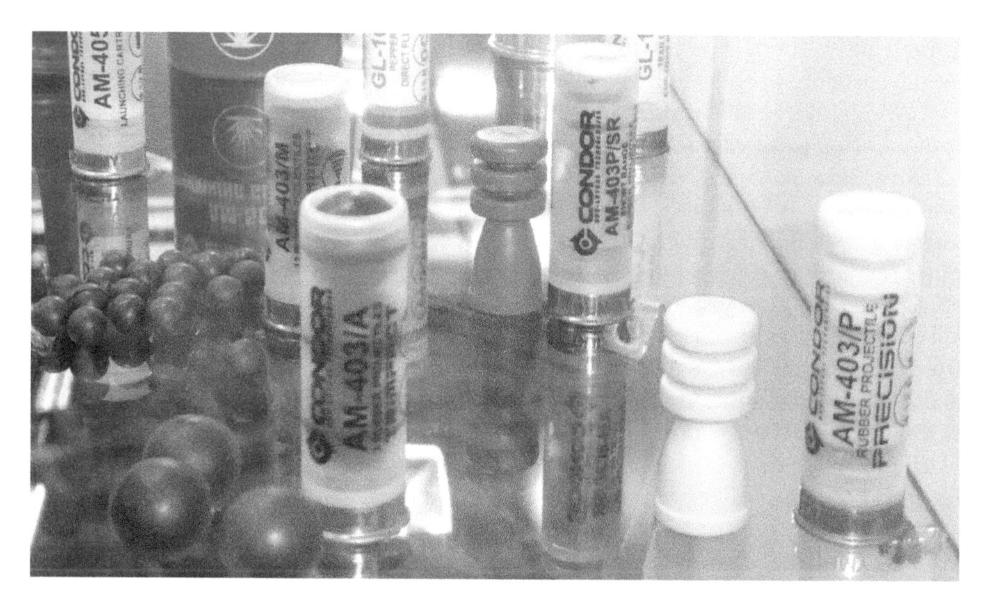

Figure 4.4 Examples of multiple projectiles (left) and single projectiles (right). *Source:* ©Omega Research Foundation. June 2018. Paris, France.

When fired from close range, kinetic impact projectiles can cause serious injury or death even when fired at the legs. For instance, in the Brazilian state of Pernambuco in 2017, a Military Police officer shot Edvaldo da Silva Alves in the thigh with a kinetic impact projectile during a peaceful protest. The projectile was fired from a distance of less than 5 m, resulting in penetration. Although the projectile was surgically removed from the victim's body, he later died of a general infection caused by his injuries (Folha de Pernambuco 2017).

Factors such as insufficient use of force reporting internally by law enforcement agencies and a lack of transparency regarding the use of force data that is compiled make it difficult to measure trends relating to the use of LLWs over time. The context-specific information that is available nonetheless seems to indicate that kinetic impact projectiles are increasingly being used for the policing of protests.

The use of kinetic impact projectiles known as 'Plastic Baton Rounds' against protesters by the Royal Ulster Constabulary and the British Army was 'the most controversial aspect of public order policing' during The Troubles in Northern Ireland (The Independent Commission on Policing in Northern Ireland 1999, para. 9.12). According to the Patten Report, between 1981 and 1999 over 56 000 plastic baton rounds were fired by the police and the army, resulting in 11 deaths in Ireland (The Independent Commission on Policing in Northern Ireland 1999, para. 9.12), of the 17 overall deaths caused by rubber or plastic impact projectiles. Over 18 years, this averages out to approximately 8.5 rounds being used every day. The year 1981 was when most plastic baton rounds were fired, with 29 695 rounds used in total (O'Halloran 1996), averaging out to 81 per day.

In recent years, law enforcement officials in several countries have used even larger numbers of kinetic impact projectiles against protesters, including in response to mass protests that occurred in Chile in late 2019. Responding to an access to information request, the *Carabineros de Chile* revealed that they fired 151288 shotgun cartridges containing 12 pellets each between 18 October and 31 December 2019 (Weibel Barahona and Jara 2020). Thus, on average, 2017 rounds were fired per day during this period. Similarly, in recent years, French police have responded to Yellow Vest protests with large quantities of LLWs. The Council of Europe Commissioner for Human Rights reported that from 17 November 2018, when the protests began, to 4 February 2019, police fired 12122 40mm kinetic impact projectiles (Mijatović 2019b, para. 16), an average of 152 per day. French police were also reported to have used 4942 stingball grenades, exploding grenades with multiple small hard rubber pellets, which are expelled at high velocity in a random pattern. Although France and Chile are both much larger countries than Northern Ireland, it is still striking that use of kinetic impact projectiles in both countries so clearly exceeded the use during even the most difficult times during the troubles.

Although such data is often not collected or made publicly available, the Omega Research Foundation's experience monitoring the use of force indicated that the use of large quantities of kinetic impact projectiles to police protests is not limited to these two cases. Assuming that these cases are, to an extent, representative of other examples, it is useful to look more closely at the type of projectiles being used in Chile and France, how they are being used, and the consequences of their use.

During the 2019–2020 protests in Chile, police used 12-gauge shotgun cartridges containing 12 small balls or pellets each. The National Human Rights Institute reported that 460 protesters suffered ocular injuries, with 35 people suffering either burst eyeballs or losing an eye (Instituto Nacional de Derechos Humanos 2020). A medical study into the cases of ocular trauma treated at the Eye Trauma Unit in Hospital del Salvador, Santiago, between 18 October and 30 November 2019, found that 182 (70.5%) of 259 patients had been injured by kinetic impact projectiles, all of which were multiple projectile rounds (Rodríguez et al. 2020). Of the 182 patients whose injuries were caused by kinetic impact projectiles, 33 had total blindness in the affected eye, and 90 had severe visual impairment or were blind at first examination (Rodríguez et al. 2020, p. 3).

As reports of large numbers of eye injuries emerged, there was international condemnation and also intense domestic scrutiny as to the nature and use of the projectiles. For instance, a study carried out by the Faculty of Physical and Mathematical Sciences of the University of Chile revealed that the projectiles contained just 20% rubber, with the other 80% made up of silica, barium sulfate, and lead (Jorquera and Palma 2019). According to Viviana Meruane, Director of the Department of Mechanical Engineering, this composition provides added hardness and energy, 'which significantly increases the damage caused by the projectiles' (Mechanical Engineering Department of the Faculty of Physical and Mathematical Sciences of the University of Chile 2019).

In response, the Carabineros issued a public statement, stating that they had carried out their own study in the Carabinero's Crime Laboratory following the publication of the independent one (Rozas Córdova 2019). The results of this second study were not published, but the public statement indicated that they also showed that the composition of the projectiles differed from the manufacturer's datasheet. The statement announced that the use of this ammunition would be suspended for crowd control. Such a statement suggests that the Chilean authorities were not previously aware of the composition of the projectiles they were using, because they had not carried out their own tests, independent of the manufacturer. UN Guidance, however, requires states to carry out tests prior to procuring LLWs, and these testing should be 'based on impartial legal, technical, medical and scientific expertise and evidence' (United Nations 2020, para. 4.2.2).

It subsequently became apparent that at least some testing had, in fact, been carried out by Chilean authorities. A classified 2012 study commissioned by the *Carabineros* was leaked to the media, showed that authorities were aware that kinetic impact ammunition containing multiple projectiles was both inaccurate and risked causing serious injury, including ocular injuries (Carabineros de Chile 2012). The study recommended that its use be limited to distances greater than 30 m and that the point of aim should always be the lower half of the body. At the same time, the study demonstrated the inaccuracy of these projectiles, with just two of the 12 projectiles striking the target when fired from 30 m.

It is important to underline the fact that the inaccuracy of these projectiles does not preclude the possibility that they were also used unlawfully. The high number of ocular injuries during the Chilean protests suggests that the upper part of the body was repeatedly targeted, running counter to the recommendations made in the 2012 study. In such cases, kinetic impact ammunition was not being used to merely disperse protesters but also to inflict additional harm.

During the Yellow Vest (Gilets Jaunes) protests, French police used various types of kinetic impact ammunition, but the use of one particular kind, launched 40 mm projectiles, was the cause of widespread public anger. The *Lanceur de balles de défense* (LBD), also called the GL06 by the Swiss manufacturer B&T, is the 40-mm launcher most commonly used by French law enforcement. Although B&T markets a range of 40 mm ammunition (namely, SIR [safe impact round], SIR-X [safe impact round – extended range], dye marking, impact CS, impact OC, rubber shot, muzzle blast OC, ballistic smoke, ballistic CS, and muzzle blast CS), this was not adopted by the French Government when the launcher was introduced, and French law enforcement mainly use 40-mm ammunition manufactured by the French company SAE Alsetex (part of the Etienne Lacroix Group).

Unlike the shotgun ammunition used in Chile, the 40-mm ammunition used in France contains single projectiles. These 40-mm kinetic impact projectiles are designed to impact head on, with the foam nose absorbing the energy upon impact and thereby reducing injuries, but the prevalence of serious injuries presented may indicate that they do not always do that.

The types of ammunition in use are designed to be launched from different ranges, with the longer range ammunition containing more propellant. The French Ombudsperson (Défenseur des Droits) has called for the use of this weapon system to be prohibited during public gatherings, in part due to the difficulty of determining the range and assessing the risk of hitting bystanders, as people are usually grouped and mobile in such a context (Toubon 2019). As with other kinetic impact projectiles, there is an increased risk of serious injury or death when rounds are used from close range or impact sensitive parts of the body (e.g. head or upper body). In January 2019, the Ombudsperson reported that use of the LBD was the cause of or implicated in 18 of the 45 complaints he had received of 'extremely serious injury' or 'permanent mutilation'.

Stun grenades

Stun grenades, also known as 'flashbangs' and 'distraction' or 'disorientation' devices, are explosive devices that can be either hand thrown or weapon launched. Upon detonation, they emit an extremely loud noise and/or bright flash(es) of light. The flash causes temporary blindness for several seconds by activating photoreceptor cells in the eye (International Network of Civil Liberties Organizations (INCLO) and Physicians for Human Rights (PHR) 2016). The loud blast can cause temporary loss of hearing and loss of balance as well as a sense of panic (International Network of Civil Liberties Organizations (INCLO) and Physicians for Human Rights (PHR) 2016). Some stun grenades have ridges to limit their movement after detonation, while other types move around and have multiple explosions, causing loud blasts and bright flashes. The casing on some stun grenades has several circular cutouts, such as those shown in Figure 4.5, to allow the sound and light from the explosion through. Some types also disperse chemical irritants.

Some such grenades can produce high-velocity fragments when they explode, which can have sufficient energy to cause serious injury or death. Safer grenades split, rather than fragment. As the proximity of a person or group of people to the explosion increases, so too does the risk of serious injury or death. The UN Guidance states

Figure 4.5 Hand-thrown stun grenades. *Source:* ©Omega Research Foundation. November, 2017. Paris, France.

that 'the use of pyrotechnic flash-bang grenades directly against a person would be unlawful' (United Nations 2020, para. 6.1.4). The concussive blast of the detonation also carries a risk of causing injuries, including burns, hearing damage, eye injuries, and psychiatric trauma, and grenades can also start fires, particularly when used in enclosed spaces. The intensity of the noise and flash emitted varies from manufacturer to manufacturer. Some manufacturers market grenades with rubber casing as suitable for indoor use (see, for example Condor Tecnologias Nao Letais 2021). Others claim that their grenades emit less heat upon detonation, lowering the fire risk, or are designed not to move upon detonation (see Pacem Defense 2020).

The height at which stun grenades may explode at is also a factor in the injuries they may cause. The risk is lowered when grenades are rolled along the ground, rather than thrown or launched. This was tragically evident in the death of Remi Fraisse, who was killed during a protest against the construction of a dam in France in October 2014. A stun grenade thrown by the French Gendarmerie reportedly exploded near his back, severing his spinal cord and killing him instantly (Désarmons-les 2018).

Originally designed as a training aid to simulate explosions, stun grenades were initially adopted for use by military special forces units and later, by law enforcement special weapons and tactics (SWAT) teams, during room clearance or hostage situations. In some countries, stun grenades have become more widely used by law enforcement personnel for crowd-control purposes. When used in the context of a public gathering, the aim is to disorientate those present and cause them to disperse. Use in this context carries with it an increased risk of secondary injuries from falls, as the detonation of stun grenades risks causing panic or stampedes among large groups of people. States may need to reassess their use of stun grenades during public gatherings in light of the recent UN Human Rights Committee General Comment on the right of peaceful assembly, which states 'when [less lethal] weapons are used, all reasonable efforts should be made to limit risks, such as causing a stampede or harming bystanders' (United Nations Human Rights Committee 2020, para. 87).

Stun grenades were used in the law enforcement response to large protests that erupted against the contested results of the Belarusian presidential election in August 2020. The UN High Commissioner for Human Rights Michelle Bachelet denounced the 'heavy crackdown' this prompted, with police using 'unnecessary and excessive force against largely peaceful protesters' (United Nations Office of the High Commissioner of Human Rights 2020). Large numbers of stun grenades were reportedly used during the first days of the protests, leading to many serious injuries (Boika et al. 2020).

The Belarussian security forces used several types of stun grenades, including the Russian-manufactured SV-1319 Viyushka (AbraxasSpa 2020), which, upon detonation, emits a loud bang, a bright flash of light, and 750 rubber pellets with a diameter of 7.5 mm. According to company information, these grenades produce a sound not less than 130 dB measured at a distance of 10 m from the explosion (Rosoboronexport 2020).

In Belarus, stun grenades were reportedly used directly against crowds of people, exploding at thigh height or lower (Boika et al. 2020). Serious injuries reportedly included broken bones in the finger, foot, and calf. In one instance, a man's foot was reportedly ripped off when a stun grenade exploded near him (Boika et al. 2020). Many protesters were injured by fragments of exploded stun grenades (Karmanau 2020), including serious injuries such as pneumothorax (Boika et al. 2020), and some people also suffered eardrum damage (Mardilovic 2020).

Stun grenades were previously also used against large crowds of peaceful protesters in Yerevan, Armenia, on 29 July 2016. Police reportedly both launched and threw stun grenades into groups of peaceful protesters, with grenades exploding and emitting thick smoke and loud blasts, resulting in injuries to dozens of protesters and several journalists (Human Rights Watch 2016). Eleven people injured during the protests spoke to Civil Society Platform monitors; of those, seven had been injured by stun grenades, including one protester who suffered second-degree burns to 16% of her skin, including her face (International Partnership for Human Rights 2016, p. 11). The use of stun grenades led to injuries including first- and second-degree burns and fragmentation wounds, and one protester reportedly lost an eye after being hit by fragments (Human Rights Watch 2016).

Human Rights Watch reported on the impact of a stun grenade exploding near a man's feet shortly after he joined the protest (Human Rights Watch 2016):

> *He was briefly blinded by thick smoke and felt severe pain in his head. As he struggled to flee, he saw that his pant legs were almost entirely burned and his legs were covered in blood. [The individual] has 30 lacerations and first- and second-degree burns covering both legs. Doctors removed five plastic fragments from the stun grenade from his legs. He was not able to walk normally at the time of the interview.*

Other less-lethal weapons

There are a wide range of other LLWs and devices in use, although perhaps not as commonly observed as those covered above, all would merit thorough examination, but a lack of space prevents it here. Water cannons are becoming more prevalent globally, due in part to lower cost arising from manufacture in China and South Korea (see, for example Jino Motors 2021) and also from more modern designs (see, for example Albert Ziegler GmbH Co. KG 2021). Direct injury from the water jets is rare but includes loss of the eyes (Grimley 2014). Secondary effects include being knocked over, pushed off high ground, or objects being entrained in the water stream impacting the body. Dyes and/or tear gas are often added to the water jet. Acoustic and optical devices are also increasingly encountered. Acoustic devices were introduced for warning or as loud hailers; however, some have 'warning tones' at high frequency or sound pressure levels, which can lead to hearing loss (Amnesty International USA 2014, p. 14). Optical devices that interfere with an individuals' vision are designed to warn or to dazzle (B.E. Meyers & Co. Inc. (2021), range of GLARE products). They have the potential to cause irreversible sight loss.

Any use of firearms against peaceful protesters is unlawful under international human rights law and policing standards. The use of firearms to disperse a crowd is unlawful in all circumstances (United Nations Human Rights Committee 2020). Law enforcement officers in a number of states, including Israel (United Nations Commission of Inquiry on the 2018 protests in the Occupied Palestinian Territory 2019, para. 294) and India (United Nations Office of the High Commissioner for Human Rights 2019), assert that their use of firearms to launch bullets or pellets is a 'less-lethal' use of firearms. However, in these cases, safer alternatives are available which would not result in such devastating, life-long, and life-limiting punitive injuries, and so the claim that firearms are being used in a less-lethal manner is spurious and such use amounts to unlawful use of force.

Conclusion

The vast majority of public assemblies pass off peacefully, and many do not even require the presence of law enforcement officials. Where police or other law enforcement officials are present, their primary duty is to facilitate, respect, and protect the exercise of fundamental rights, enabling the assembly to take place as intended and minimising the potential for injury to any person or damage to property (United Nations Human Rights Committee 2020, paras. 74–76). Nonetheless, in many countries policing of assemblies is politicised, and there is also a tendency to label entire assemblies as violent and use LLWs to disperse them. In cases of non-peaceful assemblies where force against those who are violent may be justified, there is often no attempt to distinguish between individuals engaged in unlawful conduct and others exercising their right to peaceful assembly. The obligation to prioritise non-violent means of conflict resolution is frequently ignored in such circumstances.

There are various factors that contribute to law enforcement officials continuing to use excessive and arbitrary force to police assemblies, but one crucial issue is impunity. Impunity is, in turn, facilitated by a series of interrelated causes. These include unprofessional and permissive police cultures, weak use of force protocols, inadequate training, an absence of effective, independent oversight mechanisms, allegations of police abuse being investigated by the same agency, overly close relationships between police and prosecutors, and a lack of knowledge among judges and other legal professionals concerning international use of force standards and the technical characteristics of the weapons used by law enforcement agencies.

Reliable forensic evidence is of fundamental importance to addressing impunity and achieving accountability for the excessive or arbitrary use of force. As such, we encourage medical and forensic professionals to ensure they have up-to-date knowledge of the characteristics of LLWs used in the jurisdictions where they practice, the techniques used for their employment, and the risks associated with their use. Professional bodies representing forensic professionals should offer training and support on these matters, complementing other training on the implementation of the Istanbul and Minnesota Protocols.

Acknowledgement

Figure 4.1 (CC BY 2.0.) and Figure 4.2 (CC BY-SA 3.0) are provided by permission of Creative Commons licence. Figure 4.3 is reproduced by kind permission of the photographer, Doug Brown, ACLU of Oregon. Figures 4.4 and 4.5 are from the Omega Research Foundation.

References

AbraxasSpa (2020). Russian less-lethal grenades used by Alpha, SOBR and other units in Minsk. [*Twitter*] 12 August. https://twitter.com/AbraxasSpa/status/1293485090283159552 (accessed 24 November 2020).

Albert Ziegler GmbH Co. KG (2021). PSV 9000 https://www.ziegler.de/de/produkte/sonderfahrzeuge/Spezielle%20Fahrzeuge/psv (accessed 12 July 2021).

American Thoracic Society (2020). Tear gas use during COVID-19 pandemic irresponsible; Moratorium needed, says American Thoracic Society. https://www.thoracic.org/about/newsroom/press-releases/journal/2020/tear-gas-use-during-covid-19-pandemic-irresponsible-moratorium-needed,-says-american-thoracic-society.php (accessed 2 July 2021).

Amnesty International (2019). Iraq: Iranian tear gas grenades among those causing gruesome protester deaths. https://www.amnesty.org/en/latest/news/2019/10/iraq-gruesome-string-of-fatalities-as-new-tear-gas-grenades-pierce-protesters-skulls/ (accessed 5 July 2021).

Amnesty International (2020a). Belarus: mounting evidence of a campaign of widespread torture of peaceful protesters. www.amnesty.org/en/latest/news/2020/08/belarus-mounting-evidence-of-a-campaign-of-widespread-torture-of-peaceful-protesters/ (accessed 2 July 2021).

Amnesty International (2020b). Tear gas: an investigation. https://teargas.amnesty.org/ (accessed 5 July 2021).

Amnesty International (2020c). Tunisia: authorities must investigate excessive use of force in tataouine. https://www.amnesty.org/download/Documents/MDE3027472020ENGLISH.pdf (accessed 5 July 2021).

Amnesty International USA (2014). On the streets of America: Human Rights Abuses in Ferguson.https://www.amnestyusa.org/reports/on-the-streets-of-america-human-rights-abuses-in-ferguson/ (accessed 12 July 2021).

B.E. Meyers & Co. Inc. (2021). https://bemeyers.com/ (accessed 12 July 2021).

Barrett, B. (2020). The Pentagon's hand-me-downs helped militarize police. Here's how. Wired. https://www.wired.com/story/pentagon-hand-me-downs-militarize-police-1033-program/ (accessed 16 April 2021).

Benedek, W. (2020). OSCE Rapporteur's report under the Moscow Mechanism on Alleged Human Rights Violations related to the Presidential Elections of 9 August 2020 in Belarus. https://www.osce.org/files/f/documents/2/b/469539.pdf (accessed 2 July 2021).

Boika, A., Litavrin, M., Skovoroda, Y. et al. (2020). "Brutalised Minsk: how Belarusian police beat protesters", *Mediazona*, 3 November 2020. https://mediazona.by/article/2020/11/03/minsk-beaten-en (accessed 16 November 2020).

Breeden, A. (2020). France to stop using TNT-loaded tear gas grenades. New York Times. https://www.nytimes.com/2020/01/27/world/europe/france-tear-gas-grenades.html (accessed 5 July 2021).

Carabineros de Chile (2012). Disparos con escopeta antidisturbios, con empleo de cartuchería con perdigón de goma y sus efectos en la superficie del cuerpo humano. Dirección de Investigación Delictual y Drogas, Departamento de Criminalística. https://ciperchile.cl/wp-content/uploads/INFORME-CARABINEROS_compressed.pdf.

Carothers, T. and Press, B. (2020). The global rise of anti-lockdown protests—and what to do about it. World Politics Review. https://www.worldpoliticsreview.com/articles/29137/amid-the-covid-19-pandemic-protest-movements-challenge-lockdowns-worldwide (accessed 16 April 2021).

Case of Abdullah Yaşa and Others v. Turkey (2013). (Application no. 44827/08), JUDGEMENT, STRASBOURG, 16 July 2013, FINAL, 16/10/2013, European Court of Human Rights.

Case of izci v. Turkey (2013). (Application no. 42606/05), Judgement, Strasbourg 23 July 2013, European Court of Human Rights.

Case of Oya Ataman v. Turkey (2007). (Application no. 74552/01), JUDGMENT, STRASBOURG, 5 December 2006, FINAL 05/03/2007, European Court of Human Rights.

Centro de Estudios Legales y Sociales (2018). The Internal War: How the fight against drugs is militarizing Latin America. 1a ed. Ciudad Autónoma de Buenos Aires: Centro de Estudios Legales y Sociales CELS.

Condor Tecnologias Nao Letais (2021). GB-707 Granada luz e som – Indoor. https://www.condornaoletal.com.br/gb-707/ (accessed 9 July 2021).

Cordall, S.S. (2021). 'Things are getting worse': Tunisia protests rage on as latest victim named. The Guardian. https://www.theguardian.com/global-development/2021/jan/27/things-are-getting-worse-tunisia-protests-rage-on-as-latest-victim-named (accessed 5 July 2021).

Council of Europe (2018). Journalists detained and assaulted at inauguration protests. Platform to promote the protection of journalism and safety of journalists. https://www.coe.int/en/web/media-freedom/detail-alert?p_p_id=sojdashboard_WAR_coesojportlet&p_p_lifecycle=2&p_p_cacheability=cacheLevelPage&p_p_col_id=column-1&p_p_col_count=1&_sojdashboard_WAR_coesojportlet_alertPK=35676217&_sojdashboard_WAR_coesojportlet_cmd=get_pdf_one (accessed 2 July 2021).

Désarmons-les (2018). Rémi Fraisse, tué par les gendarmes à Sivens le 26 octobre 2014. https://desarmons.net/index.php/2018/01/09/retour-sur-la-mort-de-remi-fraisse-le-26-octobre-2014/ (accessed 24 November 2020).

Faculty of Forensic & Legal Medicine (2021). Irritant sprays: clinical effects and management, Recommendations for Healthcare Professionals (Forensic Physicians, Custody Nurses and Paramedics). https://fflm.ac.uk/wp-content/uploads/2021/02/Irritant-sprays-clinical-effects-and-management-Dr-J-McGorrigan-Prof-J-Payne-James-Jan-2021.pdf (accessed 2 July 2021).

Folha de Pernambuco (2017). Polícia conclui que Caso Itambé foi homicídio. https://www.folhape.com.br/noticias/policia-conclui-que-caso-itambe-foi-homicidio/29761/ (accessed 8 July 2021).

Gershkovich, E. (2020). Those guys with whips? They're cossacks meant to keep you safe at Russia's World Cup. Moscow Times. https://www.themoscowtimes.com/2018/05/28/those-guys-with-whips-theyre-cossacks-meant-to-keep-you-safe-at-russias-world-cup-a61599 (accessed 2 July 2021).

Greiner, A., Laviana, A., Stewart, A.W. et al. (2020). Open letter advocating for an anti-racist public health response to demonstrations against systemic injustice occurring during the COVID-19 pandemic. https://drive.google.com/file/d/1Jyfn4Wd2i6bRi12ePghMHtX3ys1b7K1A/view (accessed 2 July 2021).

Grimley, N. (2014). Blinded German man urges Boris Johnson not to bring water cannon to London. BBC News. https://www.bbc.co.uk/news/uk-26226926 (accessed 12 July 2021).

Hoz, S.S., Aljuboori, Z.S., Dolachee, A.A. et al. (2020). Fatal penetrating head injuries caused by projectile tear gas canisters. *World Neurosurg* 138: e119–e123. https://www.sciencedirect.com/science/article/abs/pii/S1878875020303259# (accessed 5 July 2021).

Human Rights Watch (2016). Armenia: excessive police force at protest. https://www.hrw.org/news/2016/08/01/armenia-excessive-police-force-protest (accessed 16 November 2020).

Human Rights Watch (2020a). Nigeria: crackdown on police brutality protests. https://www.hrw.org/news/2020/10/16/nigeria-crackdown-police-brutality-protests (accessed 16 April 2021).

Human Rights Watch (2020b). "Kettling" protesters in the bronx. Systemic police brutality and its costs in the United States. https://www.hrw.org/report/2020/09/30/kettling-protesters-bronx/systemic-police-brutality-and-its-costs-united-states (accessed 2 July 2021).

Institute for Economics & Peace (2020). COVID-19 and peace. http://visionofhumanity.org/reports (accessed 16 April 2021).

Instituto Nacional de Derechos Humanos (2020). Reporte general de datos sobre violaciones a los derechos humanos: Datos desde 17 de octubre de 2019 e ingresados hasta el 13 de marzo de 2020. https://www.indh.cl/bb/wp-content/uploads/2020/04/Reporte-INDH-19-de-marzo-de-2020.pdf (accessed 8 July 2021).

International Network of Civil Liberties Organizations (INCLO) and Physicians for Human Rights (PHR) (2016) Lethal in disguise: the health consequences of crowd-control weapons. https://www.inclo.net/issues/lethal-in-disguise/ (accessed 16 November 2020).

International Partnership for Human Rights (2016). Beaten, burnt and betrayed: Armenians awaiting accountability for police violence. https://iphronline.org/wp-content/uploads/2016/09/Beaten-Burned-and-Betrayed-Armenia-report-Sept-2016.pdf (accessed 16 November 2020).

Jino Motors (2021). Titan water cannon vehicles. http://www.jinomotors.com/sub/product/list.php?idx=1 (accessed 12 July 2021).

Jorquera, E.P. and Palma, H.R. (2019). Estudio de Perdigón. Faculty of Physical and Mathematical Sciences of the University of Chile. https://www.uchile.cl/noticias/159315/perdigones-usados-por-carabineros-contienen-solo-20-por-ciento-de-goma (accessed 9 July 2021).

Karmanau, Y. (2020). Police break up protests after Belarus presidential vote. AP. https://apnews.com/article/alexander-lukashenko-belarus-international-news-elections-riots-d310c4f9811b31e44fa5536bb826cc5e (accessed 24 November 2020).

Kiai, M. (2012). Report of the UN special rapporteur on the rights to freedom of peaceful assembly and of association. UN Doc. A/HRC/20/27.

Kiai, M. and Heyns, C. (2016). Joint report of the Special Rapporteur on the rights to freedom of peaceful assembly and of association and the Special Rapporteur on extrajudicial, summary or arbitrary executions on the proper management of assemblies. UN Doc. A/HRC/31/66.

Kraft, C.A. (2018). AAP Statement in response to tear gas being used against children at the U.S. southern border. American Academy of Paediatrics. https://www.aap.org/en/news-room/news-releases/aap/2018/aap-statement-in-response-to-tear-gas-being-used-against-children-at-the-us-southern-border/ (accessed 17 December 2018).

Mardilovic, A. (2020). Комбинированная травма. Минчанка, получившая осколочное ранение от светошумовой гранаты, стала подозреваемой по уголовному делу. Mediazona. https://mediazona.by/article/2020/09/08/kombinirovannaya-travma (accessed 24 November 2020).

Mechanical Engineering Department of the Faculty of Physical and Mathematical Sciences of the University of Chile (2019). Investigación U. de Chile comprueba que perdigones usados por Carabineros contienen solo 20 por ciento de goma. https://www.uchile.cl/noticias/159315/perdigones-usados-por-carabineros-contienen-solo-20-por-ciento-de-goma (accessed 9 July 2021).

Mijatović, D. (2019a). Human rights comment: shrinking space for freedom of peaceful assembly. Council of Europe. https://www.coe.int/en/web/commissioner/-/shrinking-space-for-freedom-of-peaceful-assembly?inheritRedirect=true (accessed 8 July 2021).

Mijatović, D. (2019b). Memorandum on maintaining public order and freedom of assembly in the context of the "yellow vest" movement in France. https://www.coe.int/en/web/commissioner/-/maintaining-public-order-and-freedom-of-assembly-in-the-context-of-the-yellow-vest-movement-recommendations-by-the-council-of-europe-commissioner-for (accessed 8 July 2021).

Northwest Center for Alternatives to Pesticides, Williamette Riverkeeper, Cascadia Wildlands et al. v. U.S. Department of Homeland Security, Chad Wolf, in his capacity as Acting Secretary, U.S. Department of Homeland Security (2020). Complaint for declaratory and

injunctive relief. https://aclu-or.org/sites/default/files/field_documents/nepa_federal_teargas_lawsuit.pdf (accessed 5 July 2021).
Nowell, C. (2020). Protesters say tear gas caused them to get their period multiple times in a month. Teen Vogue. https://www.teenvogue.com/story/protestors-say-tear-gas-caused-early-menstruation (accessed 15 April 2021).
O'Halloran, M. (1996). 6,000 plastic bullets fired since Drumcree stand-off. Irish Times. https://www.irishtimes.com/news/6-000-plastic-bullets-fired-since-drumcree-stand-off-1.67516 (accessed 8 July 2021).
Olewe, D. (2020). Coronavirus in Africa: whipping, shooting and snooping. BBC. https://www.bbc.co.uk/news/world-africa-52214740 (accessed 2 July 2021).
Omega Research Foundation (2020). Lowering the risk - curtailing the use of chemical irritants during the COVID-19 pandemic. https://omegaresearchfoundation.org/sites/default/files/uploads/Publications/Position%20Paper%20with%20logo.pdf (accessed 2 July 2021).
Omega Research Foundation (2021). Coronavirus: mapping cases of excessive force by law enforcement. https://omegaresearchfoundation.org/covid.php (accessed 31 August 2021).
Omega Research Foundation and Amnesty International (2017). Tackling the trade in tools of torture and execution technologies. https://omegaresearchfoundation.org/sites/default/files/uploads/Publications/Ending%20the%20torture%20trade%20WEB%20version.pdf (accessed 2 July 2021).
Omega Research Foundation and Amnesty International (2020). Ending the torture trade: the path to global controls on the 'tools of torture'. https://omegaresearchfoundation.org/sites/default/files/uploads/Publications/Ending%20the%20Torture%20Trade%20-%20The%20Path%20to%20Global%20Controls%20on%20the%20%27Tools%20of%20Torture%27.pdf (accessed 2 July 2021).
Omega Research Foundation and OSCE Office of Democratic Institutions and Human Rights (2021). Guide on law enforcement equipment most commonly used in the policing of assemblies. https://www.osce.org/odihr/491551 (accessed 5 July 2021).
Organisation for Security and Cooperation in Europe (2016). *Human Rights Handbook on Policing Assemblies*. Warsaw, Poland: Office for Democratic Institutions and Human Rights (ODIHR).
Organisation for the Prohibition of Chemical Weapons (2019). Report of the OPCW on the implementation of the convention on the prohibition of the development, production, stockpiling and use of chemical weapons and on their destruction in 2018. https://www.opcw.org/sites/default/files/documents/2019/12/c2404%28e%29.pdf (accessed 2 July 2021).
Pacem Defense (2020). GR-60. https://www.lesslethal.com/products/tld005-detail (accessed 13 November 2020).
PepperBall (2021). VXR Live-X. https://www.pepperball.com/products/vxr-live-x/ (accessed 2 July 2021).
Physicians for Human Rights (2012). Weaponizing tear gas: Bahrain's unprecedented use of toxic chemical agents against civilians. https://phr.org/our-work/resources/weaponizing-tear-gas/ (accessed 3 April 2021).
Planned Parenthood North Central States (2021). Tear gas and reproductive health study. https://www.plannedparenthood.org/planned-parenthood-north-central-states/about-ppncs/research/tear-gas-and-reproductive-health-study (accessed 15 April 2021).
Reddy, M. and Allison, S. (2020) Police use sjamboks and rubber bullets to enforce Hillbrow lockdown. Mail and Guardian. https://mg.co.za/article/2020-03-31-police-use-sjamboks-and-rubber-bullets-to-enforce-hillbrow-lockdown/ (accessed 2 July 2021).
Rodríguez, Á., Peña, S., Cavieres, I. et al. (2020). Ocular trauma by kinetic impact projectiles during civil unrest in Chile. Eye. https://www.nature.com/articles/s41433-020-01146-w?utm_source=feedburner&utm_medium=feed&utm_campaign=Feed%3A+eye%2Frss%2Fcurrent+%28Eye+-+Issue%29 (accessed 9 July 2021).

Rosoboronexport (2020). Viyushka: combined effect hand grenade. http://roe.ru/eng/catalog/special-weapons-and-ammunitions/non-lethal-grenades/vyushka/ (accessed 24 November 2020).

Rozas Córdova, M.A. (2019). Declaración Pública del Director General de Carabineros. https://twitter.com/Carabdechile/status/1196950642159476737/photo/1 (accessed 9 July 2021).

Samayoa, M. (2020). Lawsuit accuses Homeland Security of violating environmental law with Portland tear gas deployment. Oregon Public Broadcasting. https://www.opb.org/article/2020/10/21/lawsuit-accuses-homeland-security-of-violating-environmental-law-with-portland-tear-gas-deployment/ (accessed 5 July 2021).

Slisco, A. (2020). Tear gas may have led to abnormal menstrual cycles in Seattle and Portland. Newsweek. https://www.newsweek.com/tear-gas-may-have-led-abnormal-menstrual-cycles-seattle-portland-1529912 (accessed 15 April 2021).

Stunson, M. (2020). Protesters complain of unexpected side effect from tear gas: period changes. Miami Herald. https://www.miamiherald.com/news/nation-world/national/article244212707.html (accessed 15 April 2021).

SudOuest.fr (2020). Loi sécurité globale: un manifestant a eu les doigts arrachés par une grenade samedi à Paris. https://www.sudouest.fr/politique/gerald-darmanin/loi-securite-globale-un-manifestant-a-eu-les-doigts-arraches-par-une-grenade-samedi-a-paris-1642105.php (accessed 5 July 2021).

The Independent Commission on Policing in Northern Ireland (1999). A new beginning: policing in Northern Ireland.

The Wassenaar Arrangement (2020). The wassenaar arrangement on export controls for conventional arms and dual use goods and technologies, Munitions List. https://www.wassenaar.org/app/uploads/2020/12/Standalone-Munitions-List_2020.pdf (accessed 24 April 2021).

Toubon, J. (2019). Décision du Défenseur des droits n°2019-029. https://juridique.defenseurdesdroits.fr/doc_num.php?explnum_id=18403 (accessed 9 July 2021).

United Kingdom Home Office - Centre for Applied Science and Technology (2014). CAST standard for police chemical irritant sprays: CS and PAVA. https://assets.publishing.service.gov.uk/government/uploads/system/uploads/attachment_data/file/337910/standard-police-irritant-sprays-2314.pdf (accessed 17 April 2021).

United Nations (1980). Code of conduct for law enforcement officials. https://digitallibrary.un.org/record/10639?ln=en (accessed 3 April 2021).

United Nations (1990). Basic principles on the use of force and firearms by law enforcement officials. https://digitallibrary.un.org/record/93796?ln=en (accessed 3 April 2021).

United Nations (2004). Manual on the effective investigation and documentation of torture and other cruel, inhuman or degrading treatment or punishment. https://digitallibrary.un.org/record/535575?ln=en (accessed 8 July 2021).

United Nations (2017a). The minnesota protocol on the investigation of potentially unlawful death. https://www.ohchr.org/Documents/Publications/MinnesotaProtocol.pdf (accessed 4 April 2021).

United Nations (2017b). Resource book on the use of force and firearms in law enforcement. https://www.ohchr.org/Documents/ProfessionalInterest/UseOfForceAndFirearms.pdf (accessed 2 July 2021).

United Nations (2020). UN human rights guidance on less-lethal weapons and related equipment in law enforcement. https://www.ohchr.org/Documents/HRBodies/CCPR/LLW_Guidance.pdf (accessed 16 April 2021).

United Nations Commission of Inquiry on the 2018 protests in the Occupied Palestinian Territory (2019). Report of the detailed findings of the independent international Commission of inquiry on the protests in the Occupied Palestinian Territory. UN Doc. A/HRC/40/CRP.2. https://www.un.org/unispal/wp-content/uploads/2019/06/A.HRC_.40.CPR_.2.pdf (accessed 12 July 2021).

United Nations Committee against Torture (2017). Concluding observations on the second and third periodic reports of Bahrain. UN Doc. CAT/C/BHR/CO/2-3. https://digitallibrary.un.org/record/1306835?ln=en (accessed 4 April 2021).
United Nations Human Rights Committee (2020). General comment No. 37 on the right of peaceful assembly (article 21). UN doc. CCPR/C/GC/37. https://digitallibrary.un.org/record/3884725?ln=en (accessed 4 April 2021).
United Nations Office of the High Commissioner for Human Rights (2019). Update of the situation of human rights in Indian-administered Kashmir and Pakistan-administered Kashmir from May 2018 to April 2019. https://www.ohchr.org/Documents/Countries/IN/KashmirUpdateReport_8July2019.pdf (accessed 12 July 2021).
United Nations Office of the High Commissioner of Human Rights (2020). Bachelet condemns violent response of Belarus to post-electoral protests. https://www.ohchr.org/EN/NewsEvents/Pages/DisplayNews.aspx?NewsID=26162&LangID=E (accessed 16 November 2020).
US National Institute of Justice (2019). Criminal justice restraints standard NIJ standard 1001.00. Revison A. https://www.ojp.gov/pdffiles1/nij/253876.pdf (accessed 5 April 2022).
Weibel Barahona, M. and Jara, M. (2020). Carabineros revela que disparó 104 mil tiros de escopeta en las primeras dos semanas del estallido social. CIPER. https://ciperchile.cl/2020/08/18/carabineros-revela-que-disparo-104-mil-tiros-de-escopeta-en-las-primeras-dos-semanas-del-estallido-social/ (accessed 8 July 2021).

5 Non-fatal strangulation

J. Jason Payne-James
Norfolk & Norwich University Hospital, Norwich, UK
William Harvey Research Institute, Queen Mary University of London, London, UK
Forensic Healthcare Services Ltd., Southminster, UK

Introduction

A useful practical definition of strangulation is *'Strangulation is external pressure to the neck, by any means, that impedes airflow, blood flow, or both'* (California District Attorneys Association 2020). Such pressure results in reduced oxygen availability to the brain. This may result in unconsciousness, and if the pressure is extreme or prolonged or both can result in death. Strangulation in which death is the outcome (absent of any other factors) is 'fatal strangulation' (FS). Survivors of strangulation have experienced 'non-fatal strangulation' (NFS).

NFS is widely recognised in intimate-partner violence settings. In recent years it has become recognised that it may well be a precursor to fatal assaults. The seriousness of such assaults has resulted in a substantial body of opinion that NFS should be recognised as a separate and specific criminal offence, separate from other forms of assault. In a number of jurisdictions, legislation and prosecuting authorities have already recognised this and have introduced protocols to ensure that the more appropriate charging standards (and thus sentences) apply.

This chapter explores the findings after NFS and the documentation, management, and further investigation of those subject to such assaults. It will also provide some of the background behind the recognition of the implications of such assaults.

Non-fatal strangulation and intimate-partner violence

Studies have shown that women are vulnerable to both intimate-partner violence and homicide from both non-marital and former partners as well as from current husbands (Moracco et al. 2003; John Hopkins School of Nursing 2022). Glass et al. (2008) examined NFS by an intimate partner as a risk factor for major assault and attempted or completed homicide for women. A case–control design was used to compare NFS among complete homicides and attempted homicides (n =506) and an otherwise abused control group (n =427). NFS was reported in 10% of abused

Current Practice in Forensic Medicine, Volume 3, First Edition. Edited by John A.M. Gall and J. Jason Payne-James.

controls, 45% of attempted homicides, and 43% of homicides. Prior NFS was associated with greater than sixfold odds (odds ratio [OR] 6.70 and 95% confidence interval [CI] 3.91–11.49) of becoming an attempted homicide and over sevenfold odds (OR 7.48 and 95% CI 4.53–12.35) of becoming a completed homicide, confirming NFS as an important risk factor for homicide of women. They concluded that it underscored the need to screen for NFS when assessing abused women in the emergency department settings.

Monckton-Smith (2020a), in a summary of a survey by Stand Up to Domestic Abuse, identified three reasons why NFS should be considered a separate offence in England and Wales. Firstly, intimate-partner homicide (IPH) has a strong relationship to domestic abuse and coercive control, and there are certain 'high-risk markers' in such abuse that are strongly associated with future homicide and serious harm (Monckton-Smith 2020b). NFS is one of the strongest markers because people who use strangulation are more dangerous, not just because of the risk of 'accidental' fatality. Secondly, it is a traumatic and effective way to exert the ultimate control and leave the victim in no doubt that their life has been threatened. NFS should not be considered a spontaneous and angry assault, but it is more likely to be a controlled and determined threat. Some jurisdictions (e.g. within the United States) now require NFS to be charged as a felony (indictable) offence and not a misdemeanour (summary) offence to emphasise the dangerousness of the perpetrator and the severe trauma and injury NFS causes. The California District Attorneys' Association published a document on 'The Investigation and Prosecution of Strangulation Cases' in 2020 in which it is stated:

As prosecutors, we have a responsibility to do something about it. We cannot continue to hear the words "he choked me" and treat these types of assaults like we would a slap or a punch. The difference between life and death in most strangulation assaults is only a matter of seconds. Prosecutors have a unique opportunity to stop most stranglers before they kill, but we must seize that opportunity in order to prevent homicides.

We must learn to more effectively investigate and prosecute near and non-fatal strangulation assaults as felony offenses even with little or no external visible injury. We must pursue these complicated cases even without victim participation or testimony. We must work in Family Justice Centers (FJCs) and with multi-disciplinary teams (MDTs) to effectively hold offenders accountable and provide victims with the medical advocacy and support needed by survivors. Every time we hold a strangler accountable, we reduce the likelihood of a homicide, and we send a message to stranglers: We see you and will not let you commit life-threatening and often brain-damaging assaults with impunity.

Legal status of non-fatal strangulation

Strack et al. (2001) described 300 strangulation cases reviewed by the San Diego City Attorney's Office. The predominance of cases were by manual strangulation. The means of strangulation were multiple and included one hand, two hands, chokeholds, being strangled from the front, from the rear, pinned against the wall, or while being straddled on the floor or bed. The importance of a forensic medical assessment by a trained forensic clinician is clear as the study showed that the nature of training of police may have led them to overlook symptoms of strangulation (including pain, difficulty in breathing, and difficulty in swallowing) described by victims and also by not recognising that absence of symptoms (67%) and signs (50%) did *not* exclude strangulation. Although this study was from the United States, these concerns will apply globally. The lack of recognition that 'absence of evidence' is not 'evidence of absence' means that opportunities for higher level (and appropriate) criminal prosecutions were missed. The authors emphasised that though this study did not look at fatality, victims experienced a serious and potentially lethal form of physical violence.

In England and Wales, concerns centre on the challenges to prosecute NFS under existing offences such as actual bodily harm (ABH) because injuries may not be visible, and substantial political and professional pressure, which this author strongly supported, has now made NFS a separate offence of NFS under the new Domestic Abuse Act 2021 with a potential prison sentence of 5 years (The Guardian 2021). The definitions used are broad – '*Strangulation or suffocation. . .[if a person] intentionally strangles another person. . .or. . . does any other act . . .that. . .affects [the other person's' ability to breathe*'. These definitions will thus include the application of a ligature and means of externally occluding air passages.

The details of the specific legislation in England and Wales are shown in Table 5.1.

Non-fatal strangulation and assault

Although there is now much greater awareness of NFS, its importance has been recognised for decades. In a review in 1984, Iserson (1984) considered the nature of injuries after ligature, manual, and postural neck compression. Although most of the review focussed on fatal outcome, Iserson emphasised that in the person who survived, injuries may be different to those who died and may require periods of treatment and observation in hospital. Iserson concluded that '*in the ambulatory patient arriving with the complaint of near-strangulation, but with few, if any, signs or symptoms, care must be taken to evaluate the possibility that no strangulation had taken place*'.

In the forensic setting and in court, the issue of the severity of strangulation may become a key element in the case. This is a difficult area, as fatality may occur after strangulation in the absence of visible injury. Plattner et al. (2005) suggested a classification with four categories embracing signs and symptoms (Table 5.2). This classification relates to the clinical examination findings and is a useful model if one accepts that it does not apply to all cases of NFS. Plattner et al., however, suggest that Category 1 and 2 represent non-life-threatening (light and moderate) strangulation and that Category 3 and 4 are life-threatening. This classification may be inapplicable for some cases as all those experienced in the assessment of such cases recognise that any pressure to the neck has the potential to be life-threatening without any external evidence of injury or major symptomatology, and sometimes minor visible lesions may be the only indicator of severe force pressure.

As the significance of NFS is better recognised, a number of studies have noted the incidence or presence of evidence of NFS.

McQuown et al. (2016) reviewed the prevalence of strangulation in patients surviving sexual assault (SA) and domestic violence (DV), identified the presence of lethality risk factors in intimate-partner violence, and assessed differences in strangulation

Table 5.1 Extract from the England and Wales Domestic Abuse Act 2021.

70 Strangulation or suffocation

(1) In Part 5 of the Serious Crime Act 2015 (protection of children and others), after section 75 insert—

'Strangulation or suffocation

75A Strangulation or suffocation

(1) A person ('A') commits an offence if—

(a) A intentionally strangles another person ('B'), or

(b) A does any other act to B that—

(i) affects B's ability to breathe and

(ii) constitutes battery of B.

(2) It is a defence to an offence under this section for A to show that B consented to the strangulation or other act.

(3) But subsection (2) does not apply if—

(a) B suffers serious harm as a result of the strangulation or other act, and

(b) A either—

(i) intended to cause B serious harm or

(ii) was reckless as to whether B would suffer serious harm.

(4) A is to be taken to have shown the fact mentioned in subsection (2) if—

(a) sufficient evidence of the fact is adduced to raise an issue with respect to it, and

(b) the contrary is not proved beyond reasonable doubt.

(5) A person guilty of an offence under this section is liable—

(a) on summary conviction—

(i) to imprisonment for a term not exceeding 12 months (or 6 months, if the offence was committed before the coming into force of paragraph 24(2) of Schedule 22 to the Sentencing Act 2020), or

(ii) to a fine, or both;

(b) on conviction on indictment, to imprisonment for a term not exceeding 5 years or to a fine, or both.

between SA and DV populations; 351 of the overall study group reported strangulation in 23% ($n = 1542$) during their assaults. The prevalence of strangulation was 38% with DV and 12% with SA. Most of the intimate-partner encounters with strangulation had significant risk for lethality, and they found that those patients presenting who were survivors of DV were more likely than SA patients to sustain strangulation.

Zilkens et al. (2017) reviewed 1163 women attending a Sexual Assault Referral Centre (SARC) in Western Australia over a 6-year period from 2009 to 2015. They noted an increased injury severity in women assaulted by intimate partners and had previously noted (Zilkens et al. 2016) that women sexually assaulted by their intimate partner, particularly those aged in their 30s, have a very high risk of NFS. Of the 1163 subjects in the 2017 study, 15 (1.3%) had physical signs of NFS and a history of NFS but were not referred to hospital. These were classified as moderate injuries.

Table 5.2 A classification of NFS presentation according to signs and symptoms.

Category	Signs or symptoms	Examples	Further notes
1	Superficial skin lesions	Hyperaemia, abrasions, and intracutaneous haemorrhage	These may be caused by both assailant and victim or both. The terms abrasions should be expanded to include grazes and fingernail scratches or fingertip bruising. Ligature marks may be evident
2	Subjective symptoms indicative of pharyngeal and laryngeal haematomas or swelling	Painful palpation, hoarseness, and sore throat	Symptoms may also include painful neck and difficulty in swallowing. Tongue biting or bruising may be present
3	Distinctive petechiae	Petechial haemorrhages on conjunctivae, mucosal surfaces, and facial skin	Petechiae should be specifically sought for on all skin and mucosal surfaces (including in the ears, nose, mouth, eyes, and the scalp). Petechiae may rapidly coalesce and present as larger haemorrhages or bruises
4	Subjective symptoms indicative of cerebral hypoxia	Loss of consciousness or loss of urine	Memory loss may also be apparent

Source: Modified from Plattner et al. (2005).

Thirteen (1%) with severe NFS findings required referral to hospital (including neck/throat pain, pain and/or difficult in swallowing, vocal changes, shortness of breath, linear neck abrasions, non-petechial bruising neck, and subconjunctival haemorrhages). In the 2016 study, the authors found 79 (7.4%; $n = 1064$) alleged NFS during the SA. The prevalence of NFS varied significantly by age group and assailant type. Of women assaulted by an intimate partner, 22.5% gave a history of NFS compared to less than 6% of women assaulted by other assailant types. Of all SAs with NFS, intimate partners were the assailant in 58.2% of cases, whereas in SA cases without NFS, intimate partners were the assailant in 15.9% of cases.

A retrospective review of 158 forensic assessments of women subject to non-fatal violence from their husbands between 2010 and 2015 showed that 19 (8%) sustained bruises or abrasions to their neck, although the authors observed that it was unclear if these occurred in the setting of attempted strangulation or if they were sustained randomly as part of a general assault. This emphasises the need to ensure that any forensic medical assessment accurately documents the complainant's account of how injuries were sustained (Abedr-Rahman et al. 2017).

In Taiwan, of the 220 adult femicide victims recruited, 114 were killed by intimate partners and 106 were killed by non-intimate-partner offenders (Fong et al. 2016).

The most common site of injuries in the intimate-partner group and the non-intimate-partner group was the neck and the upper limbs, respectively. The most common cause of death after sharp force injury in both groups was strangulation (27.2% and 22.6%, respectively).

Strangulation is well represented as a cause of violent death in women. Unal et al. (2016) in a series of autopsy studies from Istanbul established that strangulation represented 8.4% of deaths and that the majority were perpetrated by spouse or boyfriend in the home. Data from Portugal showed that 4.8% of fatal intimate-partner violence against women in Portugal was by manual strangulation (Pereira et al. 2012).

Cannon et al. (2020) retrospectively reviewed prevalence, trends, and characteristics of sexual assault cases involving NFS from one academic hospital in the Midwestern United States from 2002 to 2017. Of 856 cases. In total, 5.1% of cases involved NFS. NFS was more common when current/former partners (18.9%) or strangers were assailants (16.6%), compared to family members (11.1%) or friends/acquaintances.

A meta-analysis of the risk factors for IPH showed that IPH is most likely (when compared with other homicides) to occur in the victim's home or shared resided and with a white weapon or sharp object or strangulation (Matias et al. 2020).

Symptoms and signs of non-fatal strangulation (acute and longer term)

Direct pressure to the neck can cause visible damage to the skin surface and also directly and indirectly to underlying structures which may present immediately or sometime later. In general, the greater the surface area of the object applying the pressure, the less likelihood of focal external, visible injuries such as bruises at the site of the applied pressure. This may not be the case if other objects or materials (e.g. clothing, watches, and bracelets) are present between the neck and the object applying the force.

There are many ways of applying pressure with hands or limbs to the neck, and the California District Attorneys Guidance for investigators provides a useful diagram to try and assist victims in describing the method by which force was applied. The diagram is reproduced in Figure 5.1.

This diagram only addresses manual and upper limb techniques. Ligature application is another common means of causing NFS.

Physical injury (or its absence) can be documented at the time of assessment. Bichard et al. (2021) undertook a systematic review reviewing pathological, neurological, cognitive, psychological, and behavioural outcomes of NFS in domestic and sexual violence. They identified 30 empirical, peer-reviewed studies which met the inclusion criteria. Pathological changes included arterial dissection and stroke. Neurological consequences included loss of consciousness, indicating at least mild acquired brain injury, seizures, motor and speech disorders, and paralysis. Psychological outcomes included post-traumatic stress disorder (PTSD), depression, suicidality, and dissociation. However, they also determined that cognitive and behavioural sequelae, which included memory loss, increased aggression, compliance, and lack of help-seeking, were described less frequently. This may be because no studies used formal neuropsychological assessment, and

Supplemental report for strangulation assaults

Which of the below hold(s) best describe how you were strangled
(Have the victim circle and initial the type of strangulation hold used)

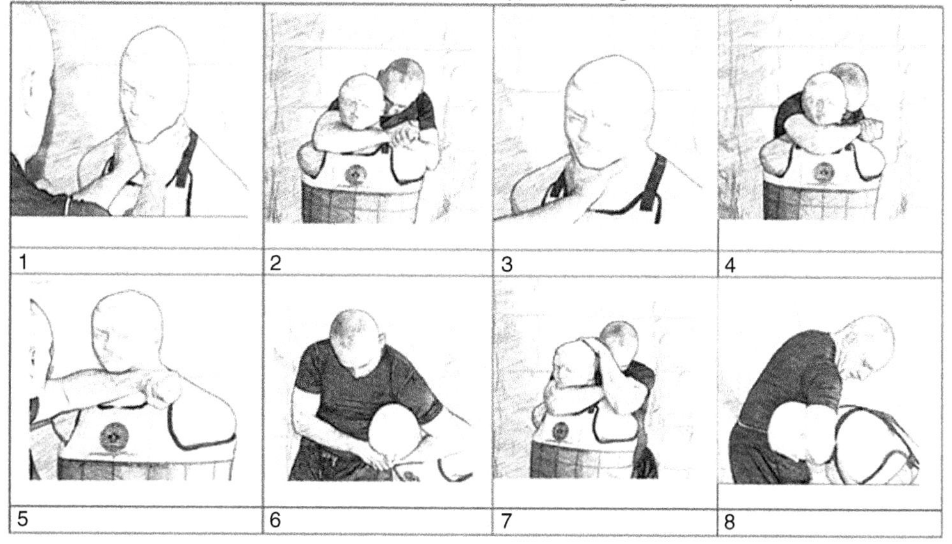

Figure 5.1 Supplemental report for strangulation assaults. *Source:* Photos by Det. Alex Smith, Los Angeles County Sheriff Department.

this emphasises the need for further research, in particular, to determine what the neuropsychological implications are and how to identify those who will need subsequent neuropsychological evaluation.

White et al. (2021) undertook a study to identify the prevalence of NFS in patients presenting to the SARC, for an acute forensic medical examination (FME) after a report of rape or SA, and explore the characteristics of patients reporting NFS compared to those who did not. Additionally, the study explored the prevalence of various symptoms and signs associated with NFS.

A retrospective analysis of 2196 adults (≥18 years old) who attended between 1 January 2017 and 31 December 2019 was undertaken. Of those, 204 (9.28%) were NFS cases. The study found that the prevalence of NFS was 18.9% where the alleged perpetrator was a partner or ex-partner. For NFS cases, 96.6% (n = 197) of the patients were female, and the alleged perpetrator was male in 98% (n = 200) of the NFS cases; 40% of the NFS cases had been strangled in their own homes. In 27% (n = 55) of the NFS cases, the patient said that the alleged perpetrator had also strangled them on a previous occasion; 46.6% had an injury to the neck or above attributable to the NFS, 15.7% (n = 32) of the NFS cases reported loss of consciousness, 8.8% (n = 18) were incontinent of urine, and 2% (n = 4) incontinent of faeces as a result of the NFS. Over one-third of the patients (36.6%) thought that they were going to die during the NFS. The authors concluded that NFS in SA is a gendered crime, with most victims female, and most assailants male. Visible NFS injuries are not the norm, yet fear of death is not uncommon. Over 1 in 6 (15.7%) reported loss of consciousness, suggesting that they were victims of a near lethal assault.

Less commonly, life-changing injury may be caused by NFS. Rarer complications of NFS may (i) result in failure to recognise that the complication is related to NFS or (ii) present late (Anscombe and Knight 1996). Examples include adult respiratory distress syndrome after attempted strangulation which has been described (Murphy et al. 1993). The patient had presented with cerebral irritability with multiple sites of bruising and petechial haemorrhage to head, face, and neck and bruises and linear scratches to the neck consistent with manual strangulation. She developed non-cardiogenic pulmonary oedema and adult respiratory distress syndrome which was prolonged.

Clarot et al. (2005) described two cases of carotid dissection, both not initially recognised as being due to strangulation, but both of which resulted in devastating neurological sequelae. The incidence of strangulation related to carotid artery dissection remains unknown and may be underreported, in part because of underreporting of NFS and because it is also recognised that spontaneous carotid artery dissection can occur. The authors emphasise that sudden neurological deficit in young adults without pathological history should always include a carefully detailed history of previous violence, strangulation or even if the strangulation is only a single attempt, and that NFS should be considered in the differential diagnosis of stroke in at-risk patients. Although not describing cases of NFS, Blanco Pampin et al. (2002) emphasised that even apparently trivial neck trauma can give rise to this complication.

Table 5.3 shows signs and symptoms and their presence in some published studies related to NFS. It will be noted that there is a range of terminology used to describe assorted signs and symptoms, and where possible these has been clarified in the table. Sommers et al. (2012) suggested that the community of scientists concerned about Intimate Partner Violence and sexual violence develop a more rigorous injury classification system to improve the quality of forensic evidence proffered, and decisions made throughout the criminal justice system. This would be helpful, but because of the range and variety of professionals undertaking such examinations (e.g. health professionals, scientists, legal, criminological, and sociological) this is unlikely to be achievable, in particular when linguistic challenges often cloud definitions and translations. There is a tendency to complicate simple ideas and introduce technical terms (particularly from a medical point of view) that do not advance understanding. It is, however, appropriate that in published studies researchers clearly outline the definitions or interpretations they are using or referring to.

A comparison of signs and symptoms that published guidance documents on NFS advise might be observed is shown in Table 5.4. There is some variation between the guidance, but there is considerable overlap, reflecting the range of effects of NFS rather than inadequacies in guidance and reinforces the requirement for detailed history taking and physical examination.

Whatever description is used, this should be supplemented by appropriate standard forensic photography. However, the use of photographs for assessment and interpretation of injury without both colour scales and rules (i.e. determine nature, size, colour, or relative colour) is inappropriate (Payne-James et al. 2012), and extreme caution should be exercised in making any definitive interpretation in the absence of appropriate medical assessment combined with appropriate quality of photography (Gall and Payne-James 2011). Certain standards are advised for photographic documentation of injuries in order to minimise the risks of misinterpretation (Evans et al. 2014; Faculty of Forensic & Legal Medicine 2020).

Table 5.3 Examples of studies recording signs and symptoms in NFS cases and noting differences in terminology used.

Study		Strack et al. (2001) ($n = 300$)	Plattner et al. (2005) ($n = 134$)	Shields et al. (2010) ($n = 102$)	Zilkens et al. (2016) ($n = 79$)	McQuown et al. (2016) ($n = 351$)
Symptoms						
	Neck/throat pain				46.8	
	Neck tender on palpation				34.2	
	Pain and/or difficulty in swallowing				19	3
	Pain with swallowing					31
	Difficulty with swallowing			25		27
	Vocal changes				15.2	
	Hoarseness			21		
	Difficult speaking					
	Shortness of breath		20	14	8.9	47
	Loss of consciousness		11	38	8.9	
	Difficult breathing			39		
	Felt dizzy/faint		10	13	8.9	
	Blurred vision				2.5	
	Urinary incontinence		3	6	1.3	
	Faecal incontinence			3		

(*Continued*)

Table 5.3 (Continued)

Study		Strack et al. (2001) ($n = 300$)	Plattner et al. (2005) ($n = 134$)	Shields et al. (2010) ($n = 102$)	Zilkens et al. (2016) ($n = 79$)	McQuown et al. (2016) ($n = 351$)
	Involuntary defaecation/ urination					8
	Pain on talking				1.3	
	Loss of memory			Undetermined		10
	Loss of voice or voice change					
	Persistent throat pain					
At least one symptom					67.1	
No symptoms			5.22			
Signs	Linear abrasions – neck				31.7	
	Abrasions			40		
	Petechiae – site not specified		21	15		
	Petechial bruising – upper neck/facial				21.5	
	Non-petechial bruising – neck				17.7	

Study		Strack et al. (2001) ($n = 300$)	Plattner et al. (2005) ($n = 134$)	Shields et al. (2010) ($n = 102$)	Zilkens et al. (2016) ($n = 79$)	McQuown et al. (2016) ($n = 351$)
	Bruising			54		
	Subconjunctival haemorrhage				3.8	
	Petechiae in conjunctivae				2.5	
	Soft tissue swelling – neck			18	1.3	
	Injury to neck					57
	Redness/discoloration			26		
At least one sign					50.6	
No signs			5.22	13		

Table 5.4 Examples of signs and symptoms that need to be sought and documented from some NFS assessment guidance (where terminology differs, the best equivalent has been used).

Guideline	Sign or symptom (these may be observed at the time of the assault or subsequently)	Royal College of Pathologists of Australasia: Clinical Forensic Assessment and Management of NFS – January 2018	Faculty of Forensic and Legal Medicine of the Royal College of Physician: Non-fatal Strangulation: in physical and sexual assault – March 2020	California District Attorneys Association (CDAA) and the Training Institute on Strangulation Prevention Investigation and Prosecution of Strangulation Cases – 2020
Symptoms				This list is taken from the table which provides a summary of the locations on the body where investigators may find signs of strangulation and/or suffocation, and the table below provides a summary of what symptoms to look for when trying to identify internal injury symptoms on a victim who has reported being strangled or who is believed to have been strangled.
Direct	Neck pain	Y		Y
	Dysphagia	Y	Y	Y
	Cough	Y		Y
	Hoarseness/hoarse voice/'hot potato voice'	Y	Y	Y
	Loss of consciousness	Y	Y	
	Unable to speak			Y
	Dribbling/drooling		Y	Y
	Vomiting/nausea		Y	Y

Guideline	Sign or symptom (these may be observed at the time of the assault or subsequently)	Royal College of Pathologists of Australasia: Clinical Forensic Assessment and Management of NFS – January 2018	Faculty of Forensic and Legal Medicine of the Royal College of Physician: Non-fatal Strangulation: in physical and sexual assault – March 2020	California District Attorneys Association (CDAA) and the Training Institute on Strangulation Prevention Investigation and Prosecution of Strangulation Cases – 2020
Indirect	Urinary incontinence	Y	Y	Y
	Faecal incontinence	Y	Y	Y
	Dizziness/ lightheaded	Y	Y	Y
	Headache	Y	Y	Y
	Memory loss	Y	Y	Y
	Auditory change	Y	Y	
	Visual change	Y	Y	
	Behavioural change			Y
	Agitation/ combativeness			Y
	Seizure	Y	Y	
Respiratory	Difficulty in breathing	Y	Y	Y
	Shortness of breath	Y	Y	
	Hyperventilation		Y	Y
	Stridor		Y	
	Cannot breathe			Y

(*Continued*)

Table 5.4 (Continued)

Guideline	Sign or symptom (these may be observed at the time of the assault or subsequently)	Royal College of Pathologists of Australasia: Clinical Forensic Assessment and Management of NFS – January 2018	Faculty of Forensic and Legal Medicine of the Royal College of Physician: Non-fatal Strangulation: in physical and sexual assault – March 2020	California District Attorneys Association (CDAA) and the Training Institute on Strangulation Prevention Investigation and Prosecution of Strangulation Cases – 2020
Signs				
	Subconjunctival haemorrhage	Y		Y
	Petechiae in skin above the level of compression/ application of force	Y	Y	Y
	Petechiae in mucosa above the level of compression/ application of force	Y	Y	
	Swelling of face	Y		
	Swelling of tongue	Y		
	Swelling of lips			Y
	Mouth and tongue mucosal trauma		Y	

Guideline	Sign or symptom (these may be observed at the time of the assault or subsequently)	Royal College of Pathologists of Australasia: Clinical Forensic Assessment and Management of NFS – January 2018	Faculty of Forensic and Legal Medicine of the Royal College of Physician: Non-fatal Strangulation: in physical and sexual assault – March 2020	California District Attorneys Association (CDAA) and the Training Institute on Strangulation Prevention Investigation and Prosecution of Strangulation Cases – 2020
	Reddening (hyperaemia) to the neck	Y	Y	Y
	Reddening to the face			Y
	Bruises	Y	Y	
	Abrasions (grazes)	Y	Y	
	Scratches (fingernails of assailant or victim)	Y	Y	Y
	Ligature marks	Y		Y
	Vaginal bleeding		Y	
	Epistaxis		Y	Y
	Fractured nose			Y
	Bleeding from external auditory meatus			Y
	Reduced Glasgow coma score		Y	
	Subcutaneous (surgical) emphysema		Y	

(*Continued*)

Table 5.4 (Continued)

Guideline	Sign or symptom (these may be observed at the time of the assault or subsequently)	Royal College of Pathologists of Australasia: Clinical Forensic Assessment and Management of NFS – January 2018	Faculty of Forensic and Legal Medicine of the Royal College of Physician: Non-fatal Strangulation: in physical and sexual assault – March 2020	California District Attorneys Association (CDAA) and the Training Institute on Strangulation Prevention Investigation and Prosecution of Strangulation Cases – 2020
	Carotid bruits		Y	
	Hyoid bone, thyroid, and laryngeal cartilage injury		Y	
Red flags (necessitating referral to ED or ENT specialist)				
	Reduces Glasgow coma score		Y	
	Use of ligature	Y		
	Loss of consciousness	Y	Y	
	Loss of control of bowel or bladder	Y	Y	
	Difficulty or pain in swallowing	Y	Y	
	Voice change	Y	In	
	Difficulty breathing	Y	Y	
	Neck pain	Y		

Examples of findings and descriptions of NFS assaults

Images below give examples of features of NFS and the accounts given by the victims at the time, but such features are very variable and their appearance changes rapidly which is why early and detailed assessment and documentation is essential. The complainants will be referred to as victims, as in each case the events described were corroborated by others or the accounts and findings were accepted in court. The extreme range of descriptions of what the victims experience is notable. The importance of forensic quality images is reflected in some of these images where details may be unclear and unambiguous due to, for example poor lighting, focus, positioning, and failure to observe basic forensic photographic rules. The visual evidence may be very subtle. Often, complainants and witnesses (including law enforcement personnel) take images with smart phones or other devices. Sometimes, they may be adequate, but most often they are not if basic forensic photography principles are not applied.

The first example (Figure 5.2a,b) shows separate areas of bruising to the right side of the neck. Some might describe these as intradermal (although realistically that cannot be established on clinical examination). These may be consistent with finger marks. The smaller areas may represent petechiae or coalesced petechiae. The description given by the 33-year-old female victim in the interview was (interviewers' questions in square parentheses):

> *[And, erm, did, did he strangle you with one hand, or two?] I, I don't remember. I think it was one. I was sat down on the sofa and that's when he strangled me. [Okay, which I know that we've photographed. How did it feel when he was strangling you?] [Did you, at any point, struggle to breathe?] A bit. Not, not a lot.*

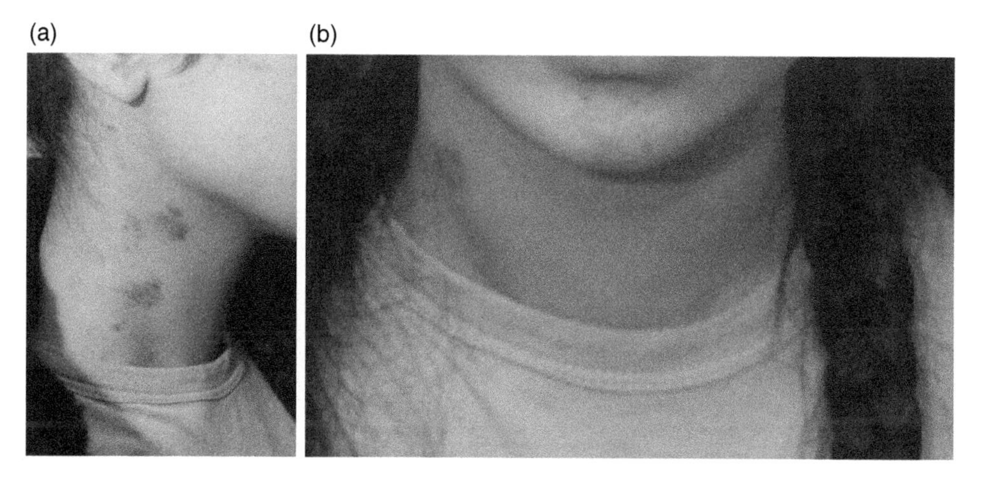

Figure 5.2 Separate areas of bruising to the right side of the neck. These may be consistent with finger marks.

The second example (Figure 5.3a,b) shows extensive bilateral bruising to the neck extending on the right to the submandibular region. The assault had occurred three days earlier. The description given by the 24-year-old female victim was:

'he put his hand over my throat again with his right hand, and then with his left hand he put his hand over my mouth and was pushing down really hard and I couldn't breathe and I literally thought I was gonna die. . .just went fuzzy and like noises were ringing. . ..second time was quite a long while, I couldn't tell you how long. . .if I didn't somehow find the strength to like quickly push him off my face so I can just get that little bit of air. . .had a bit of a sore throat but the doctor said it's fine she said it was swelling from the bruising'.

Example 3 shows an illustration of poor-quality imaging (Figure 5.4a,b) but which identifies petechial haemorrhage on the scalp skin behind the right ear (Figure 5.4a) and possible bruising below the ear lobe near the angle of the mandible (Figure 5.4b) which were the only visible signs of the choke hold. The image was taken 4 hours after the assault. This example reflects the need to ensure that appropriate imaging is undertaken and the need to ensure that all areas of skin and mucosa are examined above the level of compression. The description given by the victim and witness who found the victim was:

'He grabbed her around the throat in a choke hold whilst he was behind her. . .She could not recall if she passed out and only remember waking up in hospital. . .disclosed to officers that she thought X had raped her. . ..She remembers going to woods and not much else. . .only remembers waking up in hospital and spoke to police. . ..she had pain on the left side of her neck. . ..said she couldn't remember anything. . ..asked if I could take some photos of her neck as there appeared to be bruising and tiny red dots behind her right ear. . .took a series of photos on my work mobile phone. . .photos . . .weren't as clear as my view of

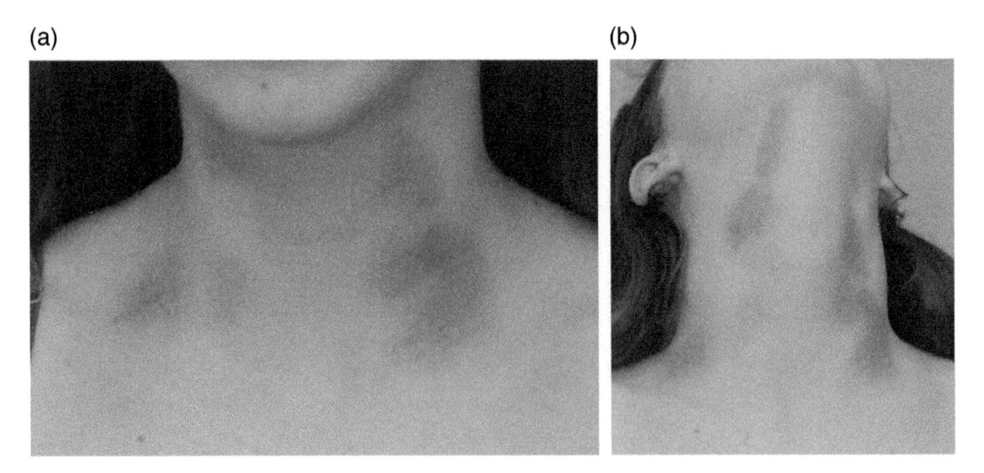

Figure 5.3 Extensive bilateral bruising to the neck extending on the right to the submandibular region.

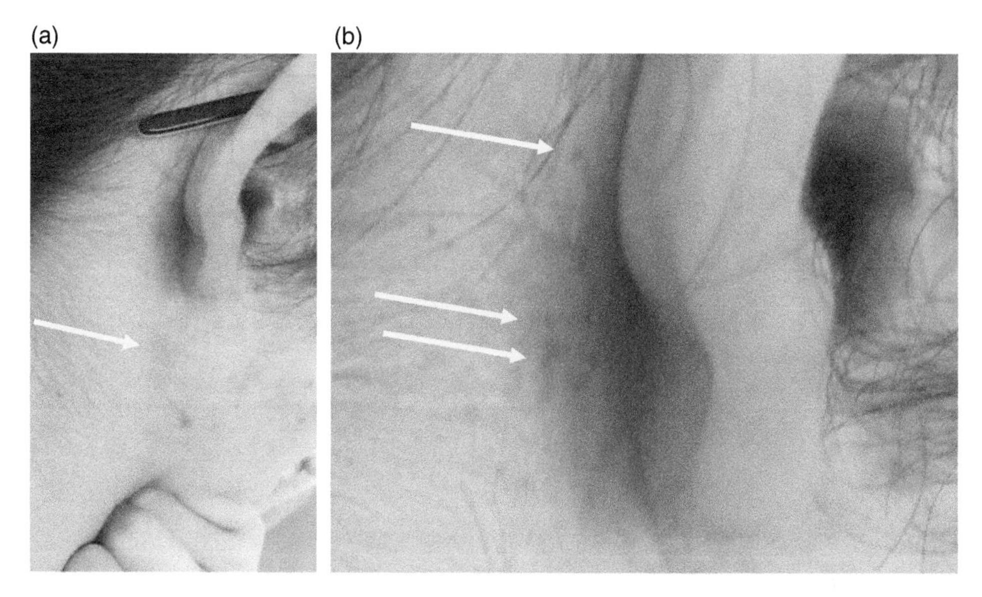

Figure 5.4 (a) Arrow points to bruising at the angle of the jaw. (b) Arrows identify petechial bruising behind the right ear.

them. . .would describe the dots as about 20 to 30 pin prick size red dots. . .about 2cm wide and 4cm long extending down towards the neck. . .had to move her hair. . .small swelling in the same place. . .some red marks around the other ear. . .she stated she remembers Jamie trying to strangle her. . .described him being behind her with his arm around her neck'.

Example 4 reinforces the point that visible evidence of contact may be present posteriorly as well as the anterior and lateral sides of the neck. The images (Figure 5.5a,b) – taken within 3 hours of the assault – show (within the bounds of the photographic quality) no apparent injury to the front of the neck; however, there are small linear areas of reddening (erythema) visible below the hairline to the left of the midline posteriorly and just above the left trapezius border.

Her account was: '*she screamed out and the male began to strangle her with both hands squeezing her neck preventing her from talking, breathing or screaming*'.

Example 5 emphasises the need to take a detailed history and not make assumptions based on the visible evidence alone. This 23-year-old female who was a complainant of rape referred to the two marks to her neck, which had been assumed to be fingerprint bruises and was very clear that no strangulation or blunt force trauma had been applied. Her account was: '*.next thing I remember is Y kissing my face and neck and giving me love bites*'. The visible similarity between love bites (hickeys) and fingertip bruises is clear (Figure 5.6). They cannot always be distinguished.

Example 6 is a 37-year-old female who was the victim of manual compression to the neck. The visible evidence (shown 24 hours later in the image – Figure 5.7) was of a mottled linear bruise, with petechial elements running down the anterior border of the left sternocleidomastoid muscle. Her account and the account of a witness who found her were: (victim)

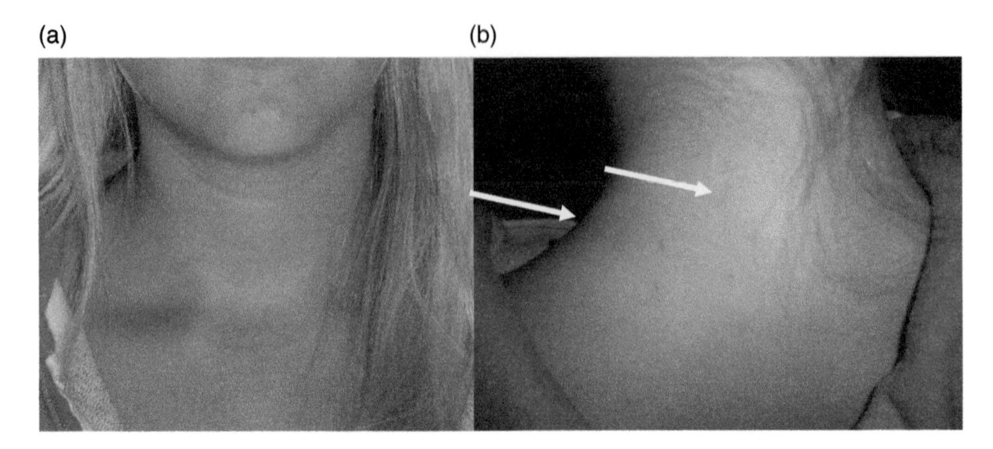

Figure 5.5 Arrows identify areas of erythema below the hairline.

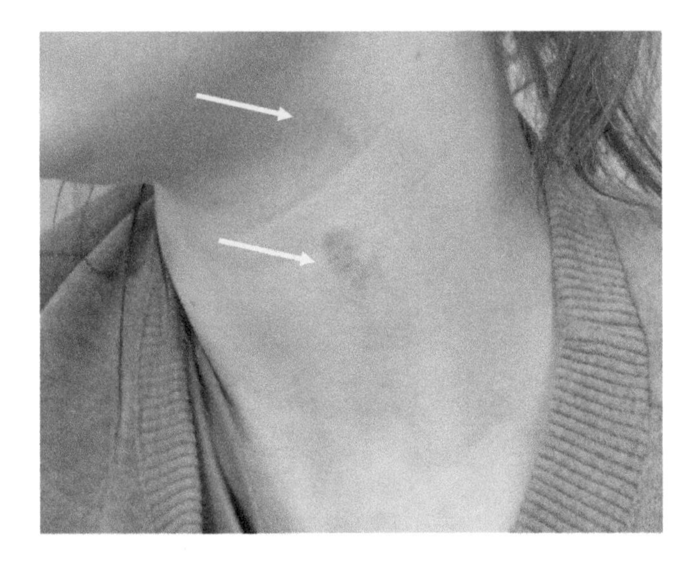

Figure 5.6 Arrows identify 'hickeys' or 'lovebites' on the neck.

> *'.Z then put his left hand up to my face/throat and I could feel the palm of his hand on the underside of my chin. . .his thumb pressing very hard on my windpipe and he continued pressing very hard. . .within a few second I had lost consciousness as I had tried to push him away and then I was unconscious. . .when I came around I remember gasping and Z was pushing my head. . .noticed that I had urinated whilst being unconscious as I did not urinate through fear' and (witness) '.her voice seemed quite hoarse and it seemed as if she was struggling for breath. . .her eyes were bloodshot. . .couldn't see any bruising or marks on the right hand side. . .did have a red mark. . .what looks like a long love bite. . .it was red in colour. . .left hand side'.*

Example 7 is a 31-year-old female who describes being strangled (with no specific detail of nature). The image (Figure 5.8) showing left scleral haemorrhage which initially (2 hours after the event) presented as four petechial haemorrhages in the

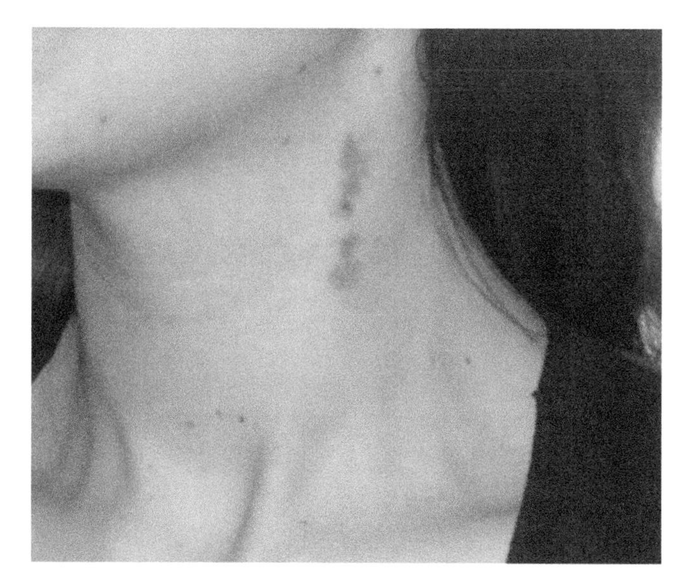

Figure 5.7 Mottled linear bruise, with petechial elements running down the anterior border of the left sternocleidomastoid muscle.

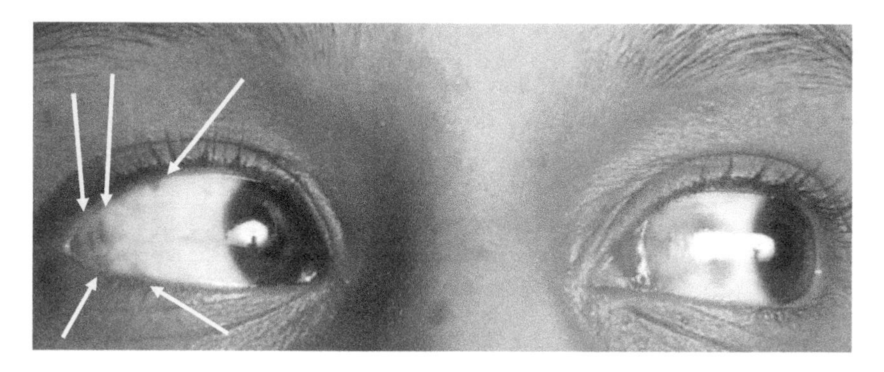

Figure 5.8 Left scleral haemorrhages (arrows).

medial aspect of the left eye, which, within 8 hours (when the image was taken), had coalesced into the visible haemorrhage seen. In the right eye, ~5 petechial haemorrhages are seen. Her account was:

> *'he came at me so quickly. . .it was like he was crazy. . ..acted so forcefully and that was when he strangled me. . . could barely breathe and that was when I passed out. . .don't know what happened after that. . .I woke up. . .I was having breathing difficulties at this point'.*

Example 8 emphasises two points: not all victims are female, and the nature of NFS may include other forms of blunt injury. The injury to the eye shown is due to neck compression, and although the 40-year-old male victim was punched to the face, it was not to the eye region (Figure 5.9a,b). The circumorbital and scleral bruising was caused by coalescence of multiple petechial haemorrhages – the image being

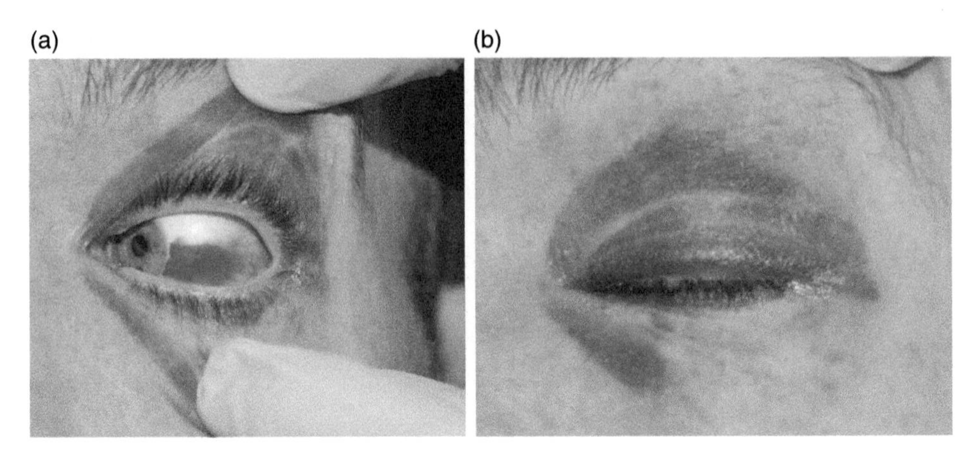

Figure 5.9 Circumorbital and scleral bruising was caused by coalescence of multiple petechial haemorrhages.

taken about 12 hours after the incident. The assault was accompanied by faecal incontinence. The victim gave the following account:

> *'.gripped me by the throat. . .punched me in the face. . . I slid onto the floor. . .I was on my knees with him crouched down in front of me. . .gripped me round the throat in a neck lock with both his arms squeezing extremely tightly. . .felt like my neck was being squeezed in a vice. . .struggled to get him off but could not break free. . .remember choking and seeing stars but I think I lost consciousness. . ..I woke up on the opposite side of the room. . .was shaking. . ..stood up and felt stuff coming down my leg. . .saw shit on my hand'. A witness to the assault stated: 'could see that P had a tight grip of S by the neck. . .his whole arm was wrapped around S's neck and he dragged S backwards into the living room floor. . . sat on top of him still gripping his throat. . .all the while S was trying to kick out with his legs to try and get P off him but P is a lot bigger than S. . .P would not release his grip. . .in the end S's body just seemed to go floppy. . .face started to bulge and turn blue. . ..eyes started to roll to the back of his head. . .the worst thing that happened was when S actually poo'd himself it was like he had lost control of all his body functions'.*

Example 9, a male of 46 years who was strangled with a ligature which he attempted to resist. The ligature marks, extensive and circumferential, may be seen where a victim has attempted to remove the ligature (Figure 5.10a–c). The arrows identify areas of irregular linear abrasion, consistent with fingernail abrasions from the victim's own nails as he tried to remove the ligature whilst struggling with his assailant. His description was:

> *'.all of a sudden M was holding a thin black electric wire around my neck . . .at the start I thought he was joking. . ..it was getting tighter and tighter. . .I put my hands up to the lead to try and get it off my neck. . ..as the wire was getting tighter and tighter I tried to pull it away but I was becoming weak. . ..could not speak, the wire was too tight around my throat. . .could feel his hands holding the wire at the back of my neck. . ..was so tight it was impossible for me to get my fingers*

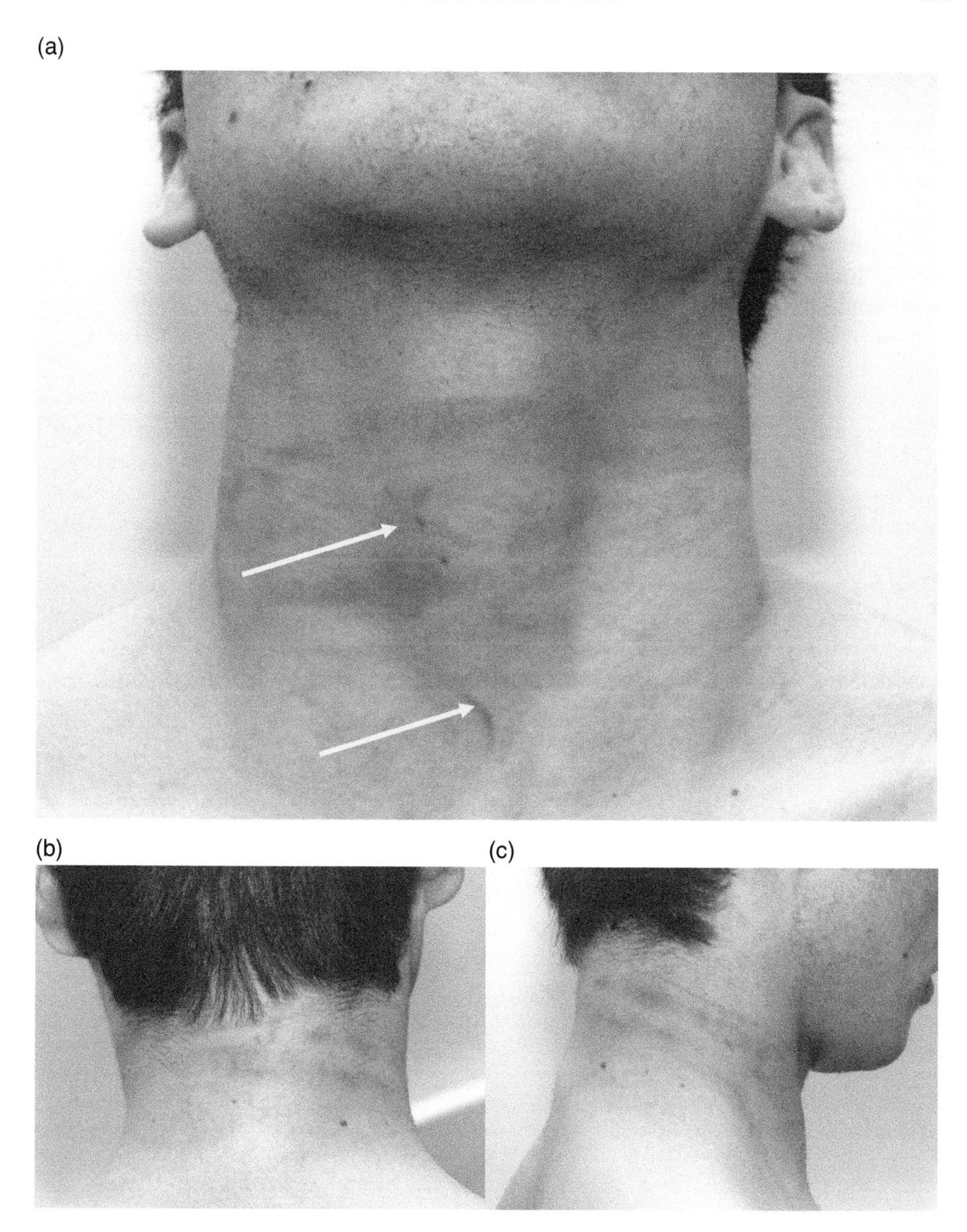

Figure 5.10 Ligature marks. The arrows identify areas of irregular linear abrasion, consistent with fingernail abrasions from the victim's own nails.

between the wire and my neck to try and save myself. . ..I couldn't stop him. . .could not breath[e], everything started going hazy, I had no feelings left. . .next thing I remember is N shaking my shoulder telling me to wake up. . ..my head felt very heavy. . ..was in a lot of pain in my neck. . ..couldn't breathe[e] properly or speak. . ..realised I had fainted'.

Management of non-fatal strangulation

Educational strategies are required to ensure that complainants of NFS report what has happened, and that healthcare and law enforcement professionals recognise the significance and the need for an appropriate forensic medical assessment. Jansen et al. (2021) observed that patients often do not seek medical treatment after NFS, and this may result in late presentation of life-threatening injuries. Thus, educational strategies must include public health messages reaching out directly to potential victims and their families and friends. Patch et al. (2017), recognising that US estimates suggest higher prevalence of strangulation of women than of men, reviewed the existing literature related to women's injuries and their subsequent experiences in seeking health care after surviving intimate-partner strangulation. They found that non-fatal intimate-partner strangulation was associated with multiple negative physical and psychological outcomes for women, but a wide variation in reporting to healthcare with between 5% and 69% of women subject to NFS seeking assistance. The authors recommended further research to support the scientific base for screening and treatment protocols as well as to understand the long-term consequences of NFS.

The management of cases of NFS should be divided into two main parts, both of which overlap. From a medical perspective, it is essential to exclude current or potential long-term complications of the neck injury and any accompanying other injury. This requires a full history and detailed examination with any injuries documented in written, body diagram, and appropriate image capture (still and moving) form. This may be done with an app such as ForensiDoc™ using a smart device. It is recommended that all those undertaking such examinations should have forensic photographic training if their access to professional police or clinical photographers is limited. From a legal perspective, it is essential to ensure that all relevant evidential material is collected at the time. This may include documentation of symptoms and signs, requesting appropriate investigations which may corroborate findings and undertaking evidential biological sampling.

Otorhinolaryngological or emergency department referral may be relevant in those with hoarseness or dysphagia or any of the other red flags shown in Table 5.4, for further investigation. Virtual computed tomography (CT) endoscopy may have a role (Becker et al. 2014).

Radiological imaging in non-fatal strangulation

From a clinical perspective, it is important to determine whether further investigations or treatment are required in addition to documentation of injury. Visible evidence of manual strangulation can and should be documented in written, body diagram, and photographic/moving image form. In addition to concerns about possible structural anatomical damage, absence of visible injury with the presence of neck symptoms may also require further investigation. In fatal cases, neck dissection as part of the post-mortem may provide further information (e.g. bruising to strap muscles), but such information obviously will not be available for NFS. Consideration should therefore be given to the use of other imaging modalities to identify and corroborate accounts where there is no external sign of injury.

CT is the most widely used modality for assessing the extent of neck injuries from blunt trauma to the neck, and this can be supplemented by CT angiograph to assess vascular injury. CT, however, is most often used in the therapeutic setting. Christe et al. (2010) reviewed 56 NFS cases comparing clinical forensic medical findings with magnetic resonance imaging (MRI) findings when attempting to determine whether the NFS was life-threatening or non-life-threatening. They concluded that MRI had a role to play in identifying life-threatening NFS when compared with clinical assessment and was better than CT for diagnosing soft tissue injury. Bruguier et al. (2019) reported on 112 victims who were clinically examined after suspected strangulation. Eleven underwent an MRI examination of the neck. Eighty-four presented objective lesions during the clinical examination, with eight showing signs of both petechiae and bruising. Neck-MRI was performed in four of these eight and three of them showed lesions visible in MRI. Of 76 with bruising as the only finding, 66 described clinical symptoms. Of those 66, seven were examined by MRI and two demonstrated lesions in MRI. When MRI was performed, relevant findings were detected in 45% of the cases. They concluded that 'This leads to the suspicion that many more findings could have been detected in the other victims, if an MRI had been performed in those cases. Our results lead us to the conclusion that an MRI examination of victims of suspected strangulation is useful, and strict indications for its application should be established'.

Gascho et al. (2020) emphasised that the radiation exposure of CT may not be justified in the NFS setting if symptoms are not severe. MRI may be more appropriate in the forensic or medicolegal setting to document non-externally visible injury. In a study of 195 cases of NFS subject to MRI, a variety of abnormalities were found – see Table 5.5.

Table 5.5 MRI findings in 195 cases of NFS.

	Number of subjects	% of subjects
Superficial tissue zone		
Intracutaneous haemorrhage	22	11.3
Subcutaneous haemorrhage	56	28.7
Platysma haemorrhage/swelling	29	14.9
Middle tissue zone		
Lymph node haemorrhage	36	18.5
Intramuscular haemorrhage	29	14.9
Deep tissue zone		13.3
Perilaryngeal fluid accumulation	26	13.3
Rare findings		
Oedema/haemorrhage of the salivary glands	12	6.2
Parapharyngeal haemorrhage	9	4.6
Glottis/vocal cord oedema	7	3.6
Thyroid haemorrhage	2	1
Blood serum level in the glottis space	1	0.5
Fluid accumulation in the vessel/nerve sheath	1	0.5
Laryngeal fracture	1	0.5

Source: Adapted from Gascho et al. (2020).

At present, the role of MRI is unclear in the forensic and medicolegal setting. A reasonable approach might be to consider MRI when there is an absence of external visible findings in cases of NFS, to discuss with radiology teams and for a joint radiology and forensic medical opinion to be provided summarising those findings. Prospective studies in this area will more clearly define when an MRI or any other imaging modality should be undertaken.

Conclusion

The significance of NFS is now well recognised. It is important that all healthcare professionals (e.g. forensic practitioners, SARC practitioners, emergency department practitioners, and primary care practitioners), law enforcement professionals, safeguarding professionals, and any who come into contact with instances of NFS understand (i) its implications, (ii) the importance of accurate and unambiguous documentation of the signs and symptoms present, (iii) in the absence of visible signs recognise when further investigations such as MRI be helpful, and (iv) know to whom to refer to ensure that relevant evidence is captured and interpreted appropriately.

References

Abedr-Rahman, H., Salameh, H.O., Salameh, R.J. et al. (2017). Role of forensic medicine in evaluating non-fatal violence against women by their husbands in Jordan. *Journal of Forensic and Legal Medicine* 49: 33–36.

Anscombe, A.M. and Knight, B.H. (1996). Case report. Delayed death after pressure on neck: possible causal mechanisms and implications for mode of death in manual strangulation discussed. *Forensic Science International* 78: 193–197.

Becker, M., Leuchter, I., Platon, A. et al. (2014). Imaging of laryngeal trauma. *European Journal of Radiology* 83: 142–154.

Bichard, H., Byrne, C., Saville, C.W.N., and Coetzer, R. (2021). The neuropsychological outcomes of non-fatal strangulation in domestic and sexual violence: a systematic review. *Neuropsychological Rehabilitation* 1–29. https://doi.org/10.1080/09602011.2020.1868537.

Blanco Pampin, J., Morte Tamayo, N., Hinojal Fonseca, R. et al. (2002). Delayed presentation of carotid dissection, cerebral ischaemia and infarction following blunt trauma: two case. *Journal of Clinical Forensic Medicine* 9: 136–140.

Bruguier, C., Genet, P., Zerlauth, J.-B. et al. (2019). Neck-MRI experience for investigation of survived strangulation victims. *Forensic Sciences Research* 5: 1–6.

California District Attorneys Association (CDAA) (2020). Investigation and prosecution of strangulation cases.

Cannon, L.M., Bailey, J.M., Ernst, S.D. et al. (2020). Examining trends in non-fatal strangulation among sexual assault survivors seeking Sexual Assault Nurse Examiner care from 2002 to 2017. *International Journal of Gynaecology and Obstetrics* 149 (1): 106–107. https://doi.org/10.1002/ijgo.13058.

Christe, A., Oesterhelweg, L., Ross, S. et al. (2010). Can MRI of the neck compete with clinical finding in assessing danger to life for survivors of manual strangulation? A statistical analysis. *Legal Medicine* 12: 228–232.

Clarot, F., Vaz, E., Papin, F., and Proust, B. (2005). Fatal and non-fatal bilateral delayed carotid artery dissection after manual strangulation. *Forensic Science International* 149: 143–150.

Evans, S., Baylis, S., Carabott, R. et al. (2014). Guidelines for photography of cutaneous marks and injuries: a multi-professional perspective. *Journal of Visual Communication in Medicine* 37 (1–2): 3–12. https://doi.org/10.3109/17453054.2014.911152).

Faculty of Clinical Forensic Medicine Committee: Royal College of Pathologists Australasia (2018). Clinical Forensic Assessment and Management of non-fatal strangulation.

Faculty of Forensic & Legal Medicine. (2019). PICS Working Group: Guidelines on photography. PICS-Working-Group-Guidelines-on-Photography-Dr-Will-Anderson-Dec-2019.pdf (fflm.ac.uk) (accessed 13 April 2022)

Faculty of Forensic and Legal Medicine (2020). Non-fatal strangulation: in physical and sexual assault.

Fong, W.-L., Pan, C.-H., Lee, J. et al. (2016). Adult femicide victims in forensic autopsy in Taiwan: a 10 year retrospective study. *Forensic Science International* 266: 80–85.

Gall, J. and Payne-James, J. (2011). Injury interpretation – common errors. In: *Current Forensic Medicine* (ed. J. Gall and J.J. Payne-James). Wiley.

Gascho, D., Heimer, J., Thali, M.H., and Flach, P.M. (2020). The value of MRI for assessing danger to life in nonfatal strangulation. *Forensic Imaging* 22: 200398.

Glass, N., Laughon, K., Campbell, J. et al. (2008). Non-fatal strangulation is an important risk factor for homicide of women. *The Journal of Emergency Medicine* 35 (3): 329–335. https://doi.org/10.1016/j.jemermed.2007.02.065.

Iserson, K.V. (1984). Strangulation: a review of ligature, manual and postural neck compression injuries. *Annals of Emergency Medicine* 12: 179–185.

Jansen, J.H., Yi, J.M., and Capps, A.E. (2021). Posterior neck hematoma formation after strangulation injury. *Visual Journal of Emergency Medicine* 22: 100924.

John Hopkins School of Nursing (2022). Danger assessment. https://www.dangerassessment.org/ (accessed 1 January 2022).

Matias, A., Goncalves, M., Soeiro, C., and Matos, M. (2020). Intimate partner-homicide: a metaanalysis of risk factors. *Aggression and Violent Behaviour* 50: 101358.

McQuown, C., Frey, J., Steer, S. et al. (2016). Prevalence of strangulation in survivors of sexual assault and domestic violence. *American Journal of Emergency Medicine* 34: 1281–1285.

Monckton-Smith, J. (2020a). Intimate partner femicide: using foucauldian analysis to track an eight stage progression to homicide. *Violence Against Women* 26 (11): 1267–1285. https://doi.org/10.1177/1077801219863876.

Monckton-Smith J. (2020b). Non-fatal strangulation. A summary report on data collected from SUTDA Survey. https://sutda.org/wp-content/uploads/Non-fatal-strangulation-word.docx (accessed 1 January 2022).

Moracco, K.E., Runyan, C.W., Butts, J.D. et al. (2003). *Journal of the American Medical Women's Association (1972)* 58 (1): 20–25.

Murphy, P.G., Jackson, E., Kirollos, R. et al. (1993). Adult respiratory distress syndrome after attempted strangulation. *British Journal of Anaesthesia* 70: 583–586.

Patch, M., Anderson, J.C., and Campbell, J.C. (2017). Injuries of women surviving intimate partner strangulation and subsequent emergency healthcare seeking: an integrative evidence review. *Journal of Emergency Nursing* 44: 384–393.

Payne-James, J.J., Hawkins, C., Bayliss, S., and Marsh, N. (2012). Quality of photographic images for injury interpretation: room for improvement? *Forensic Science, Medicine, and Pathology* 8 (4): 447–450. https://doi.org/10.1007/s12024-012-9325-2.

Pereira, A.R., Vieira, D.N., and Magalhaes, T. (2012). Fatal intimate partner violence against women in Portugal: a forensic medical national study. *Journal of Forensic and Legal Medicine* 20: 1099–1107.

Plattner, T., Bolliger, S., and Zollinger, U. (2005). Forensic assessment of survived strangulation. *Forensic Science International* 153: 202–207.

Shields, L.B., Corey, T.S., Weakley-Jones, B., and Stewart, D. (2010). Living victims of strangulation: a 10-year review of cases in a metropolitan community. *The American Journal of Forensic Medicine and Pathology* 31 (4): 320–325. https://doi.org/10.1097/paf.0b013e3181d3dc02.

Sommers, M.S., Brunner, L.S., Brown, K.M. et al. (2012). Injuries from intimate partner and sexual violence: significance and classification systems. *Journal of Forensic and Legal Medicine* 19: 250–263.

Strack, G.B., McClane, G.E., and Hawley, D. (2001). A review of 300 attempted strangulation cases Part I: criminal legal issues. *The Journal of Emergency Medicine* 21 (3): 303–309.

The Guardian (2021). https://www.theguardian.com/society/2021/mar/01/non-fatal-strangulation-to-carry-five-years-in-prison-under-reforms (accessed 1 January 2022).

Unal, E.O., Koc, S., Unal, V. et al. (2016). Violence against women: a series of autopsy studies from Istanbul, Turkey. *Journal of Forensic and Legal Medicine* 40: 42–46. https://doi.org/10.1016/j.jflm.2015.11.025.

White, C., Martin, G., Schofield, A.M., and Majeed-Ariss, R. (2021). I thought he was going to kill me': analysis of 204 case files of adults reporting non-fatal strangulation as part of a sexual assault over a 3 year period. *Journal of Forensic and Legal Medicine* 29: 102128.

Zilkens, R.R., Phillips, M.A., Kelly, M.C. et al. (2016). Non-fatal strangulation in sexual assault: a study of clinical and assault characteristics highlighting the role of intimate partner violence. *Journal of Forensic and Legal Medicine* 43: 1–7. https://doi.org/10.1016/j.jflm.2016.06.005.

Zilkens, R.R., Smith, D.A., Kelly, M.C. et al. (2017). Sexual assault and general body injuries: a detailed cross-sectional Australian study of 1163 women. *Forensic Science International* 279: 112–120.

https://victimscommissioner.org.uk/news/non-fatal-strangulation-joint-statement-from-the-victims-commissioner-and-domestic-abuse-commissioner/

https://www.strangulationtraininginstitute.com/

6 DNA: current developments and perspectives

Denise Syndercombe-Court
King's College London, London, UK

Introduction

Short tandem repeat (STR) DNA analysis has been the mainstay of forensic DNA analysis since the 1990s and remains so today. Seventy countries, worldwide, amongst the 194 INTERPOL member countries, report having a DNA database (INTERPOL 2019), including all 27 EU countries. China has the largest DNA database (more than 80 million profiles, representing less than 1% of the population); the United States reports 14.3 million profiles, representing 4.5% of the population; and the United Kingdom is the next largest, with 6.6 million profiles (Gov.UK 2020), representing around 9% of the population. As a result of the S and Marper judgement (Case of S and Marper v The United Kingdom 2008) in which the indefinite retention of DNA profiles from individuals acquitted, or having charges dropped, was declared a violation of the right to privacy under the European Convention on Human Rights, the United Kingdom introduced the Protection of Freedoms Bill in 2011, which led to the removal and destruction of around 1.38 million samples and associated profiles from the UK National DNA Database (NDNAD) and a law that limits future retention of profiles. Nevertheless, the match rate of the database continues to increase (Amankwaa and McCartney 2019).

The usefulness of DNA as a tool in forensic science was highlighted in the 2016 report from the President's Council of Advisors on Science and Technology (PCAST) (Executive Office of the President: President's Council of Advisors on Science and Technology 2016) in the document Forensic Science in the Criminal Courts: Ensuring Scientific Validity of Feature-Comparison Methods, which found that DNA analysis of single-source samples, or simple mixtures of two individuals, had strong foundational validity, in contrast to many other methods that they considered.

The efficacy of a database that represents a significant proportion of the population can be shown in the UK statistics that provide match rates to a named individual from crime scene profiles of sufficient quality to be loaded onto the database of over 65%. Even so, it must be remembered that around 15% of profiles are thought to be duplicates, and many crimes do not result in DNA being of sufficient quality, or are even recovered from a scene, resulting in an estimate of only 0.3% of crimes being linked to DNA (UK Office of the Biometrics Commissioner 2017),

---

Current Practice in Forensic Medicine, Volume 3, First Edition. Edited by John A.M. Gall and J. Jason Payne-James.

although its importance is more obvious in homicides in which the rate increases to over 8%.

There has been, however, a continuum of improvement in STR typing over time as manufacturers responded to increased desire for greater certainties of prediction from samples of even lower quantity and quality and the desire for faster analysis speeds and improving cost-effectiveness through on-site testing technologies. New STR kits containing more loci, targeted to use, for example kinship, or legal decisions made on locus choice in different legislations, have been produced, and more rapid methods developed. With increased sensitivity, achieved through changes in chemistry, comes greater detection of mixtures. The PCAST 2016 approach found that techniques being used at the time to examine mixtures were subjective and had the potential to produce erroneous results, and they recommended that probabilistic methods be developed. Significant efforts have been made over the past few years to provide statistical tools to help with the increased complexity of mixture interpretation.

Since the establishment of a database of small-scale nucleotide variants (dbSNP) (NCBI Insights 2019) in 1999, the number of single-nucleotide polymorphisms (SNPs) recorded has grown exponentially, and scientists began to employ these variants to help describe the phenotypes of individuals as early as 2008 (Sulem et al. 2008). The associated reduction in the cost of analysis has put this technology in the hands of many working in forensic science, and new intelligence tools have been developed in the intervening years. Although potentially helpful, these techniques disappoint when compared with the high predictions of individualisation achieved through STR analysis.

Significant reduction in genetic sequencing costs and the popularisation of genetic genealogy through direct-to-customer approaches, with its associated public sharing of genetic information for kinship information, has allowed law enforcement agents, principally from the United States, to utilise this information to solve historic crimes. Such approaches, however, raise significant ethical issues which will need to be considered before they are introduced into routine forensic provision.

STR improved autosomal multiplexes used for criminal justice

The potential for the analysis of SNP markers to replace STR loci has been the subject of consideration over the past decade (Butler et al. 2007); while SNPs have some benefit and advantages where DNA is fragmented, the replacement of such a successful methodology as STR multiplexing is thought very unlikely, at least in the near future for two main reasons: the multiple alleles associated with each STR enable higher match statistics and easier detection and interpretation of DNA mixtures with fewer loci, in comparison with biallelic SNPs; the considerable investment in global STR databases could not easily be replaced and would require significant efforts in dual processing to build up an as yet not fully tested methodology, over many years, without the promise that newer approaches might not overtake what is possible today.

Commercial kits typically cover the range of 15–22 autosomal STR loci in order to cover the core loci specified in Europe or the United States. Although many countries were working with multiplexes with more loci, seven core loci were confirmed in a resolution of the European Council in 2004 (Corte-Real 2004). In 2005, the Treaty of Prüm envisaged mass cross-border exchange of DNA profiles which significantly increased the chances of adventitious matches, made more relevant by the significant portion of partial crime scene profiles in individual nations' databases. Increasing the core loci within Europe therefore became a priority, and a meeting of the European Network of Forensic Science Institutes (ENFSIs) and the European DNA Profiling Group (EDNAP) in 2005 set out priorities for the development of new multiplexes based on the following criteria:

- Including loci available in the current STR multiplexes that also encompass the core 7 loci
- Including new loci with small amplicon sizes around 150 base pairs (bps) (mini STRs)
- Including established highly polymorphic loci with relatively small amplicon sizes.

In addition, they called for manufacturers to provide increased sensitivity when designing new multiplexes.

Five additional loci were added to the original 7 core loci in 2008 (Council of the European Union, Brussels (2009)), bringing the total to 12 core European Standard Set (ESS) loci. Table 6.1 defines these loci.

In the United States, there were similar thoughts. The combined DNA index system (CODIS) 13 core loci were defined in 1997, and in 2010, the FBI Core Loci Working Group identified three reasons why they should further increase the CODIS core loci (Hare 2012):

- To reduce adventitious matches
- To increase international compatibility
- To increase discrimination power.

In 2015, seven additional loci were added to the core, bringing the total CODIS loci to 20 in number (Hare 2015) (Table 6.1). The Chinese national DNA database includes an additional locus: D6S1043.

STR multiplexes are provided by a number of commercial companies. The main three manufacturers serving forensic science institutes are Promega, Qiagen, and Thermo Fisher Scientific. The amount and quality of the DNA template will influence the success of any amplification, but varying primer design and different amplification chemistries will modify the performance of any multiplex. Some multiplexes are further designed to enable fast analysis, or direct amplification from

Table 6.1 Comparison of European ESS (2004) and US CODIS STR core loci (1997), highlighted in blue, and expansion of these sets in 2008 (ESS) and 2015 (CODIS), highlighted in green.

Locus	European ESS	United States CODIS
CSF1PO		
D3S1358		
D5S818		
D7S820		
D8S1179		
D13S317		
D16S539		
D18S51		
D21S11		
FGA		
TH01		
TPOX		
VWA		
D1S1656		
D2S441		
D2S1338		
D10S1248		
D12S391		
D19S433		
D22S1045		

paper matrix punched stains, or to analyse core (ESS) loci as mini-STRs, to provide quality information or additional loci for kinship analysis (not shown here).

All include primers to target a 6-bp deletion on the X chromosome to enable the presence of both X and Y chromosomes to be identified and used for determination of sex. Deletions of part of the Y chromosome that result in failure of male designation in the amelogenin sex test (AMELY null) have, however, been reported on many occasions with a frequency range from 0.018% (Steinlechner et al. 2002) to over 8% in a South Asian population (Santos et al. 1998). This has led manufacturers to include additional Y markers in some of their autosomal multiplex kits for more reliable identification of sex.

Table 6.2 details the main loci used for forensic DNA analysis. Incorporating additional loci into a multiplex is limited by the desirability in forensic analysis to keep the amplicons as small as possible (preferably under 400 bp) and for the markers not to overlap in the 100–400-bp space. In addition to spatial separation, different coloured fluorescent dyes are used to differentiate markers, and over time, the number of dyes used by some manufacturers in their kits has increased from four to six in order to provide better separation.

The total number of desired markers is also limited by the need for them to be inherited independently (unlinked, or in linkage equilibrium) so that the statistical assessment of occurrence of a profile can be calculated through multiplication of the individual allele frequencies. In commonly used multiplexes, the closest loci are D12S391 and VWA which are separated by 11.94 centimorgans (cM) and have a recombination rate of 0.117. One cM is defined as the distance between the loci at which there is a 1% chance (recombination rate of 0.01) that they will be separated from each other in a single generation (meiosis). Loci that are completely unlinked have a recombination rate of 0.5. It is believed that the distance between these two loci is sufficient to avoid linkage disequilibrium at the population level and unlikely to affect match probabilities, but linkage should be considered if close relatives are being considered (Bright et al. 2014). Ignoring linkage can result in match probabilities being overstated by up to around 10%, depending on the closeness of the relationship and population substructure.

Rapid DNA

As early as 2012, there was a vision for a 'lab that fits in your hand' completing an entire DNA process in less than two hours with no human intervention, minimal operator education, and using a single instrument with no liquid handling. Time-saving developments focused on the polymerase chain reaction (PCR) process with new polymerases that were more resistant to inhibitors and more efficient. Fast-processing alternative multi-mixes are now a routine offer to the laboratory for many of the multiplexes described in Table 6.2.

Microfluidic solutions offered a way to miniaturise the process of DNA amplification and profiling, and instruments began to be developed as early as 2010 (Hopwood et al. 2010). In the same year, in the United States the FBI established its Rapid DNA technology programme (FBI 2019) to develop this technology for use by law enforcement with a goal of producing a CODIS core loci DNA profile within two hours. The intended purpose was to be able to profile arrestees booked into police custody to search unsolved crimes. The Rapid DNA Act of 2017 authorised the development and testing of this process, but the system is not currently approved, and significant challenges exist before this technology can be used for crime scene samples.

There are two instruments that have been approved for use in the US National DNA index system (NDIS) which holds the CODIS database:

- ANDE 6C Rapid DNA System using the FlexPlex 27 typing kits (the loci covered include all those in column C in Table 6.2, with the substitution of SE33 to replace Penta D)
- Applied Biosystems RapidHIT Rapid DNA system using the GlobalFiler typing kit.

Table 6.2 Locus components of the main forensic multiplexes from three different manufacturers: Promega (blue); Qiagen (green); ThermoFisher Scientific (pink).

Locus	A	B	C	D	E	F	G	H	I	j
D1S1656										
D2S1338										
D2S441										
D3S1358										
D5S818										
D6S1043										
D7S820										
D8S1179										
D10S1248										
D12S391										
D13S317										
D16S539										
D18S51										
D19S433										
D21S11										
D22S1045										
CSF1PO										
FGA										
Penta D										
Penta E										
SE33	*									
TH01										
TPOX										
VWA										
Amelogenin										
Y indel										
DYS391										
DYS570										
DYS576										
Quality 1										
Quality 2										

ANDE 6C

This is a microfluidic instrument that incorporates a specific ANDE swab, consumable analytical chip, and an automated expert system. There are two chips: the A chip can analyse five samples and has NDIS approval, and the I chip has been designed as a single-use item for use in forensic casework and disaster victim identification (DVI) situations and incorporates a DNA concentration module (Carney et al. 2019).

RapidHIT ID

This instrument, from Thermo Fisher Scientific, was originally developed by IntegenX and consists of single-use microfluidic cartridges for use in a 150-sample consumables cartridge housed in the instrument and RapidLink software to allow sample result viewing and matching with local databases. The system uses a range of compatible swabs, and separate cartridges can accommodate extracted and quantified DNA extracts to improve sensitivity (Salceda et al. 2017).

As part of a blinded assessment of the technology, the National Institute of Standards and Technology (NIST) reviewed the capabilities of these instruments to produce automated and manually reviewed output to provide genotyping for the CODIS loci. The results gave an automated average success of between 77% and 85%, depending on the instrument, typical of first pass expert systems within crime laboratories in the United States, increasing to 90% on manual review (Romsos et al. 2020). A similar review of the RapidHIT ID system in the United Kingdom revealed fully concordant results with similar success rates (85%) to the US study but below the expected US operational success rate of 95% (Shackleton et al. 2019).

This technology has much promise for the future, but whether it will be fully embraced, even after being accredited for use, will depend on a cost–benefit analysis, comparing likely the high analytical costs with savings on police custody and the extent of delay in analysing reference profiles that will vary between countries. Instruments generally consume the whole of a swab, and the risk of failures may mean that duplicates need to be retained. Although the ability to analyse extracts, available in one of the instruments, is helpful, this process is not one that is likely to be done outside a forensic laboratory setting, and there may be better alternatives.

It is likely to be some time before the technology will be suitable for analysis of crime scene samples. Experiments conducted by ANDE using mock crime scene swabs with blood, semen, and saliva on a variety of substrates gave high success rates with an average first pass rate of over 90%, even from samples retained for 12 months and no in-run contamination (Turingan et al. 2020). Two person mixtures down to a 1 : 19 ratio in a total 100 ng deposit, however, revealed only the major contributor in this extreme ratio, with other more balanced mixtures being flagged for review. It seems clear that, while potentially being able to identify the presence of a mixture, analysis of mixtures in the casework that is regularly encountered in a forensic laboratory is unlikely to be possible in the future using this technology.

The ability of this instrumentation to analyse low-level DNA, recovered from touch contacts and successfully analysed in routine laboratories today, remains to be evaluated. Proposals for use of this technology in DVI situations may seem attractive but will require recovery of samples that are not comingled with additional material being made available for analytical processes that the current system cannot provide (Y and X chromosome analysis, SNP typing, and mitochondrial DNA). In addition, despite the large number of autosomal STRs being typed, these are unlikely to be useful unless family comparisons are being made with first-degree relatives. These factors, the associated high analytical costs, and the time required to recover DNA from relatives and conduct complex statistical assessments probably mean that the technology has little to offer DVI.

DNA mixtures

Improvement in the sensitivity provided with the new-generation STR multiplexes has meant that scientists are now also encountering complex mixtures of DNA from an uncertain number of contributors that need to be analysed in a fair and unbiased fashion.

Mixtures are inherently difficult, especially for small amounts of DNA, because many DNA components (alleles) will be shared between unrelated individuals by chance. The relative contributions to the mixture may be unknown, as will often the knowledge that any contributors may be related to each other. Low-level components within a mixture will need to be evaluated alongside artefactual peaks in a profile, such as stutter peaks that are always present, and the risks of both drop-in (from low-level contaminants) and drop-out (from analytical loss of particular alleles that should be in the mixture but are lost because of the low concentration of DNA and competitive amplification processes) need to be considered.

The PCAST report (Executive Office of the President: President's Council of Advisors on Science and Technology 2016) highlighted the risk of analytical error and confirmation bias associated with the subjective analysis of complex mixtures that have been used, historically. In particular, they found that the commonly used combined probability of inclusion (CPI) methods lacked foundational validity unless only used in particular circumstances, and they recommended that DNA analysis of complex mixtures should move to methods based on probabilistic methods.

Combined probability of inclusion (CPI)

The CPI method, sometimes known as random man not excluded (RMNE) method, is described here because of its widespread use across the Americas, Asia, Africa, and the Middle East. It describes the probability that an unrelated person would be included as a contributor to the observed DNA mixture.

Its popularity comes from its simplicity in its calculation and ease of explaining mixtures in court (Bieber et al. 2016). In addition, there is no need to decide how many people have contributed to the mixture or define a prominent contributor to the mixture. Its relatively simplistic analysis which only requires the recording of alleles in the mixture, however, may lead to it being used inappropriately. Calculation of the CPI requires unambiguous designation of all of the peaks present and that is often difficult when DNA levels are compromised and can lead to erroneous conclusions if appropriate guidelines for peak interpretation are not followed. Further, the method has come under increasing criticism, particularly in the United States (Brenner 2011). It also does not use all of the available genetic information and so is a deficient model.

Binary models of mixture interpretation

Binary models are not suitable for low-template mixtures and usually limited to clear two-person mixtures. Clayton and colleagues (Clayton et al. 1998) offered guidelines for a robust interpretation of simple mixtures that continues to be useful today. It considers factors such as heterozygote balance and stutter and provides stepwise guidelines to define the components of the mixture before any comparison with reference profiles.

Qualitative, or semi-continuous models of mixture interpretation

Qualitative models are suitable for mixtures with low-level DNA and multiple contributors. Normally, the relative heights of the different peaks in the profile are not considered, but if a clear major profile can be defined, this can be used to infer parameters in the model.

Several probabilistic models have been developed in software solutions available as open source or commercially. The outputs from different models, given in the form of a likelihood ratio, are unlikely to be identical because each model uses different distributions. It is, however, expected that they will produce results that are similar. Table 6.3 lists some of the software programs currently available to analyse mixtures using a qualitative approach.

Quantitative, or continuous models of mixture interpretation

Mixture interpretation software that uses information on the relative amounts of DNA in a mixture of different contributors, inferred through the heights of peaks in the electropherogram, is a natural development of the qualitative models described above. They offer significant advantages to complex mixture interpretation as they generally produce higher likelihood ratios to support or refute a hypothesis because they use more of the available information. Both open source and commercial models are available, the latter sometimes being favoured by forensic laboratories because of the large amount of training and ongoing support and updates that are part of the continuing licencing costs.

Table 6.3 Semi-continuous mixture software.

Software	Availability
FST (Mitchell et al. 2011)	Commercial
Lab Retriever (Inman et al. 2015)	Open source
LikeLTD (Balding 2013)	Commercial
LiRa (Puch-Solis and Clayton 2014)	Commercial
LRmix (Haned et al. 2012) and LRmix Studio (GitHub 2019)	Open source

All evaluate the evidence, providing likelihood ratios that compare two hypotheses, but they use several different mathematical models and require many parameters to be defined and entered into the programs. Because of this, the application of different software to interrogate identical mixture profiles produces different numerical likelihood ratios. This can cause confusion when a court is presented with different numbers to assess the same evidence, but it is expected that all of these models will not differ by more than two orders of magnitude. (An order of magnitude is a power of 10: two orders would be a ratio difference of 100.)

Evaluation of evidence using both qualitative and quantitative models requires an estimate of the number of contributors. While assessment of the minimum number of contributors to a mixture can be determined through simple counting of alleles at a locus, the actual number of contributors can be very difficult to assess, especially in circumstances where any contribution is at a low level. Additionally, as the number of contributors increases, the number of different combinations and necessary computations can overwhelm the capacity of most computers available to analysts, or computations may take weeks or longer to complete. Because of this, many software solutions will not, or it will not be practical to, analyse mixtures with more than about four unknowns.

Table 6.4 provides a list of the currently available software programs along with the mathematical model or distribution that each is based on.

Table 6.4 Continuous mixture software.

Software	Mathematical models	Availability
CEESIt (Swaminathan et al. 2016)	Normal	Open source/commercial
DNAmixtures (Cowell et al. 2015)	Gamma	Open source but requires commercial HUGIN software
DNAStatistX (Benschop et al. 2019)	Gamma	Open source
DNAView mixture solution (Brenner 2015)	Gamma	Commercial
EuroForMix (Bleka et al. 2016)	Gamma	Open source
Kongoh (Manabe et al. 2017)	MCMC	Open source
LikeLTD v 6 (Steele et al. 2016)	Gamma	Open source
LiRa v 3 (Puch-Solis et al. 2013)	Gamma	Commercial
MaSTR (Adamowicz et al. 2019)	MCMC	Commercial
STRmix (Bright et al. 2016)	MCMC	Commercial
TrueAllele (Perlin et al. 2011)	MCMC	Commercial

Guidance on probabilistic statistical analysis

Both the Scientific Working Group on DNA Analysis Methods (SWGDAM 2017) and the UK Forensic Regulator (Forensic Science Regulator 2018) have published guidance to assist in the interpretation and use of appropriate software.

The latter document offers advice to scientists as to how mixtures should be reported in circumstances where it has not been possible to provide a quantitative analysis. Several historic cases in the United Kingdom had relied on an approach in which the interpretation of a complex mixture was provided by stating the number of alleles that matched the alleles in the defendant's profile, and this approach was formalised in the Appeal Court judgement in the case of Dlugosz (Dlugosz v Regina 2013). Subsequently, however, the approach was shown sometimes to be prejudicial (Evett and Pope 2013), and its use since discouraged. The guidance also advises that qualitative opinions should only be presented as investigative opinions for intelligence purposes.

Massively parallel sequencing

While STRs have been the bedrock of forensic DNA typing for many years, developments in sequencing technology that have been embraced in the medical field are finding a place in forensic analysis.

Once the human genome project had reached completion, the need for a more efficient solution to the high cost and laborious processing offered through Sanger sequencing led to technological developments that produced the first high-throughput sequencing platform developed by Solexa (Bentley et al. 2008), subsequently taken over by Illumina. Originally referred to as next-generation sequencing (NGS) as the cost of sequencing using these new platforms opened up the technology to all, massively parallel sequencing (MPS) is now the preferred term because of the miniaturised platforms that enable sequencing of many hundreds of megabases to gigabases of nucleotides through massive parallel analysis in a single run.

Sanger sequencing, developed in the late 1970s (Sanger et al. 1977), remains a gold standard process that has developed over the years through the use of slab gels through to fluorescent dye incorporation and detection in capillary electrophoresis enabling 384-well Sanger sequencing reactions (sequencing by synthesis - SBS) to be undertaken in a single run, significantly improving the throughput. Although highly accurate, in comparison with MPS technologies, the process in comparison is slow. Sanger sequencing involves copying a DNA sequence through the PCR, but as well as introducing standard nucleotides, modified di-deoxynucleotides are also present at a low ratio. When these are incorporated into the growing chain at random, the elongation stops (chain-termination PCR) producing many random lengths of sequence. Fluorescent labelling identifies the last-added nucleotide, and ordering all of the fragments allows the sequence to be read. While Sanger methods can sequence up to 1000 bases, sequencing length using MPS is much shorter. Its main advantage, however, is its ability to bind and read millions of DNA-sequenced fragments on a single flow cell in comparison with Sanger methodology in which only a single fragment can be sequenced at the same time.

In MPS 'libraries' of fragmented DNA, use DNA ligase to add a universal sequence adaptor to each end of the fragment that allows them to bind to the complementary DNA sequences on a solid surface, such as a bead or flat glass surface, followed by clonal amplification on the substrate. This technology enables all types of nucleic acids to be sequenced, including mRNA and methylated DNA, increasing the potential applications that forensic scientist can make use of.

Although there are several different sequencing technologies that can be employed, two bench top sequencers are common in forensic genetics use: the Illumina MiSeq FGx and the Thermo Fisher Ion Torrent PGM and the newer Ion S5 instrument. Both use a sequencing by synthesis (SBS) approach in which a polymerase is used along with a signal, such as a fluorophore (used by the MiSeq instrument) and change in pH (used in the Ion Torrent instrument), to identify the incorporation of a nucleotide.

These two instruments use slightly different approaches to clonal amplification, essential to enable signals to be recognised above background noise. The Illumina instrument uses bridge amplification (Figure 6.1), creating many identical single-stranded copies of a fragment in a defined area (cluster) that are sequenced to generate a fluorophore. In contrast, the Thermo Fisher approach employs an emulsion of beads with bound primers to generate many single-stranded identical copies on each bead (Figure 6.2); sequencing is through detection of pH changes in ion proton sequencing. Massive parallelisation with single beads in millions of wells on a single silicon chip allows millions of individual SBS reaction centres to be sequenced simultaneously.

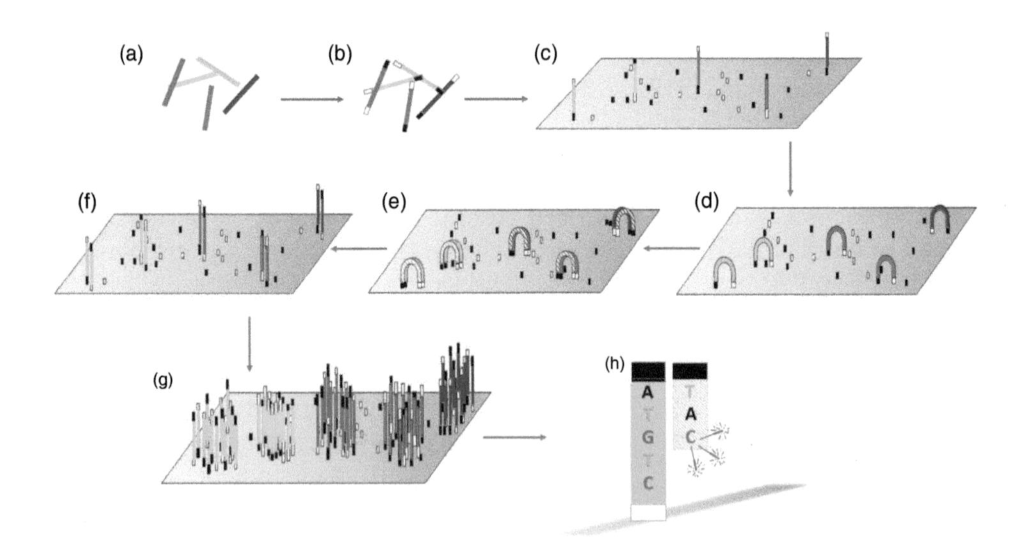

Figure 6.1 Illumina sequencing: (a) fragmentation. (b) Adapters annealed. (c) Fragments bind to glass surface loaded with complementary primers. (d) Bridge formation. (e) Bridge amplification. (f) Dissociation. (g) Cluster formed from repeated bridge formation and amplification. (h) Sequencing by synthesis and fluorescence detection.

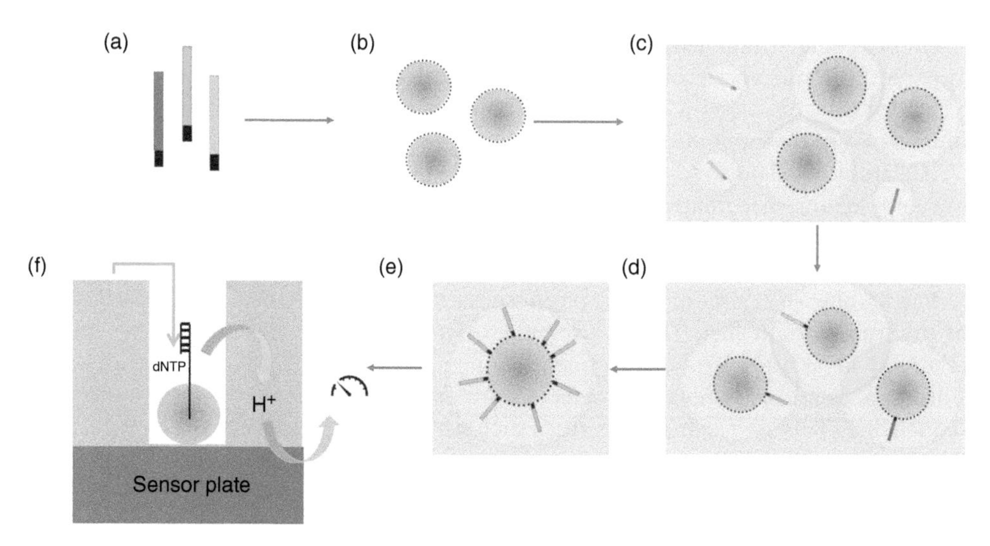

Figure 6.2 ThermoFisher Ion Torrent sequencing. (a) Adaptors annealed to fragmented library; (b) beads prepared with primers attached to surface; (c) beads and library mixed in an aqueous/oil mixture; (d) dilutions prepared so that only a single fragment anneals to a single bead; (e) standard PCR cycling on the emulsion enables clones of amplified bead-bound oligonucleotides to be developed with up to 100 000 copies on each bead; (f) detector chips with millions of wells, each holding a single bead, enables massively parallel sequencing of the bound oligonucleotides within the well. Different dNTPs are introduced to the chip in sequence and when one binds to a complementary nucleotide a hydrogen ion is released, resulting in a change of pH.

MPS has the ability to transform forensic genetics of the future. Although many of the techniques now being investigated could be analysed using the equipment that all forensic scientists will have in their laboratory, the often limited nature of forensic material puts MPS at an advantage where many different analyses can be done at the same time. Such is the high sensitivity of the current capillary electrophoresis methods for STR analysis and the success of DNA databases across the world, not forgetting the significant financial investment, a newer technology could never replace what had gone before unless it retained that ability to type the STRs that populate all worldwide national databases, and to demonstrate concordance; all MPS platforms have, however, revealed platform-specific problems with some loci (SE33, for example). Verogen, an offshoot from Illumina, has provided its ForenSeq methodology which adds SNPs, X STRs, and Y STRs to a set of autosomal STRs and accompanies these with the choice of including phenotypic and biogeographic ancestry SNPs.

Sequenced STRs

Sequencing of STRs using MPS techniques has enabled a significant increase in allelic variants to be identified (Devesse et al. 2017) and allowed examination of variation within the flanking region of the repeats.

Figure 6.3 illustrates a standard CE electropherogram and takes an example of the apparently homozygous 20 allele. The expected four-base repeat, typical of most STRs, when sequencing a collection of individuals with this allele, reveals a minimum of seven different variants.

The ability to detect increased variants provides more information when examining complex forensic mixtures. Figure 6.4 shows what appears to be two unbalanced alleles, possibly suggesting that more than one person has contributed. Sequencing confirms a clear mixture of two with a minor contribution from a third person.

The number of additional alleles seen is very variable, dependent on the locus. One of the least variant loci, TPOX, present in databases for historic reasons, remains relatively invariant even when sequenced (Figure 6.5), and D18S51, generally considered a useful polymorphic locus, shows little increase in allelic variation on sequencing. This is in contrast to loci such as D12S391 and D21S11 which more

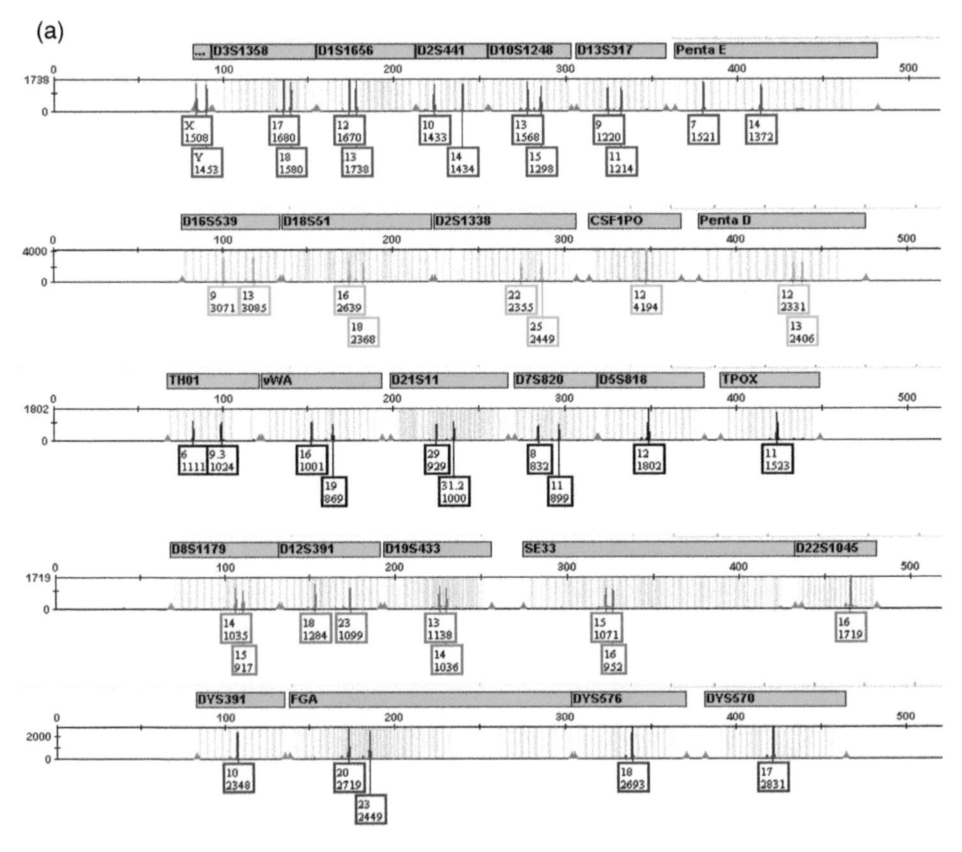

Figure 6.3 Increased allelic variation through sequencing. (a) Standard CE electropherogram revealing heterozygous and apparently homozygous alleles; (b) an allele 20 identified on CE which on sequencing of a population sample in (c) reveals five different sequence variations.

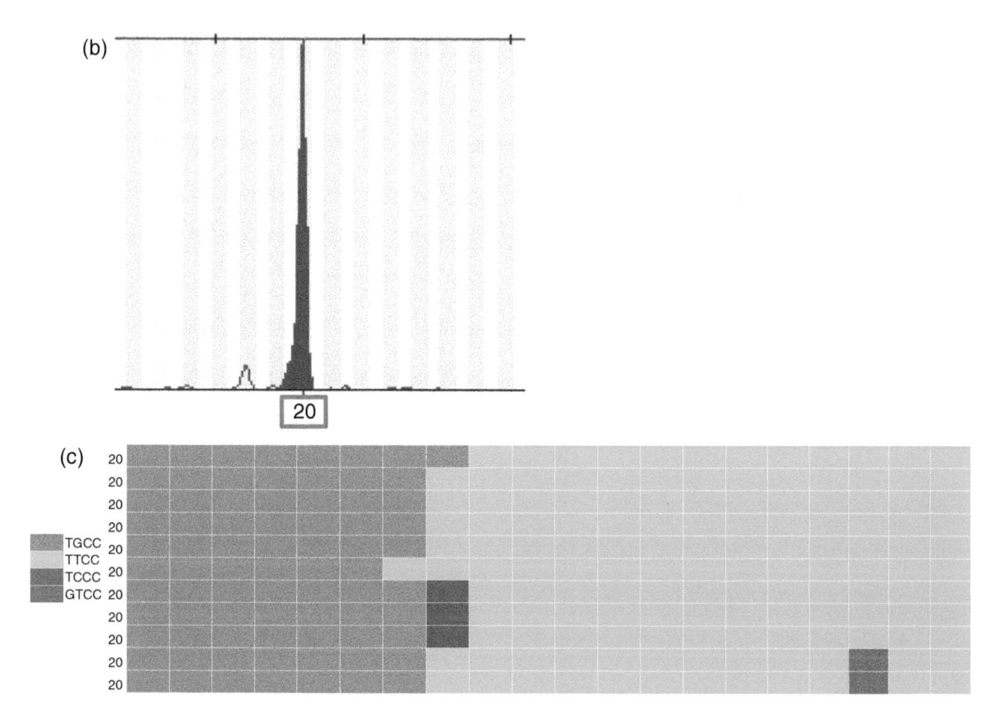

Figure 6.3 (*Continued*)

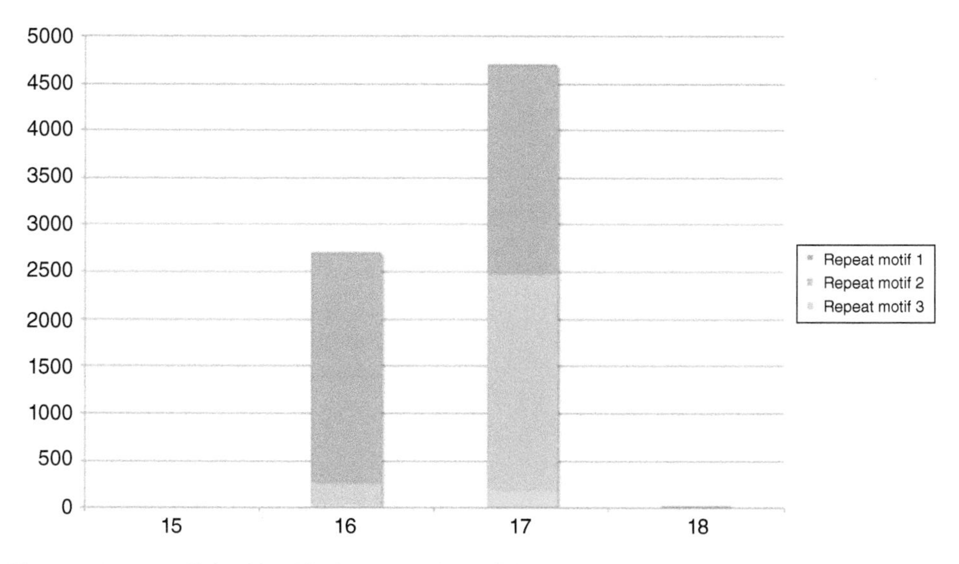

Figure 6.4 Two alleles identified as 16 and 17 alleles revealed at this locus through sequencing to comprise four alleles (a 16 allele, plus stutter, and three different 17 allelic variants).

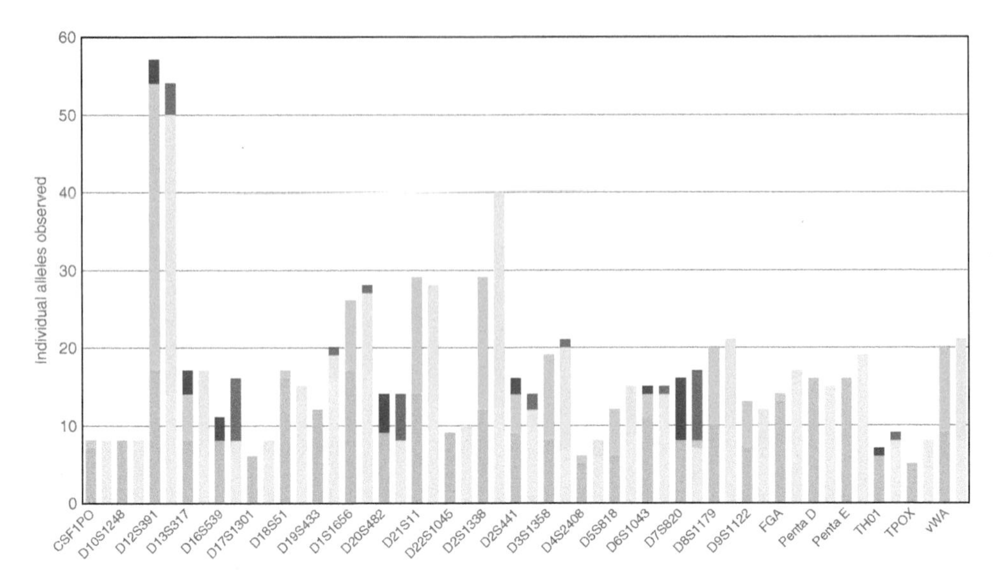

Figure 6.5 Increased sequence- based variation (in yellow) above the standard length-based variation (in blue) in two populations: European and South Asian. Additional variation in the flanking region is shown in purple.

than double the number of sequence-based alleles in the repeat region; other loci, such as D7S820, D16S539, and D20S482, have limited sequence variation in the repeat region but much higher-than-normal flanking region variation.

While the world populations typed here appear to show similar amounts of sequence variation, some alleles appear to be population specific, which may be a helpful addition in the prediction of geographic ancestry.

Forensic DNA phenotyping

In the absence of a database identification to name a potential suspect whose DNA is found at a crime scene, forensic geneticists are questioning what other information can be recovered from the DNA profile. Forensic DNA phenotyping (FDP) consists of a set of methodologies to infer information about externally visible characteristics (EVCs) of an individual but can also include the often-linked concept of someone's biogeographical ancestry and the not-so-often-included chronological age. All may offer valuable intelligence to criminal investigations and the identification of missing persons.

While none of these techniques are expected to be used in a court of law, they are framed as intelligence information to support a criminal investigation (Scudder et al. 2019). Their use is not without controversy mainly because it is perceived as reinforcing existing stigmatisation by concentrating on differences often associated with race (Skinner 2020), and some European jurisdictions already ban the use of some of these technologies because of fear of increased racial discrimination within the criminal justice system. The impact of the Human Genome Project on health-related research aiming to reduce health inequalities for all has also led to the risks posed by Duana Fullwiley in her statement 'Ancestry informative markers have emerged as a dispassionate research product that reportedly rises above subjective practices of racialising the phenotypes of others' (Fullwily 2008). While some consider the risk of FDP too great, through its inability to decouple the science from the social understandings leading to ideas of racial profiling, others point to 'visible' characteristics as not being private and that prediction of biogeographical ancestry does not depend on the former (Kayser and Schneider 2012) and argue for more education to reduce preconceptions and misinterpretation of FDP.

Biogeographic ancestry

While individuals vary considerably in their genetic composition, differences associated with the geographical locations of population groups are small (Lewontin 1972) driven by the high distance and geographical barriers that historically have prevented full interbreeding of humans. In 2002, Rosenberg et al. (2002) studied over 1000 individuals from over 52 populations collected as part of the Centre d'Etude du Polymorphisme Humain (CEPH) Human Diversity Project defining broad continental assignments: Eurasia, sub-Saharan Africa, East Asia, America, and Oceania. Limited refinements have been attained since, other than a decrease in the number of markers that are used to produce the distinction. The assumption that it is possible to define a country of origin has always been simplistic, as had been attempted by the UK Border Agency (Tutton et al. 2014), and unlikely to be attained other than in highly selected circumstances and population groups (Winney et al. 2012).

Biogeographical ancestry makes use of ancestry-informative marker (AIM) SNPs, and different collections have been developed since the first panel was developed from a framework published by Shriver et al. (1997) with the aim of being able to distinguish reliably, for forensic inference, different continental ancestries. One such panel was the 34-plex assay from the SNP*for*ID Consortium (Fondevila et al. 2013), designed for CE analysis, but with the advent of MPS better sets have been gathered, principally drawn from two collections from Kidd et al. (2014a) (differentiating African, European, East Asian, Native American, and Oceanians) and the EUROFORGEN Consortium (Phillips et al. 2014a) (differentiating African, European, and East Asians). A Eurasia-plex (Phillips et al. 2013), to differentiate European and South Asian populations, and a Pacifiplex panel (Santos et al. 2016a) for Oceanians have also been developed.

In commercial SNP assays, Verogen, in its ForenSeq DNA Signature Prep Kit, provides ancestry inference using 56 selected SNPs, and Thermo Fisher analyses 165 SNPs in its Precision ID Ancestry Panel.

Several prediction algorithms have been developed in order to assist in population differentiation.

Snipper

Snipper is a Bayesian web-based likelihood calculator for population classification (Santos et al. 2016b). It uses a set of samples of each population as training sets and assigns individuals using a maximum-likelihood calculation.

Figure 6.6 illustrates a heat map of predictions (-log likelihood) using *Snipper* predictions for two groups of Africans with self-declared West African or Somalian parental ancestries, and these numbers can be used to develop acceptance rules.

Structure

STRUCTURE (Pritchard et al. 2000*)* is a widely used Bayesian clustering approach with a graphical interface to provide cluster plots. The plots are described in a number (K) of classifications that best represent the populations being considered. Figure 6.7 shows a *STRUCTURE* plot ($K = 5$) of populations of different geographical heritage residents in the United Kingdom.

STRUCTURE analysis is also considered to be quite robust when genotypes are incomplete, as may be encountered in forensic casework, and to be relatively sensitive to undeclared admixture. While it is important not to overinterpret memberships of under about 10%, this figure reveals apparent admixture with mixed West African and African-Caribbean ancestry in individuals who self-declare their parents

	Caucasian	West African	Somalian	Chinese	Pakistan	Bangladesh
West_Africa	129.59	29.08	39.05	71.59	84.32	67.87
West_Africa	103.78	24.57	28.61	70.72	70.82	59.88
West_Africa	105.33	22.64	34.00	77.97	75.89	62.81
West_Africa	123.72	23.96	35.07	82.72	89.09	76.34
West_Africa	99.70	21.27	32.26	65.46	71.97	59.74
West_Africa	91.74	26.65	35.08	72.01	70.68	59.43
West_Africa	112.01	25.02	37.69	73.69	82.36	67.61
West_Africa	108.80	28.38	35.62	68.26	78.25	65.91
West_Africa	114.97	24.47	36.72	76.64	84.48	69.92
Somalian	81.39	38.27	29.87	61.67	59.47	50.69
Somalian	94.85	43.40	26.56	54.77	62.44	54.84
Somalian	108.20	42.23	31.40	59.19	65.17	52.78
Somalian	88.27	29.63	32.55	73.85	72.00	61.84
Somalian	70.87	36.72	27.32	64.06	52.90	48.06
Somalian	89.28	46.44	34.87	61.69	57.38	48.81
Somalian	67.13	52.22	27.47	48.78	46.13	39.76
Somalian	77.53	49.74	32.59	57.76	49.62	44.32
Somalian	85.90	49.60	31.39	59.73	59.35	53.82

Figure 6.6 Heat map of *Snipper* -log likelihood predictions according to a group of individuals' self-declared dual parental ancestry – West African or Somalian in this example. The darker orange values indicate the higher predictions according to the categories being considered here: Caucasian; West Africa; Somalia; China; Pakistan; and Bangladesh.

as from north-east Africa and also apparent admixture in a Pakistan heritage group who, historically, had significant British Caucasian presence particularly in the west of the Indian subcontinent.

Figure 6.8 shows the *STRUCTURE* plot of three individuals with declared mixed African/Caucasian ancestry making the admixture clear.

Principal components analysis

Cavalli-Sforza and Edwards (1963. p923) pioneered the use of principal components analysis and developed it in the 1990s as a way to summarise complex population data, with the principal components of variation being displayed on three axes. 3D plots are difficult to visualise, but 2D plots can illustrate population differences well when the populations are significantly divergent. *Snipper* software can also provide 2D plots.

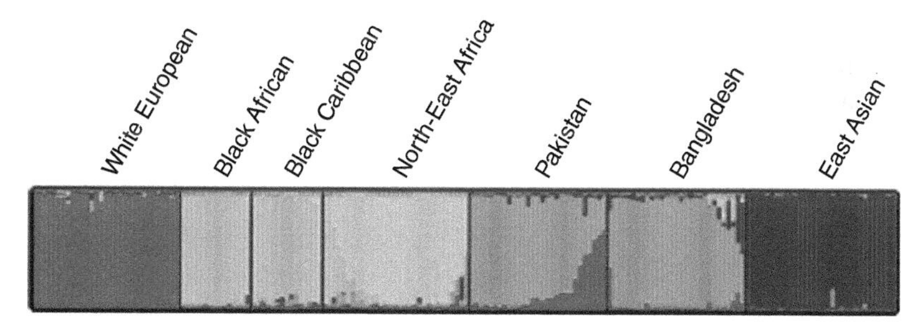

Figure 6.7 *STRUCTURE* plot (K =5) of UK residents with self-declared heritage in non-admixed individuals.

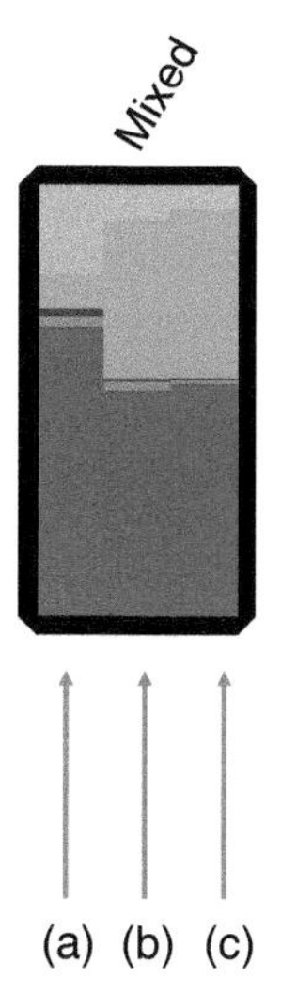

Figure 6.8 Structure plot of three individuals with mixed ancestry: (a) three Caucasian and one African grandparent; (b) and (c) two Caucasian and two African grandparents.

Figure 6.9 shows two PCA plots placing an individual in the left plot firmly within a representative population group and two individuals in the plot on the right who have known mixed ancestry. Care must be taken when inferring ancestry in any models as the true ancestry may not be represented in the plot, and this uncertainty must always be made explicit in forensic intelligence reports.

Further analyses

Although analysis of AIMs forms the basis for biogeographical analysis prediction, it is possible that use of other markers such as STRs, indels, and tri-allelic SNPs could provide additional population-specific inference.

Phillips et al. (2014b) highlighted the population-specific STRs in some populations, but they were not frequent enough to be particularly helpful. Analysis of sequenced STRs may provide additional assistance when incorporated with AIMs. Analysis has revealed several population-specific alleles. Figure 6.10 is a Sankey plot showing the large number of unique alleles that are only seen in one population; further evaluation will be needed to see if any of these can improve population discrimination. The loci D2S1338, D12S391, and D21S11 have the largest number of population unique alleles.

Microhaplotypes are short fragments of DNA with two or more closely linked SNPs producing multi-allelic haplotypes. Conceptualised by Kidd et al. (2014b) this group has already identified informative haplotypes in several populations, and Cheung et al. (2019) have demonstrated that the microhaplotypes outperform biallelic markers in the population metrics of allele frequency differences, Rosenberg's informativeness, and the fixation index.

Both tri- and tetra-allelic SNPs are more common than originally thought and reveal significant population differentiation. Some are already included in the Global-AIMs panel, and discovery of more is likely to contribute further to population differentiation.

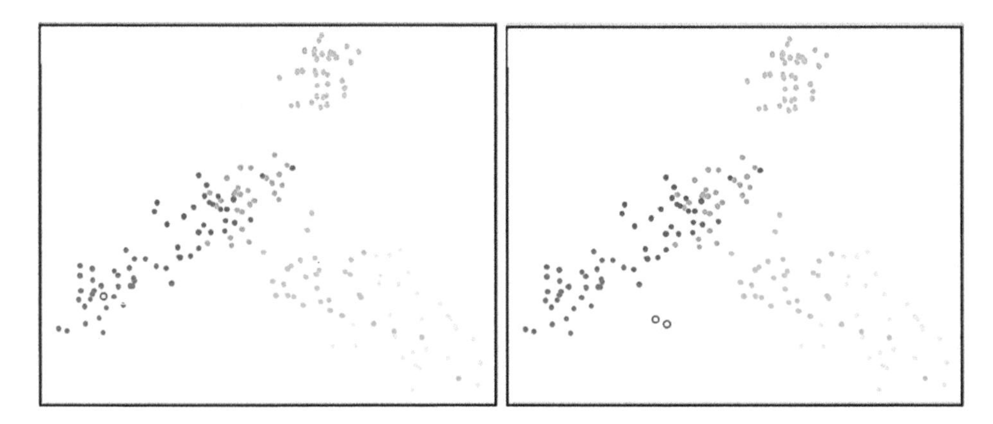

Figure 6.9 PCA plots of Caucasian (blue), Pakistan (dark pink), Bangladesh (orange), Chinese (pale pink), north-east African (green) and west African (ice-blue) populations with questioned samples (open circles) falling into a Caucasian set (top panel) and outside all sets (bottom panel).

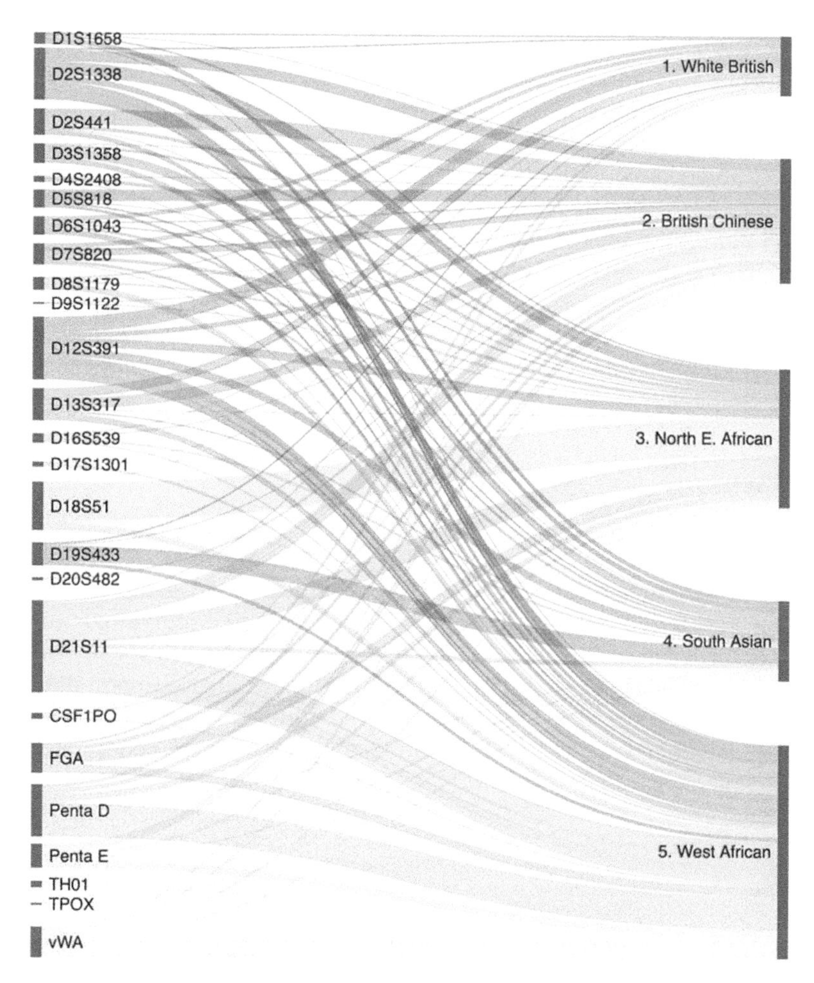

Figure 6.10 Sankey plot of sequenced STRs revealing population specific STRs; the thicker the band the higher frequency allele it represents.

Externally visible characteristics (EVCs)

The ability to produce a realistic drawing of an individual whose DNA has been found at a crime scene has long been the vision of those working with DNA. Parabon NanoLabs has been working in this area for many years, commercialising their 'Snapshot' algorithm that uses over 800 000 SNPs to predict skin colour, eye colour, hair colour, freckling, and ancestry, producing a composite facial sketch. Although the service has been used by law enforcement, critics point to its lack of peer review, and an investigation by the New York Times produced results that were not very impressive (Murphy 2015). While comparison of a DNA-generated sketch from a known individual can reveal similarities, the false positives that are likely to result from using that picture to identify an unknown raise considerable concern. In addition, the composite picture does not provide any indication of the age of the donor and requires large amounts of DNA that are less likely to be encountered at a crime scene.

Visible phenotypic characteristics, provided independently, may provide helpful intelligence if presented with information about likely error as predictions are nothing like those associated with identification of a person and may be misunderstood. Some features are easier to predict such as eye colour, but others are polygenic in nature and more difficult to predict, such as height. Much of the research has also been conducted in populations that are dominated with European subjects and for use these techniques must be validated in a range of world populations.

Human pigmentation genes (eye, hair, and skin colour) are the least complex in that they provide a high genetic contribution from a low number of genetic markers.

Eye colour

Research done since Frudakis published a series of predictive SNPs in the *OCA2* gene (Frudakis et al. 2007) led to the IrisPlex tool from Walsh et al. (2011) that provides strong prediction from only six SNPs in limited DNA samples, suitable for forensic use and tested on more than 9000 European-wide individuals to give eye colour accuracy levels of 84%, increasing to 93% when only blue or brown eye colour is considered (Walsh et al. 2012). Predictions in non-European populations with European admixture perform less well, however, with the poorer predictions being particularly associated with the rs12913832 SNP in *HERC2* in its heterozygous form.

Hair colour

Red hair colour prediction based on variants in the *MCR1* gene was first published in 2001 (Grimes et al. 2001), and work from Branicki et al. (2011) was developed into the combined HIrisPlex assay (Walsh et al. 2013) using a total of 24 SNPs to provide prediction accuracies of 87.5% (black), 80% (red), 78.5% (brown), and 69.5% (blonde), independent of the biogeographical origin. Reduced accuracy in those predicting blonde with a brown phenotype appears to be associated with the darkening effects of blonde individuals as they age.

Skin colour

Skin colour prediction has caused more difficulties because most of the genetic studies have been done within population groups, whereas the greatest differences are seen between populations. The first comprehensive study was done in 2014 by Maroñas et al. (2014), but this was limited due to the small sample size. With the aim of improving predictions across a global set, Walsh and colleagues added 36 SNPs from 16 genes to produce the HIrisPlex-S (Chaitanya et al. 2018) prediction tool, validating it in 17 different populations, providing predictions of very pale, pale, intermediate, dark, and dark-black skin colour. While sun exposure can influence the phenotype, validation in a set of individuals with the full range of skin colours produced a high correlation between predicted phenotype and skin photographs. Further validation in a global set (D. Syndercombe-Court, King's College London, unpublished results) has, however, revealed more prediction errors seen in Asian populations, and more work needs to be done in a wider population set.

Additional EVCs

A large number of genome-wide association studies in recent years have identified many additional genes associated with freckles, hair structure, male-pattern baldness, eyebrow colour, face shape, and body height, displaying varying degrees of heritability, although all models developed have a lower predictive success in comparison with HIrisPlex-S.

Height, while being highly heritable, has been a relatively disappointing trait to study. Early successes in predicting the 5% tallest Dutch (Liu et al. 2014) have not been followed with further successes to any great extent. Extensively studied, researchers have identified 697 height-significant variants in 423 genetic loci, but these only explain about 20% of height hereditability (Wood et al. 2014). Increasing the number of associated SNPs to 9500 explains about another 10%, and only by using MPS techniques would it be possible to analyse the number of SNPs in forensic casework. Although more predictive SNPs have been discovered that significantly alter height (Marouli et al. 2017), these are at low frequency in the population, and whether sufficient loci will ever be discovered to improve the prediction of height is currently unclear.

The story relating to face shape is not dissimilar but with even fewer genes being associated with face shape and suggesting that likely a large number of genetic variants are implicated, and Kayser predicts that accurate facial prediction and forensic capability will be unlikely to be achieved, if ever (Kayser 2015).

Age

Prediction of the chronological age of the donor of a crime scene stain, while being independently useful, is helpful as age also impacts on a large number of EVCs, such as hair loss, hair greying, skin wrinkles, and height.

Several biomarkers of age have been researched making use of DNA; both telomere shortening and mitochondrial DNA deletion approaches achieved poor accuracy, however (Meissner and Ritz-Timme 2010). A third DNA-based method relies on quantifying signal-joint T-cell receptor excision circles (sjTRECs), produced as T lymphocytes mature, and the amount is inversely related to age (Zubakov et al. 2010). Age prediction is also poor, limited to blood, and adversely influenced by immune responses.

Quantification of epigenetic factors, such as DNA methylation, that can change the way a person's genetics are expressed at different stages of life, has provided the most accurate and sensitive biomarkers for age to date (Aliferi et al. 2018). Because epigenetic signatures vary, not only with age but also with disease and environmental factors, markers must be carefully selected so that disease factors do not bias the determination. Environmental exposure can also impact on the estimation of age, and study samples must reflect the population of interest.

While MPS has enabled increased sensitivity and decreased cost and analysis time important for forensic casework, most of the models that have been developed have used blood as the tissue in which predictions are modelled.

One of the main advantages for the use of DNA methylation as a biomarker is its high stability with no significant change being evident even after decades of storage (Hollegaard et al. 2013) and has been successfully applied in populations of distinct biogeographical accuracy (Zaghlool et al. 2015).

DNA methylation is a chemical modification that, in mammalian cells, predominantly affects cytosines when these are followed by guanines in a 5′–3′ direction in the DNA double helix and results in the addition of a methyl group ($-CH_3$) to the 5′ carbon and are known as CpG dinucleotides.

Quantifying DNA methylation using current technologies relies on pre-treatment of the DNA molecule and bisulphite conversion (Frommer et al. 1992) of CpGs, resulting in unmethylated cytosines being preferentially converted into uracils. Amplification then replaces uracils with thymine, and sequencing can then ratio the number of cytosines (methylated status) and the number of cytosines plus thymines (total methylation) to determine the methylation proportion as detected in a single-base extension assay using MPS technologies. One of the main disadvantages of bisulphite treatment is the high template loss and requires amounts of DNA that are greater than would normally be analysed in other techniques and, similar to other forensic analyses, also relies on interrogation of a limited number of genetic areas so that analysis can be achieved with small amounts of crime scene material.

Eleven CpG candidates from 10 genes have been used to develop the blood model shown in Figure 6.11 with an average prediction error of ±3.3 years in a test set maintaining accuracy down to 5 ng DNA input for conversion. This age estimation method revealed increased accuracy for individuals under the age of 54 years, arguably more relevant in forensic casework, with an average error ±2.6 years.

While new sets of markers and models need to be defined for other tissues of forensic relevance, all need careful selection to examine for amplification bias that some markers have which can impact on model accuracy. The most successful marker discovered to date relates to CpGs in the *ELOVL2* gene, and some models have been developed using this gene only (Zbieć-Piekarska et al. 2015). The success of these markers is not just their high correlation with age but also the large methylation range that they exhibit over the human lifespan, and this characteristic is an important factor in successful design.

Forensic genealogy

On 24 April 2018, a 72-year-old former police officer, Joseph James DeAngelo, was arrested and charged with eight counts of first-degree murder, and in June 2020 he pleaded guilty to multiple counts of murder, kidnapping, and rape between 1973 and 1986. He was sentenced to life imprisonment in August 2020 (Anguiano 2020).

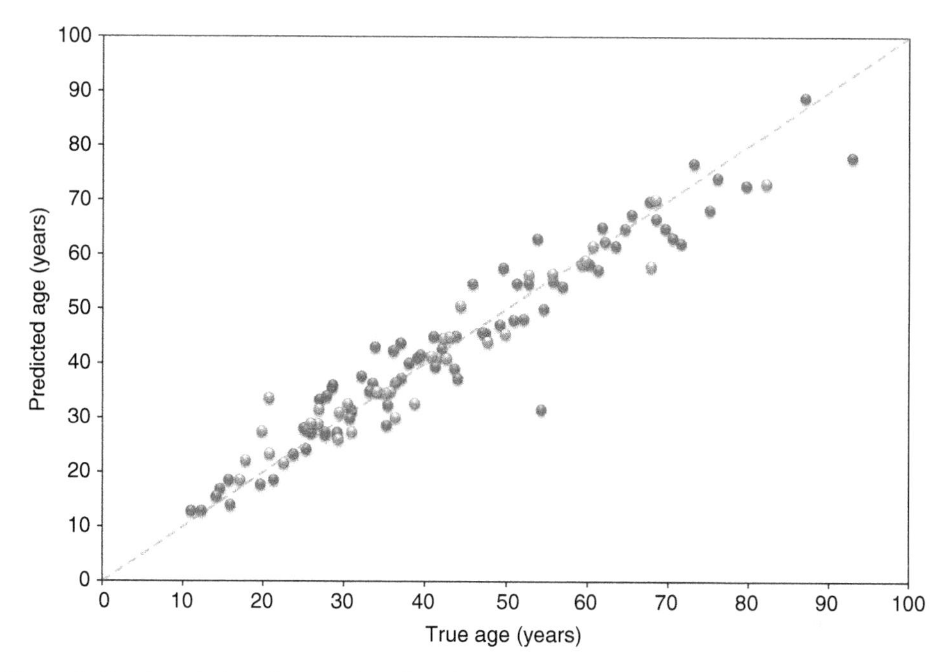

Figure 6.11 Association of true and predicted age derived from a training set (in blue) and a test set (in silver).

DeAngelo had been identified through a genetic genealogy hobbyist site used by many to identify their ancestors. Variously known as the East Coast Rapist and the Original Night Stalker crime scene DNA had, in 2001, identified that they were one and the same; in 2013, he was first referred to as the Golden State Killer.

Companies such as 23andMe, AncestryDNA, FamilyTreeDNA, and MyHeritage all provide direct-to-consumer DNA tests for genealogical purposes. Another site, GEDmatch, provided a free service that allowed anyone to load their appropriately formatted DNA to the site so that relatives who may have used different companies to test their DNA could be discovered. Databases used by the police, such as CODIS, consist of STRs and can at best identify only very close relatives if there is no direct match. In contrast, genealogy companies test a genome-wide set of SNPs with many thousands of SNPs on a single high density or whole-genome array to reveal probable third and fourth cousins that are present linked through the lengths of DNA shared on individual chromosomes. Use of the commercial services by law enforcement without a warrant was not legal in 2018, but there would be nothing to prevent a set of compatible SNPs being tested through other means and the profile put together for a GEDmatch search. Although not all genealogical ancestors are also genetic ancestors, with an estimated 950 000 profiles on the GEDmatch site in 2018 there was a good chance that possible ancestors would be found in the search for the Golden State Killer. Investigators appeared to have interrogated the GEDmatch database and identified an extensive set of around 1000 distant relatives who shared a common great-great-great grandparent, with around 20 of those related as third or fourth cousins (Zhang 2018). Subsequent genealogy followed taking many thousands of hours, but two individuals with links to areas close to the crime scenes were thought the most likely. Standard DNA profiling on a discarded item from DeAngelo produced a complete DNA match with the crime scene, and the arrest followed.

While the identification and arrest of a serial killer has been welcomed worldwide, the potential invasion of people's privacy and the use of their DNA for a purpose other than what they gave their consent for has caused considerable concern (Biometrics and Forensic Ethics Group 2020). Nevertheless, the procedure has been used subsequently to identify some 40 individuals suspected of committing serious crimes. While GEDmatch allowed police to use the database for serious crimes, when it was criticised for its use in identifying an individual for a less-serious crime, GEDmatch immediately changed its terms and conditions, opting everyone out from law enforcement matching, while allowing people to opt in if they wished; only about 170 000 did at the time and their policy has changed again since. Most other companies reinforced their conditions not to allow police to use the database without a warrant, but FamilyTreeDNA, who had already been working with the FBI, offered an 'opt-out' option to their US contributors; Europeans were automatically opted out because of data protection laws, but reportedly, few US individuals have actively removed their DNA.

What appeared to be a legal, if innovative, investigation, has now been revealed as much more concerning, ethically. The difficult ongoing genealogical investigation was overtaken through a process which involved the FBI making a search of the FamilyTreeDNA site without the site's explicit knowledge and a civilian genealogist loading the profile to the MyHeritage site (St John 2020), identifying a probably second cousin, closer than previously identified. This underhand process is only likely to increase distrust in law enforcement in the future (Callaghan 2019).

The obvious success of genetic genealogy in North America has caught the interest of law enforcement across the world, and recent cases in Canada (Premji 2020) and Sweden (Hofverberg 2020) are testament to its utility, particularly to resolve cold cases. Experience in the United Kingdom where DNA collected from a crime scene immediately provides a suspect name in about 66% of cases (Home Office – National Police Chiefs' Council 2020), and the associated costs, means that the use of genetic genealogy, even if approved, is likely to be limited to serious cold cases.

The United States Department of Justice has set out criteria for the use of forensic genetic genealogy (United States Department of Justice 2019): unsolved violent crimes; forensic samples from a putative perpetrator; and unidentified human remains, and they may also authorise its use in other situations when the circumstances surrounding the criminal act(s) present a substantial and ongoing threat to public safety or national security. Prior to use, the investigator must have loaded a standard forensic profile to the CODIS database and pursued all relevant investigative leads in the case of unidentified remains.

Considering the future potential of using the technique in other jurisdictions, Scudder et al. (2020) have looked at this from within an Australian context and have suggested that the application is considered in the context of the type of offence and prior genetic testing, including FDP, in order to understand the likely success of the technique. The practice must also include careful consideration of privacy as the practice will make use of genetic data from not only individuals who have uploaded their profiles to online databases but also of individuals who are identified by inference of being a perpetrator or a close relative and may be approached and questioned by investigators based on genetic information only (NBC News 2018).

The possibility of success in the casework has been investigated in two small experiments (Aldhous 2019; Thomson et al. 2019), with around a 50% success rate, but is likely to be lower today because the availability of the databases to law enforcement has been massively reduced. Further, if the perpetrator of a crime does not have Western European heritage, the probability is that the search will fail because of the underrepresentation of individuals from other parts of the world in searchable genealogical databases.

In 2019, Verogen acquired GEDmatch with the aim of being able to maintain the privacy of greater than one million contributors. Instead of making use of DNA-heavy SNP arrays they plan to design custom sets targeted for low-level and degraded DNA (Verogen 2020), and this approach could also mean that privacy is improved by avoiding the typing of medically sensitive sites.

Conclusion

Forensic genetics has been transformed since sequencing the human genome was completed in 2003. Not only has sensitivity of forensic DNA testing increased to the point where only a few cells can provide reliable results, so have the techniques to analyse their increasing complexity.

In the years to come DNA, both nuclear and mitochondrial, and the associated forms of RNA will provide more answers to uncertainties, such as those posed by activity-level questions as well as provision of better phenotypic information. DNA testing will become more efficient, with translation to reliable onsite investigation of crime scene material becoming a reality. Non-human genetic information will come to the fore, offering further intelligence to the criminal justice system, and not least, the enormous research effort that has already started in the area of forensic metagenomics (Mason-Buck et al. 2020).

Forensic research in all forms, however, relies on sufficient funding being available in the future, and competition for this in the light of other world priorities may slow progress for some time.

References

Adamowicz, M., Clarke, J., Rambo, T. et al. (2019). Validation of MaSTR software: Extensive study of fully-continuous probabilistic mixture analysis using PowerPlex Fusion 2 – 5 contributor mixtures. *Forensic Sci Int Genet Suppl Ser* 7: 641–643.

Aldhous, P. (2019). We tried to find 10 BuzzFeed employees just like cops did for the Golden State Killer. *BuzzFeed News*. https://www.buzzfeednews.com/article/peteraldhous/golden-state-killer-dna-experiment-genetic-genealogy (accessed 20 December 2020).

Aliferi, A., Ballard, D., Gallidabino, M. et al. (2018). DNA methylation-based age prediction using massively parallel sequencing data and multiple machine learning models. *Forensic Sci Int Genet* 37: 215–226.

Amankwaa, A. and McCartney, C. (2019). The effectiveness of the UK national DNA database. *Forensic Sci Int Synergy* 1: 45–55.

Anguiano, D. (2020). Golden state killer: Joseph DeAngelo sentenced to life in prison. *The Guardian*. https://www.theguardian.com/us-news/2020/aug/21/golden-state-killer-joseph-deangelo-to-be-sentenced-to-life-in-prison (accessed 13 July 2021).

Balding, D.J. (2013). Evaluation of mixed-source, low-template DNA profiles in forensic science. *Proc Natl Acad Sci USA* 110: 12241–12246.

Benschop, C.C.G., Hoogenboom, J., Hovers, P. et al. (2019). NAxs/DNAStatistX: development and validation of a software suite for the data management and probabilistic interpretation of DNA profiles. *Forensic Sci Int Genet* 42: 81–89.feri.

Bentley, D.R., Balasubramanian, S., Swerdlow, H.P. et al. (2008). Accurate whole human genome sequencing using reversible terminator chemistry. *Nature* 456: 53–59.

Bieber, F.R., Buckleton, J.S., Budowle, B. et al. (2016). Evaluation of forensic DNA mixture evidence: protocol for evaluation, interpretation, and statistical calculations using the combined probability of inclusion. *BMC Genet* 17: 125.

Biometrics and Forensic Ethics Group (2020). Should we be making use of genetic genealogy to assist in solving crime? A report on the feasibility of such methods in the UK. *Gov.UK*. https://www.gov.uk/government/publications/use-of-genetic-genealogy-techniques-to-assist-with-solving-crimes/should-we-be-making-use-of-genetic-genealogy-to-assist-in-solving-crime-a-report-on-the-feasibility-of-such-methods-in-the-uk-accessible-version (accessed 20 December 2020).

Bleka, Ø., Storvik, G., and Gill, P. (2016, 2016). EuroForMix: an open source software based on a continuous model to evaluate STR DNA profiles from a mixture of contributors with artefacts. *Forensic Sci Int Genet* 21: 35–44.

Branicki, W., Lui, F., van Duijn, K. et al. (2011). Model-based prediction of human hair color using DNA variants. *Hum Genet* 129: 443–454.

Brenner, C.H. (2011). The mythical 'exclusion' method for analyzing DNA mixtures – does it make any sense at all? *Proc Am Acad Forensic Sci* 17: 79.

Brenner, C.H. (2015). The DNA-view mixture solution. Poster presented at the International Society for Forensic Genetics 26th Congress, Krakow, 2015.

Bright, J.-A., Hopwood, A., Curran, J.M. et al. (2014). A guide to forensic DNA interpretation and linkage. *Profiles in DNA*. https://www.promega.co.uk/resources/profiles-in-dna/2014/a-guide-to-forensic-dna-interpretation-and-linkage (accessed 20 December 2020).

Bright, J.-A., Taylor, D., McGovern, C. et al. (2016). Developmental validation of STRmix™, expert software for the interpretation of forensic DNA profiles. *Forensic Sci Int Genet* 23: 226–239.

Butler, J.M., Coble, M.D., and Vallone, P.M. (2007). STRs versus SNPs: thoughts on the future of forensic DNA testing. *Forensic Sci Med Pathol* 3: 200–205.

Callaghan, T.F. (2019). Responsible genetic genealogy. *Science* 365: 155.

Carney, C., Whitney, S., Vaidyanathan, J. et al. (2019). Developmental validation of the ANDE rapid DNA system with FlexPlex assay for arrestee and reference buccal swab processing and database searching. *Forensic Sci Int Genet* 40: 120–130.

Cavalli-Sforza, L.L. and Edwards, A.W.F. (1963). Analysis of human evolution. *Genetics today: Proceedings of the 11th International Congress of Genetics,* The Hague, The Netherlands. New York: Pergamon. Vol. 3, 923–993.

Chaitanya, L., Breslin, K., Zuñiga, S. et al. (2018). The HIrisPlex-S system for eye, hair and skin colour prediction from DNA: introduction and forensic developmental validation. *Forensic Sci Int Genet* 35: 123–135.

Cheung, E.Y.Y., Phillips, C., Eduardoff, M. et al. (2019). Performance of ancestry-informative SNP and microhaplotype markers. *Forensic Sci Int Genet* 43: 102141.

Clayton, T.M., Whitaker, J.P., Sparkes, R. et al. (1998). Analysis and interpretation of mixed forensic stains using DNA STR profiling. *Forensic Sci Int* 7: 92–97.

Corte-Real, F. (2004). Forensic DNA databases. *Forensic Sci Int* 146S: S143–S144.

(2008). Council of Europe: European court of human rights grand chamber. Case of S and Marper v The United Kingdom. (Applications nos. 30562/04 and 30566/04). Judgement Strasbourg 4 December 2008.

Council of the European Union, Brussels (2009). 15870/09 *ENFOPOL 287 CRIMORG* 170:1–7.

Cowell, R., Graversen, T., Lauritzen, S. et al. (2015). Analysis of forensic DNA mixtures with artefacts. *J R Stat Soc Ser C* 64: 1–48.

Devesse, L., Ballard, D., Davenport, L. et al. (2017). Concordance of the ForenSeq system and characterisation of sequence-specific autosomal STR alleles across two major population groups. *Forensic Sci Int Genet* 34: 57–61.

Dlugosz v Regina (2013). England and Wales Court of Appeal [ECWA Crim 2] (Criminal Division) Crim. App. Rep 32, 30 January 2013.

Executive Office of the President: President's Council of Advisors on Science and Technology (2016). Report to the President. Forensic science in criminal courts: ensuring scientific validity of feature-comparison methods. https://obamawhitehouse.archives.gov/sites/default/files/microsites/ostp/PCAST/pcast_forensic_science_report_final.pdf (accessed 20 December 2020).

Evett, I., and Pope, S. (2013). Science of mixed results. *The Law Society Gazette*. https://www.lawgazette.co.uk/practice-points/science-of-mixed-results/5036961.article (accessed 20 December 2020).

FBI (2019). Rapid DNA. https://www.fbi.gov/services/laboratory/biometric-analysis/codis/rapid-dna (accessed 20 December 2020).

Fondevila, M., Phillips, C., Santos, C. et al. (2013). Revision of the SNP*for*ID 34-plex forensic ancestry test: assay enhancements, standard reference sample genotypes and extended population studies. *Forensic Sci Int Genet* 7: 63–74.

Forensic Science Regulator (2018). DNA mixture interpretation FSR-G-222 issue 3. https://www.gov.uk/government/publications/dna-mixture-interpretation-fsr-g-222 (accessed 20 December 2020).

Frommer, M., McDonald, L.E., Millar, D.S. et al. (1992). A genomic sequencing protocol that yields a positive display of 5-methylcytosine residues in individual DNA strands. *Proc Natl Acad Sci USA* 89: 1827–1831.

Frudakis, T., Terravainen, T., and Thomas, M. (2007). Multilocus *OCA2* genotypes specify iris human iris colours. *Hum Genet* 122: 311–326.

Fullwily, D. (2008). The biologistical construction of race: 'admixture' technology and the new genetic medicine. *Soc Stud Sci* 38: 695–735.

GitHub (2019). LRmixStudio. https://github.com/smartrank/lrmixstudio (accessed 20 December 2020).

Gov.UK (2020). National DNA database statistics. http://www.gov.uk/government/statitics/national-dna-database-statistics (accessed 20 December 2020).

Grimes, E.A., Noake, P.J., Dixon, L. et al. (2001). Sequence polymorphism in the human melanocortin 1 receptor gene as an indicator of the red hair phenotype. *Forensic Sci Int* 122: 124–129.

Haned, H., Slooten, K., and Gill, P. (2012). Exploratory data analysis for the interpretation of low template DNA mixtures. *Forensic Sci Int Genet* 6: 262–774.

Hare, D. (2012). Expanding the core CODIS loci in the United States. *Forensic Sci Int Genet* 6: e52–e54.

Hare, D. (2015). Selection and implementation of expanded CODIS core loci in the United States. *Forensic Sci Int Genet* 17: 33–34.

Hofverberg, E. (2020). FALQs: the use of DNA from genealogy sites to solve crimes in Sweden. In Custodia Legis: Law Librarians of Congress. https://blogs.loc.gov/law/2020/09/falqs-the-use-of-dna-from-genealogy-sites-to-solve-crimes-in-sweden (accessed 20 December 2020).

Hollegaard, M.V., Grauholm, J., Nørgaard-Pedersen, B. et al. (2013). DNA methylome profiling using neonatal dried blood spot samples: a proof-of-principle study. *Mol Genet Metab* 108: 225–231.

Home Office – National Police Chiefs' Council (2020). National DNA database strategy board biennial report (2018–2020). https://assets.publishing.service.gov.uk/government/uploads/system/uploads/attachment_data/file/913015/NDNAD_Strategy_Board_AR_2018-2020_print.pdf (accessed 20 December 2020).

Hopwood, A.J., Hurth, C., Yang, J. et al. (2010). Integrated microfluidic system for rapid DNA analysis: sample collection to DNA profile. *Anal Chem* 82: 6991–6999.

Inman, K., Rudin, N., Cheng, K. et al. (2015). Lab retriever: a software tool for calculating likelihood ratios incorporating a probability of drop-out for forensic DNA profiles. *BMC Bioinform* 16: 298.

INTERPOL (2019). Global DNA profiling survey results. www.interpol.int/en/How-we-work/Forensics/DNA (accessed 20 December 2020).

Kayser, M. and Schneider, P. (2012). Reply to "Bracketing off population does not advance ethical reflection on EVCs: a reply to Kayser and Schneider" by A. M'charek, V. Toom, and B. Prainsack. *Forensic Sci Int Genet* 6: e18–e19.

Kayser, M. (2015). Forensic DNA phenotyping: predicting human appearance from crime scene material for investigate. *Forensic Sci Int Genet* 18: 33–48.

Kidd, K.K., Speed, W.C., Pakstis, A.J. et al. (2014a). Progress toward an efficient panel of SNPs for ancestry inference. *Forensic Sci Int Genet* 10: 23–32.

Kidd, K.K., Pakstis, A.J., Speed, W.C. et al. (2014b). Current sequencing technology makes microhaplotypes a new type of genetic marker for forensics. *Forensic Sci Int Genet* 12: 215–224.

Lewontin, R.C. (1972). The apportionment of human diversity. *Evol Biol* 6: 381–398.

Liu, F., Hendricks, A.E.J. et al. (2014). Common DNA variants predict tall stature in Europeans. *Hum Genet* 133: 587–597.

Manabe, S., Morimoto, C., Hamano, Y. et al. (2017). Development and validation of open-source software for DNA mixture interpretation based on a quantitative continuous model. *PloS One* 12: e0188183.

Maroñas, O., Phillips, C., Söchtig, J. et al. (2014). Development of a forensic skin colour predictive test. *Forensic Sci Int Genet* 13: 34–44.

Marouli, E., Graff, M., Medina-Gomez, C. et al. (2017). Rare and low-frequency coding variants alter human adult height. *Nature* 542: 186–190.

Mason-Buck, G., Graf, A., Elhaik, E., et al. (2020). DNA based methods in intelligence – moving towards metagenomics. https://www.preprints.org/manuscript/202002.0158/v1 (accessed 20 December 2020).

Meissner, C. and Ritz-Timme, S. (2010). Molecular pathology and age estimation. *Forensic Sci Int* 203: 34–43.

Mitchell, A.A., Tamariz, J., O'Connell, J. et al. (2011). Likelihood ratio statistics for DNA mixtures offering for drop-out and drop-in. *Forensic Sci Int Genet Suppl Ser 3*: E240–E241.

Murphy, H. (2015). I've just seen a (DNA generated) face. *New York Times*. https://www.nytimes.com/2015/02/24/science/dna-generated-faces.html (accessed 20 December 2020).

NBC News (2018). DNA used in hunt for Golden State Killer previously led to wrong man. *Associated Press*. https://www.nbcnews.com/news/us-news/dna-used-hunt-golden-state-killer-previously-led-wrong-man-n869796 (accessed 20 December 2020).

NCBI Insights (2019). *National Library of Medicine*. dbSNP celebrates 20 years! https://ncbiinsights.ncbi.nlm.nih.gov/2019/10/07/dbsnp-celebrates-20-years/#:~:text=dbSNP%20was%20established%20in%20August,of%20small%20scale%20nucleotide%20variants (accessed 20 December 2020).

Perlin, M.W., Legler, M.M., Spencer, C.E. et al. (2011). Validating TrueAllele DNA mixture interpretation. *J Forensic Sci* 56: 1430–1447.

Phillips, C., Freire Aradas, A., Kriegel, A.K. et al. (2013). Eurasiaplex: a forensic SNP assay for differentiating European and South Asian ancestries. *Forensic Sci Int Genet* 7: 259–366.

Phillips, C., Parson, W., Lundsberg, B. et al. (2014a). Building a forensic ancestry panel from the ground up: the EUROFORGEN Global AIM-SNP set. *Forensic Sci Int Genet* 11: 13025.

Phillips, C., Gelabert-Beseda, M., Fernandez-Formosa, L. et al. (2014b). New turns from old STaRs: enhancing the capabilities of forensic short tandem repeat analysis. *Electrophoresis* 35: 3173–3187.

Premji, Z. (2020). How genetic geneaology identified the killer in a Toronto cold case. *CBC News*. https://www.cbc.ca/news/technology/genetic-genealogy-identifies-killer-jessop-1.5764107 (accessed 20 December 2020).

Pritchard, J.K., Stephens, M., and Donnelly, P. (2000). Inference of population structure using multilocus genotype data. *Genetics* 155: 945–959.

Puch-Solis, R., Rodgers, L., Mazumder, A. et al. (2013). Evaluating forensic DNA profiles using peak heights, allowing for multiple donors, allelic dropout and stutters. *Forensic Sci Int Genet* 7: 555–563.

Puch-Solis, R. and Clayton, T. (2014). Evidential evaluation of DNA profiles using a discrete statistical model implemented in the DNA LiRa software. *Forensic Sci Int Genet* 6: 762–774.

Romsos, E.L., French, J.L., Smith, M. et al. (2020). Results of the 2018 rapid DNA maturity assessment. *J Forensic Sci* 65: 953–959.

Rosenberg, N.A., Pritchard, J.K., Weber, J.L. et al. (2002). Genetic structure of human populations. *Science* 298: 2381–2385.

Salceda, S., Barican, A., Buscaino, J. et al. (2017). Validation of a rapid DNA process with the RapidHIT ID system using GlobalFiler Express chemistry, a platform optimized for decentralized testing environments. *Forensic Sci Int Genet* 28: 21–34.

Sanger, F., Nicklen, S., and Coulson, A.R. (1977). DNA sequencing with chain-terminating inhibitors. *Proc Natl Acad Sci USA* 74: 5463–5467.

Santos, C., Phillips, C., Fondevila, M. et al. (2016a). Pacifiplex: an ancestry0informative SNP panel centred on Australia and the Pacific region. *Forensic Sci Int Genet* 20: 71–80.

Santos, C., Phillips, C.P., Goméz Tato, A. et al. (2016b). Inference of ancestry in forensic analysis II: analysis of genetic data. In: *Methods in Molecular Biology: Forensic DNA Typing Protocols*, vol. 1420 (ed. W. Goodwin), 255–285. New York: Humana Press.

Santos, F., Panda, A., Tyler-Smith, P. et al. (1998). Reliability of DNA-based sex tests. *Nat Genet* 18: 103.

Scudder, N., Robertson, J., Kelty, S.F. et al. (2019). A law enforcement intelligence framework for use in predictive DNA phenotyping. *Aust J Forensic Sci* 51 (Suppl 1): S255–S258.

Scudder, N., Daniel, R., Raymond, J. et al. (2020). Operationalising forensic genetic genealogy in an Australian context. *Forensic Sci Int* 316: 110543.

Shackleton, D., Gray, N., Ives, L. et al. (2019). Development of RapidHIT I using NGMSElect Express Chemistry for the processing of reference samples within the UK criminal justice system. *Forensic Sci Int* 295: 179–188.

Shriver, M.D., Smith, M.W., Jin, L. et al. (1997). Ethnic-affiliation estimation by use of population-specific DNA. *Am J Hum Genet* 60: 957–964.

Skinner, D. (2020). Forensic genetics and the prediction of race: what is the problem. *BioSocieties* 15: 329–349.

Steele, C.D., Greenhalgh, M., and Balding, D.J. (2016). Evaluation of low-template DNA profiles using peak heights. *Stat Appl Genet Mol Biol* 15: 431–445.

Steinlechner, M., Berger, B., Niederstätter, H. et al. (2002). Rare failures in the amelogenin sex test. *Int J Legal Med* 116: 117–120.

St John, P. (2020). The untold story of how the Golden State Killer was found: a covert operation and private DNA. *Los Angeles Times*. https://www.latimes.com/california/story/2020-12-08/man-in-the-window (accessed 20 December 2020).

Sulem, P., Gudbjartsson, D.F., Stacey, S.N. et al. (2008). Genetic determinants of hair, eye and skin pigmentation in Europeans. *Nat Genet* 20: 835–837.

Swaminathan, H., Garg, A., Grgicak, C.M. et al. (2016). CEESIt: a computational tool for the interpretation of mixtures. *Forensic Sci Int Genet* 22: 149–160.

SWGDAM (2017). Interpretation guidelines for autosomal STR typing by forensic DNA testing laboratories. www.swgdam.org/publications (accessed 20 December 2020).

Thomson, J., Clayton, T., Cleary, J. et al. (2019). The effectiveness of forensic genealogy techniques in the United Kingdom – an experimental assessment. *Forensic Sci Int Genet Suppl Ser* 7: 765–767.

Turingan, R.S., Tan, E., Jiang, H. et al. (2020). Developmental validation of the ANDE 6C system for rapid DNA analysis of forensic casework and DVI samples. *J Forensic Sci* 65: 1056–1071.

Tutton, R., Hauskeller, C., and Sturdy, S. (2014). Suspect technologies: forensic testing of asylum seekers at the UK border. *Ethn Racial Stud* 37: 738–752.

UK Office of the Biometrics Commissioner (2017). Annual report 2016: commissioner for the retention and use of biometric material. Paul Wiles. March 2017. https://assets.publishing.service.gov.uk/government/uploads/system/uploads/attachment_data/file/644426/CCS207_Biometrics_Commissioner_ARA-print.pdf (accessed 20 December 2020).

United States Department of Justice (2019). Interim policy forensic genetic genealogical DNA analysis and searching. https://www.justice.gov/olp/page/file/1204386/download (accessed 20 December 2020).

Verogen (2020). Application note: forensic genetic genealogy with GEDmatch. https://verogen.com/wp-content/uploads/2020/12/forensic-genetic-genealogy-with-gedmatch-application-note-vd2020005-b.pdf (accessed 20 December 2020).

Walsh, S., Liu, F., Ballantyne, K.N. et al. (2011). IrisPlex: a sensitive DNA tool for accurate prediction of blue and brown eye colour in the absence of ancestry information. *Forensic Sci Int Genet* 5: 170–180.

Walsh, S., Chaitanya, L., Clarisse, L. et al. (2012). Developmental validation of the HirisPlex system: DNA-based eye and hair colour prediction for forensic and anthropological usage. *Forensic Sci Int Genet* 9: 150–161.

Walsh, S., Liu, F., Wollstein, A. et al. (2013). The HIrisPlex system for simultaneous prediction of hair and eye colour from DNA. *Forensic Sci Int Genet* 7: 98–115.

Winney, B., Boumertit, A., Day, T. et al. (2012). People of the British Isles: preliminary analysis of genotypes and surname in a UK-control population. *Eur J Hum Genet* 20: 203–210.

Wood, A.R., Esko, T., Yang, J. et al. (2014). Defining the role of common variation in the genomic and biological architecture of adult human height. *Nat Genet* 46: 1173–1186.

Zaghlool, S.B., Al-Shafai, M., Al-Muftah, W.A. et al. (2015). Association of DNA methylation with age, gender, and smoking in an Arab population. *Clin Epigenetics* 2015: 6.

Zbieć-Piekarska, R., Spólnicka, M., Kupiec, T. et al. (2015). Examination of DNA methylation status of the *ELOVL2* marker may be useful for human age prediction in forensic science. *Forensic Sci Int Genet* 14: 161–167.

Zhang, S. (2018). How a genealogy site led to the alleged Golden State Killer. *The Atlantic*. https://www.theatlantic.com/science/archive/2018/04/golden-state-killer-east-area-rapist-dna-genealogy/559070/ (accessed 20 December 2020).

Zubakov, D., Liu, F., van Zelm, M.C. et al. (2010). Estimating human age from T-cell DNA rearrangements. *Curr Biol* 20: R970–R971.

7 The utility of forensic radiology in evaluation of soft tissue injury

Curtis E. Offiah[1,2]
[1] *Department of Radiology, Royal London Hospital, Barts Health NHS Trust, London, Uk*
[2] *Cameron Forensic Medical Sciences, William Harvey Research Institute, Barts and the London School of Medicine and Dentistry, Queen Mary University London, London, UK*

Introduction

The forensic assessment of injuries in survivors of trauma – accidental or criminally inflicted – is a frequent requirement of criminal and coronial investigation. Examples include victims of road traffic collisions, deliberately inflicted blunt trauma, sharp force-penetrating trauma, projectile trauma, intimate-partner violence, elder abuse, and non-accidental injury in the paediatric population. The assessment of external soft tissue injuries in such cases has long been a standard practice for the forensic physician. However, with the significant advances in imaging technology and, specifically, clinical radiology and imaging, the majority of victims subject to injury requiring medical assessment and treatment will undergo some form of imaging which typically takes the form of plain X-ray (also called plain radiographs) and/or computed tomography (CT) scan and/or magnetic resonance imaging (MRI) scan. In particular, CT scan images and MRI scan images, although performed primarily with the aim of directing medical or surgical management of injuries of a victim, may, when interpreted from a forensic radiological perspective, assist in defining the mechanism of injury, wound profile, and potential causation. There are several advantages to the retrieval and forensic radiological evaluation of such clinical imaging:

1. Commonly (after, for example a serious assault requiring hospitalisation) it may be weeks before investigative agencies are alerted to the possibility of criminal injury requiring forensic evaluation of such injuries, by which time, healing of such injuries may be advancing, or surgically-altering treatment may have been performed on the victim, for example the victim of severe blunt trauma to the head requiring emergency craniectomy to evacuate large extra-axial haemorrhage and alleviate precipitously raised intracranial pressure. While forensic evaluation may rely on retrospective review of contemporaneous clinical documentation in medical records, such documentation will frequently omit detailed account related to the injuries because these are deemed irrelevant to emergency medical or surgical and

Current Practice in Forensic Medicine, Volume 3, First Edition. Edited by John A.M. Gall and J. Jason Payne-James.

potentially life-saving treatment. The emergency admission imaging assessments will, however, provide a radiological account of the injuries before such medical/surgical intervention on the victim has taken place.

2. Frequently, where there are 'internal injuries' requiring only conservative management (e.g. localised pulmonary contusions resulting from isolated rib fracture), forensic indication of the extent of such injury may only be objectively demonstrated on clinical imaging such as CT scans performed on presentation of the victim to hospital for diagnostic evaluation and medical management. Other examples include depth and direction of stab wound tracks where operative intervention was not required.

3. Clinical imaging such as CT scans and MRI scans performed in the medical assessment of victims presenting to hospital for diagnostic evaluation of injuries is archived indefinitely on generic software platforms called PACS (picture archiving and communicating systems) and can be retrieved and interpreted several years later by a suitable forensic radiological expert and provide an account of historical injuries that the individual may have sustained (Figure 7.1a).

4. Clinical imaging such as CT scans and MRI scans can assist in corroborating human identification by demonstrating past injuries including soft tissue injuries on historical clinical imaging – the so-called secondary identification criteria (this is in addition to the use of normal anatomical variants discernible on clinical imaging (another form of secondary identification) and primary identification techniques such as DNA profiling and dental X-rays) (Payne-James and Jones 2019; Levy et al. 2010; Schuh et al. 2013).

Clinical imaging, if available, provides a wealth of information for forensic purposes at an early and, importantly, at a later stage. An important consideration in reviewing such clinical imaging is the fact that clinical radiology reports are undertaken by clinical radiologists, and frequently, medical reports and records allied to such clinical imaging (clinical radiology) reports will omit key forensic findings as these are frequently irrelevant to the clinical/medical management and treatment of the victim. It is paramount therefore that any such clinical imaging is reviewed and evaluated by a radiologists with experience in forensic issues.

One of the significant facilities of cross-sectional radiology such as CT scanning is the ability to identify small or large foreign body fragments – not only the identification of knife blade tip fragments, glass fragments, or embedded wooden fragments but also the presence of scattered tiny particles either on and in the wound or on the skin surface of the body, characteristic of the nature of a hard blunt implement/weapon or location of the body of the victim at the time of the injury (for example, particles of grit on the body surface or wound from road surface or outdoor ground area).

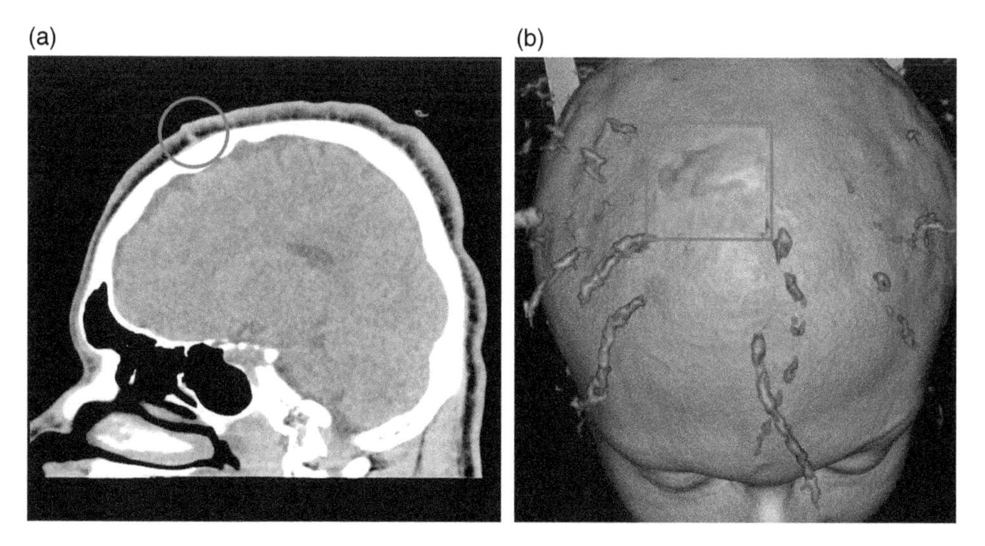

Figure 7.1 (a) Sagittal CT scan image of a 24-year-old male with stab wound to the right side of the vertex of the head which produced a divot fracture of the outer table of the underlying skull vault (not shown) – stab wound track through the scalp tissue and subperiosteal haematoma demonstrated (circled). (b) Volume-rendered reconstructed image (3D reconstruction) from the CT scan of the same victim – the stab wound on the scalp skin (magnified and annotated) is less well demonstrated on 3D reconstruction because of overlying blood and limitations of 3D reconstruction that can occur. Image postprocessed to reduce conspicuity of thick hair braids (arrowed).

Limitations

It is important to understand the limitations of radiological imaging in the evaluation of soft tissue injuries in the living in order to understand and appropriately direct the use of radiological imaging in the evaluation of such injuries. Such limitations are frequently only appreciated by the radiologist. CT scans can, if of adequate resolution, be used to create three-dimensional (3D) image reconstructions (volume- and surface-rendered reconstructions), which are a useful way of presenting soft tissue wounds to the lay person, particularly police investigation officers and, ultimately, judge and jury. It may also be viewed by some as a more 'palatable' form of exhibiting graphic injuries even though the 3D reconstructions are of the victim and of his or her injuries. However, it is the remit of the radiologist to understand that whenever source data acquired in clinical imaging such as CT scans is postprocessed or manipulated in anyway, information within such data (i.e. the source data) may be lost (Figure 7.1b) – in particular, it is the remit of the radiologist to understand the nature, severity, and impact of such artefact on any post-processed or manipulated data on the information being presented (Borowska-Solonynko and Solonynko 2015). The aforementioned limitations of source data post-processing such as 3D image reconstructions mean that certain soft tissue injuries (and even bone fractures) may not be presentable in 3D format. Soft tissue wound depiction on 3D reconstructions may be further hindered by dressings applied to heavily bleeding soft tissue wounds by medical staff prior to clinical imaging as well as by clotted blood on and around the wound and, for example in the hair of a victim with a scalp laceration or incised scalp wound (Figure 7.2a).

The inherent limitations in the resolution of clinical imaging also mean that superficial soft tissue injuries such as abrasions (grazes) may not always be visible, particularly in hair-bearing or dependent areas. Typically, the back of the body as victims are usually scanned supine on the CT scanner and dependency will compress the subcutaneous soft tissues of the dependent parts of the body and render injuries in these locations, including sharp force injuries, on the skin surface 'invisible' or poorly delineated. The presence of accumulated congealed blood on the skin surface of the wound or the interim application of pressure dressings or dressing cover to wounds (which may also subsequently become blood soaked) applied as part of initial emergency medical/surgical injury management also limits the detailed evaluation of such injuries – evaluation which is routinely undertaken by the forensic physician or forensic pathologist. While these limiting factors in the radiological discernibility of cutaneous and subcutaneous soft tissue injuries are important to be aware of and understand, cross-sectional radiological imaging (most frequently, CT scans) still frequently alerts the forensic radiologist to the location of a potential site of application of physical force and directs further radiological scrutiny of the imaged tissues in this location and elsewhere for more subtle injury (such as underlying microfracture

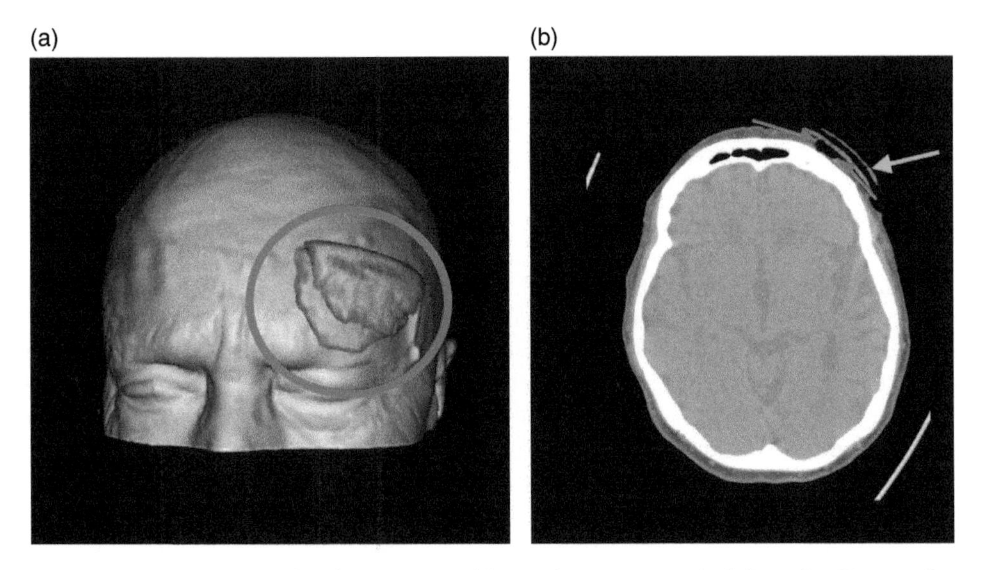

Figure 7.2 (a) Volume-rendered reconstructed image (3D reconstruction) from the CT scan of an 87-year-old male who suffered a fall following alleged assault and presented with left forehead laceration. Wound has been covered with a dressing during pre-hospital attendance (circled). (b) Source CT scan image of the same victim – the radiological appearances and morphology of the laceration can, however, still be assessed on the axial source CT images. The wound edges have been arrowed on this image. The covering dressing is also evident (green arrow).

or crack fracture, for example). In the author's experience, such radiological evaluation, although limited compared to direct clinical forensic medical evaluation, when identified and properly evaluated by the radiologist may also direct the forensic physician or forensic pathologist to areas of soft tissue injury which might have gone unrecognised (particularly scalp injuries) (Figure 7.2b) or encourage reconsideration of injuries misinterpreted as inflicted soft tissue injury but, in fact, are iatrogenic (such as thoracostomy incisions).

A final potential limitation of any cross-sectional radiological imaging relates to the way in which the scan parameters are set up and performed (sometimes referred to as *the scan protocol*): there is no standardised protocol, and scan protocol is very much directed and instructed by the radiologist. This impacts significantly on the resolution of the scan and determines the sensitivity and specificity of the information that the radiologist can extract from the scan. There is a significant difference between, for example a CT scan protocol and CT scan interpretation by a forensic radiologist compared to a clinical radiologist because the requirements of the interpretation of each are very different: the CT scan evaluation of the *forensic radiologist* will be directed at injury profiles, mechanisms of injury, potential causation, and likely life-threatening injuries and very much requires a CT scan of adequate resolution and optimised scan protocol, while the CT scan evaluation of the clinical (hospital) radiologist will be directed at the identification of life-threatening injuries or injuries requiring medical or surgical or critical care management and as such, mechanisms of injury and causation are frequently not a consideration: this is particularly the case in relation to identification of soft tissue injuries, particularly blunt force traumatic soft tissue injuries. The distinction lies in understanding the radiological imaging requirement and perspective, i.e. imaging of a *patient* versus imaging of a *potential victim*. This is similar to the differing emphasis of the role of the forensic physician vs the emergency department physician. Examples include the stab wound to the chest with resultant haemopneumothorax – identification of the specific site of a stab wound through the chest wound soft tissues or fracture of an underlying rib is not treatment defining and, therefore, is not an essential consideration necessary for the clinical radiologist – the identification of a large haemothorax and actively bleeding transected intercostal artery is the more pressing requirement of the clinical radiologist; in contrast, identification of the specific site of a stab wound through the chest wound soft tissues or fracture of an underlying rib is a critical part of the forensic radiological consideration; similarly, the acute periorbital soft tissue haematoma and old historical orbital floor fracture in the absence of the underlying acute mid-facial or acute orbital fracture on the CT scan of a patient are not essential considerations or commentary of the clinical radiologist but clearly are significant considerations on the CT scan of the victim for the forensic radiologist (Grassberger et al. 2011; Aromatario et al. 2016; Bauer et al. 2004).

Types of cross-sectional radiological imaging

In the acute trauma setting, this falls largely into two types of scan or imaging modalities – CT scanning and MRI scanning. The vast majority of cross-sectional imaging of this nature will be in the form of CT scanning – both for the surviving victim of physical injury and for the post-mortem imaging of fatalities.

The basics

CT scanning

CT is the workhorse for imaging of traumatic injuries in the twenty-first century, both for the clinical assessment of the majority of significant acute traumatic injuries and for extracting forensic radiological findings relating to those injuries where required or helpful to the forensic investigation of such injuries. There have been significant advances in CT technology (in particular multi-detector technology) over the past decade, which make CT scanning central to the clinical (and subsequently forensic) evaluation of various traumatic injuries. The technique involves volume acquisition of a stack of axial slices of the body or region of the body using X-ray tubes and an array of aligned detectors. The acquired images are translated into a grey scale. In the live patient, intravenous contrast can be injected into the patient/victim to identify significant vascular injury and active haemorrhage and increase the sensitivity and specificity of the identification of visceral injury. The inherent disadvantage of CT scanning is that it utilises ionising radiation (X-rays) with all of the potential hereditary and deterministic side effects associated with ionising radiation exposure. But where the benefits to the patient/victim surpass the potential risks, CT scanning will be undertaken. The advantages of CT scanning are that it is a fast imaging modality and is particularly sensitive and specific in the setting of trauma (Rydberg et al. 2000; Prokop 2003).

MRI scanning

MRI is an imaging modality based on the application of radiofrequency pulses to the body having placed the patient/victim in a high-strength magnet (typically 1.5 or 3 T). The radiofrequency pulses applied within the magnetic field cause movements of protons in the tissues of the body, which will subsequently emit energy when they return to their normal position when the radiofrequency pulse ceases (Pooley 2005). The energy emitted is captured (in k-space) and, using mathematical algorithms, converted to an image. The main advantages of MRI scanning are that it does not use ionising radiation and it has inherently superior contrast resolution for soft tissue interrogation than CT scanning. The disadvantage and significant limiting factors of MRI scanning in the acute trauma imaging setting are that it is slower (comprehensive scanning of a body region can take approximately 20–30 minutes); it is not as sensitive or specific for identification of acute fractures as CT scanning; and patients/victims with ferromagnetic substances within them (such as certain metals and certain types of pacemakers or bullets or knife tips) cannot be imaged in MRI scanners because of the magnetic field strength. However, in the forensic setting, MRI scanning does lend itself to more specific forensic radiological evaluation of certain types of old injuries such as stab wounds.

Types of injury

There are a number of ways of categorising soft tissue injuries, but a simple and broad division is between sharp force injury and blunt force injury (although there are other types of frequently encountered soft tissue injuries such as chemical injury and thermal/fire-related injury which do not fall into either of these two broad categories). This broad categorisation also allows useful observation of national statistics in any particular country and trends in overall violent crime. In the United Kingdom, the Crime Survey of England and Wales (Office of National Statistics) estimated approximately 1.2 million incidents of violence for the survey year ending March 2020. While a proportion of these will have been minor injuries and do not include the non-violent accidental injuries which may still require forensic investigation, it gives an indication of the likely encounters of forensic investigators with such injuries.

Blunt force injury

The most common type of fatal and non-fatal trauma remains blunt force injury. Where a blunt force is imparted to the body, or alternatively, the body strikes a blunt surface, the mechanical energy absorbed by impact site alters its anatomical integrity and creates a wound by compression, traction (tension), or shear or torsional forces. The occurrence and extent of injury depends on the amount of force sustained. The amount of force applied can (arbitrarily) be described as mild/weak, mild to moderate, moderate, moderate to severe, and severe. In addition, the higher the amount of kinetic energy associated with, the greater the potential for injury (however, therein, a number of variables such as the nature of the surface impacted and tissue elasticity then become significant). A helpful forensic perspective of blunt force trauma is to consider the outcome of the trauma (Di Maio and Di Maio 2001; Payne-James and Jones 2019):

1. no injury;
2. tenderness;
3. pain;
4. reddening (or erythema);
5. swelling (or oedema);
6. bruising (or contusion and haematoma);
7. abrasion (or graze);
8. lacerations; and
9. fractures.

Routine radiological imaging can identify some but not all of these changes – plain X-rays may identify bone fractures and may identify significant soft tissue swelling; CT scans will identify fractures and may identify swelling and bruising and frequently lacerations and sometimes abrasions with significantly increased sensitivity compared to plain X-rays; MRI scans may identify reddening, swelling, abrasions, lacerations with increased sensitivity compared to CT scans, and fractures, but MRI scan is very much dependent on the scan protocol directed by the radiologist and because of the inherent restrictions associated with MRI (which is beyond the scope of this chapter) is not a suitable imaging modality of choice (or 'workhorse') in the setting of the emergency trauma imaging assessment. While the sensitivity of CT imaging is limited in relation to cutaneous soft tissue wound assessment compared to the direct forensic examination undertaken by the forensic physician or the forensic pathologist, the utility of the radiological imaging assessment and forensic radiological interpretation is in demonstrating the extent of the underlying soft tissue (Figure 7.3a,b) and non-soft tissue injury beneath the skin surface and potentially inferring the amount of force associated with the trauma as well as the potential mechanism of the trauma (Schuh et al. 2013) (Figure 7.4a–d); a cutaneous soft tissue injury is frequently an external sign of severe underlying visceral injury (for example, abdominal cutaneous soft tissue contusion over the upper abdomen associated with high-grade blunt force liver injury) in keeping with significant absorption of mechanical energy at that location (Mason and Purdue 2000; Levy et al. 2010; Malli et al. 2013).

Patterned injuries cannot be evaluated on routine acute cross-sectional radiological imaging (CT scanning), but the forensic radiologist will be familiar with the subtle *geographical* distribution of 'intradermal' soft tissue contusion which makes stamping injuries, for example (particularly from the shod foot) a potential mechanism of the entire radiologically discernible injuries identified on CT scan.

CT scanning has exquisite sensitivity and significant specificity in identifying and localising foreign bodies within soft tissue wounds, particularly metallic or glass foreign body fragments, which not only aids emergency medical/surgical management of such wounds but provides valuable forensic information relating to the mechanisms of such injuries (Figure 7.5).

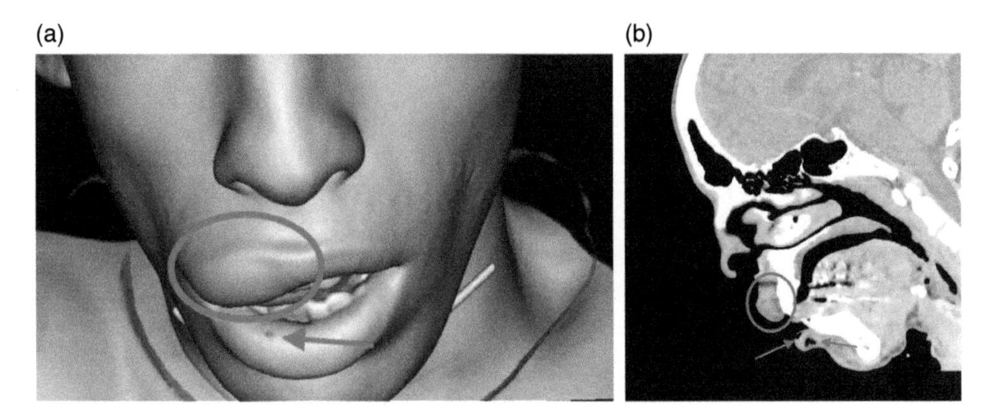

Figure 7.3 (a) A 24-year-old male victim of a punch injury to the face – volume-rendered reconstructed image (3D reconstruction) from the CT scan performed on hospital attendance demonstrated haematoma of the right upper lip (fat lip) (oval annotation) and small area of superficial-looking injury to the right lower lip caused by puncture by a tooth of the victim (arrow). (b) A 24-year-old male victim of a punch injury to the face – CT image (sagittal) from axial source CT dataset shows the haematoma of the right upper lip producing swelling (oval annotation) but demonstrates deeper puncture of the soft tissues of the right lower lip by a tooth of the victim than would have been discernible from the 3D-reconstructed image (arrows).

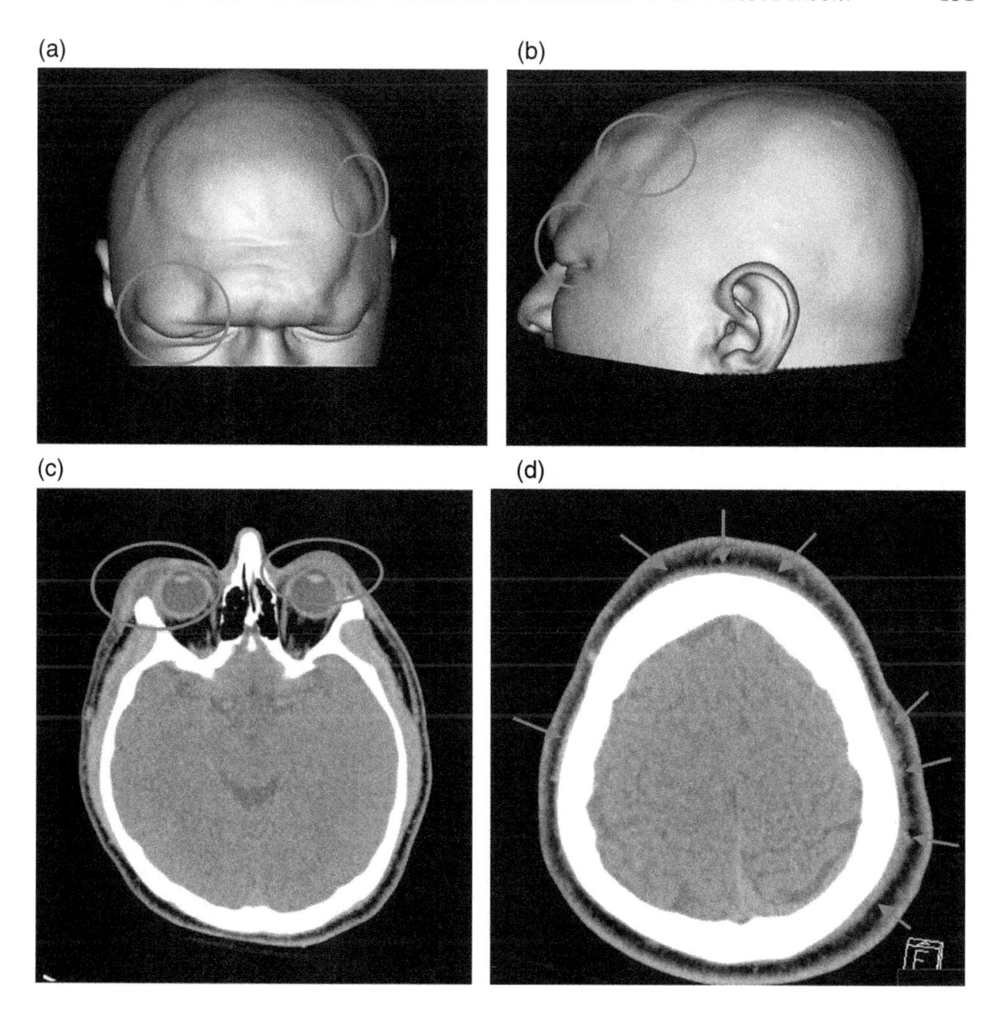

Figure 7.4 (a, b) A 21-year-old male victim of alleged assault of males with 'knuckledusters' – frontal and left lateral volume-rendered reconstructed images (3D reconstructions) from the CT scan performed on attending the emergency department. Right lateral brow and periorbital soft tissue haematoma and left frontal scalp haematoma (annotated). The normal soft tissue appearances can make identifying the extent and additional areas of acute soft tissue injury difficult to identify. (c, d) Axial source data CT scan images of the same victim confirm that the degree of acute soft tissue injury is more extensive with bilateral (as opposed to only right sided) periorbital haematoma (oval annotation) as well as more extensive subgaleal scalp soft tissue haematoma present on the left and frontally and also shallower subcutaneous soft tissue injury on the right (arrows) than evident on 3D reconstruction.

While the degree of contusion can reflect the severity of the blunt force trauma (including inflicted trauma), a number of modifying factors need to be borne in mind by the forensic radiologist when detailing the soft tissue injury profile identified on radiological imaging as these falsely 'upgrade' the derived impression of the degree of trauma sustained. The most significant of these are medications currently being taken by the victim such as antiplatelet therapy (such as aspirin and, in particular, clopidogrel)

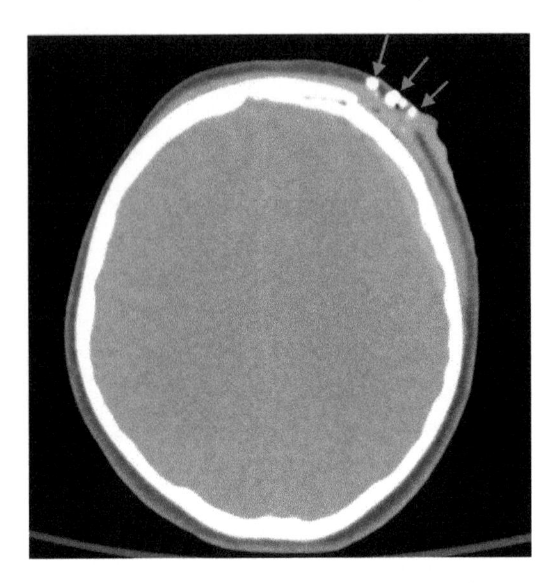

Figure 7.5 Axial CT image of an 11-year-old male occupant of a vehicle involved in collision with a lorry – severe left frontal scalp laceration and subgaleal and subperiosteal scalp haematoma (associated with underlying skull fracture and intracranial trauma) – the typical appearances of multiple vehicular glass foreign body fragments (arrows) in the left frontal scalp wound.

or various forms of oral anticoagulation therapy such as warfarin or directly acting oral anticoagulants and medical conditions associated with bleeding diathesis. A particular example is liver cirrhosis frequently associated with and frequently an indicator of chronic alcohol excess in the victim (a demographic already predisposed to accidental and inflicted blunt force trauma). The elderly should also be recognised as more prone to bruising because of dermal atrophy with resulting senile ecchymoses a complicating issue: bruising and soft tissue haematoma can be extensive in the elderly on radiological imaging but not necessarily reflective of a severe degree of blunt force trauma (Di Maio and Di Maio 2001; Ko and Dang 2010; Payne-James and Jones 2019). The soft tissue bruising also extends and evolves much more rapidly in this group where serial cross-sectional radiological imaging is performed for any reason and is therefore not necessarily indicative of interval new blunt force trauma nor of a direct impact site because of the tendency of blood to track through atrophic dermal layers. Nevertheless, the recognition of elder abuse has increased significantly, particularly with the growing elderly population, and radiological imaging evaluation of this entity has an ever-increasing role to play and should be considered by the radiologist when evaluating imaging in a similar vein to non-accidental injury in the infant population (Wong et al. 2017; Lee et al. 2018). Because of the loose elastic delicate cutaneous architecture, the neonatal and infant population bruise more readily, and while this can assist in the forensic radiological evaluation of abusive head trauma in cases of non-accidental injury, it is also relevant where an accidental causation of bruising in a child is identified radiologically. Obesity also enhances bruising and can be conspicuous on radiological imaging where the skin remains integral such that the subcutaneous blood cannot evacuate; however, in sharp force injury, such as stab wounds, the bleeding in breached generous subcutaneous fat layers present in the obese can be much more subtle.

An essential part of the forensic assessment by a forensic physician and forensic pathologist, ageing of cutaneous soft tissue injuries, is limited on radiological cross-sectional imaging such as CT scanning to lacerations and sometimes sizeable abrasions and, by and large, to classification of recent, non-recent/subacute, or old (healed and historical). However, this crude classification can provide an indication of potentially non-contemporaneous blunt traumatic injuries (for example, to the scalp) and possible mechanisms of the injuries.

Sharp force injury

In the United Kingdom, according to the Office of National Statistics, the Crime Survey for England and Wales ending March 2020 determined that the most common method of killing, for both male and female victims, was by a sharp instrument. The volume of victims of sharp force trauma encountering radiological imaging, typically CT scanning, is considerable, particularly in the major trauma centres – these victims may or, sadly, may not survive, but therein exists a wealth of forensically extractable information in the emergency hospital admission CT scans undertaken on victims of inflicted sharp force trauma. These scans, where of adequate (albeit, not always optimised) CT protocols, can form a valuable piece of the forensic assessment when interpreted by the forensic radiologist. Not infrequently, where contemporaneous wound photographs are not available in victims surviving inflicted sharp force trauma, the forensic pathologist or forensic physician is reliant on clinical details recorded in the medical records of survivors including the clinical radiology reports; however, in the author's opinion, such CT scans should be reviewed for expert opinion by a forensic radiologist where forensically relevant detail is imperative to the overall forensic investigative process. Forensically relevant detail is frequently omitted (because frequently, it is not clinically relevant to the emergency medical/surgical management of the victim) on the clinical radiological evaluation and reports of such scans (Tarani et al. 2017).

Incised wounds (cuts) (Figure 7.6a–c) and stab wounds (Figure 7.7a) are examples of sharp force injuries. It is the depth of the wound track relative to the length of the cutaneous wound which distinguishes a stab wound from an incised wound: the wound track of the stab is greater than the length of the wound on the skin surface, i.e. the wound is deeper than it is long (Mason and Purdue 2000; Di Maio and Di Maio 2001; Schmidt and Pollak 2006; Bohnert et al. 2006; Payne-James and Jones 2019). A slash wound is essentially an incised wound caused by a sharp-edged instrument such as a knife moving across the skin surface. Any instrument that has an edge and a point such as a knife can produce an incised wound or a stab wound, whereas an instrument that has a point but not an edge such as a pin or a knitting needle can produce a stab wound but not typically produce an incised wound. While distinction is typically made between sharp force injury and blunt force injury, some injuries overlap and exhibit a combination of the two such as machete or axe injuries (chop injuries) and sometimes 'glassing' injuries.

With the advances in multi-detector (multi-slice) CT imaging, the forensic radiological evaluation of sharp force injuries has expanded significantly in addition to the clinical diagnostic evaluation (Figure 7.7a,b). In addition, in the 'cold' injury

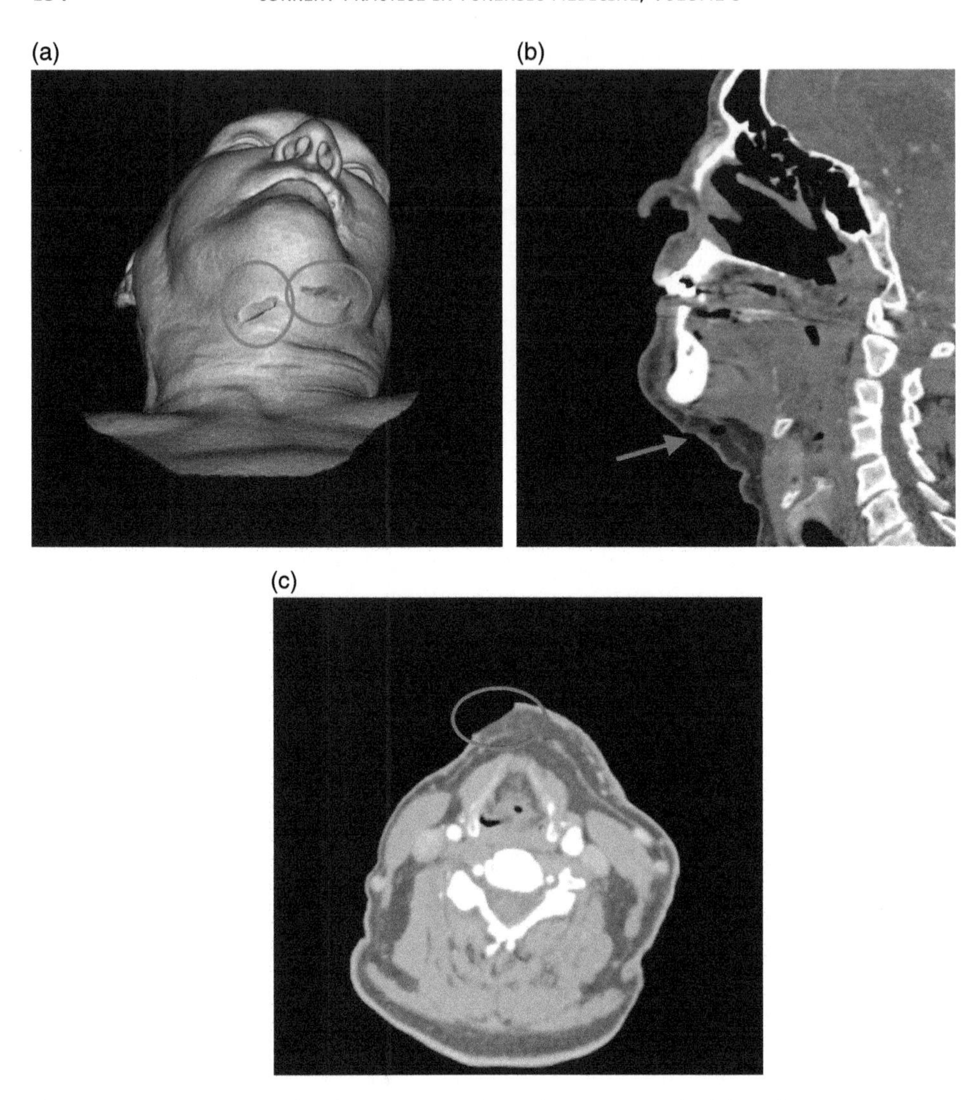

Figure 7.6 (a) Volume-rendered reconstructed image (3D reconstruction) from the CT scan performed on a 72-year-old male admitted with sharp force injury to the neck – two incised wounds to the anterior neck (annotated) with typical features of self-infliction – close proximity 'hesitation'-type relatively horizontal appearances. (b, c) CT images from axial source CT dataset show the relatively superficial depth of the incised wounds on the left (b) (arrows) and on the right (c) (oval annotation) and give better indication of the depth of each wound than the 3D-reconstructed imaging (with no breach of platysma).

evaluation such as confirming and ageing historical sharp force injuries, imaging utilising MRI scanning has provided an assist to forensic investigation (Yen et al. 2004, 2009; Woźniak et al. 2015; Aromatario et al. 2016). In the acute sharp force injury setting, CT scanning of the victim is frequently undertaken as part of medical/surgical management, and stab wound tracks as well as tissue and organ injury can be evaluated as part of forensic wound profiling of directionality as well as the life-threatening nature

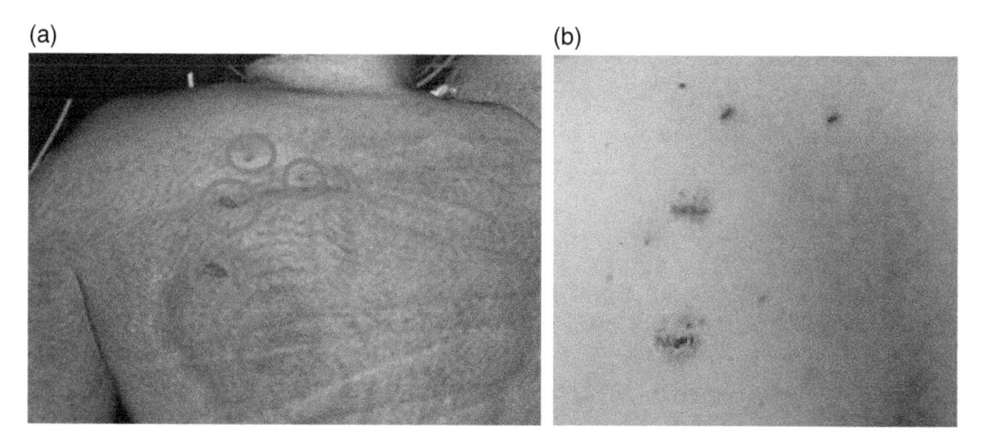

Figure 7.7 (a) Volume-rendered reconstructed image (3D reconstruction) from the CT scan performed on a 30-year-old female victim of intimate-partner violence admitted to the emergency department with acute multiple stab wounds inflicted to the chest and arms (not shown) and to the back (annotated). (b) Actual photograph of the stab wounds to the back of the same victim taken after treatment and hospital discharge and healing.

of the stab wound (Humphrey et al. 2017). Wound tracks can be depicted in the live victim by assessing not only the characteristic skin surface wounds but also haematoma along the wound track which may be accompanied by soft tissue emphysema extending along the wound track and active bleeding from small or critical blood vessels captured on contrast-enhanced CT scans performed on the live victim.

The amount of force required to inflict a stab wound is related to the shape and sharpness of the weapon (Mason and Purdue 2000; Di Maio and Di Maio 2001; Hugar et al. 2012; Bohnert et al. 2006; Payne-James and Jones 2019). It is the overlying skin which is often the most resistant portion of the body to the penetrating stab wound, and once this has been breached, very little additional force is needed to penetrate the weapon further (although this does not necessarily mean that little force was used).

The directionality or trajectory and the length of the wound track on CT imaging can be estimated and inferred, but definitive extrapolation of the length of the offending blade cannot be reliably assessed and, in the author's opinion, should be avoided for a number of reasons: the victim is typically scanned in a physiological position (i.e. supine on the CT scanner), which, in a similar vein to the autopsy of a deceased victim on an autopsy table, is rarely an accurate reproduction of the position of the victim at the point the injury was inflicted; the organs of a victim while standing or sitting position will assume a different anatomical position and configuration than in the relatively controlled supine position of the CT scanner; there is variable distensibility of the soft tissues, particularly the skin in different parts of the body; the depth of thrust of the blade of the weapon may be subtotal, total, or beyond the heel or even to the bolster; and in addition, the infliction of sharp force trauma such as a stab injury is a very dynamic process where there is movement of the assailant (in non-accidental injury), movement of the penetrating instrument or weapon, and movement of the victim (typically in an attempt to evade or fend the stab injury) – this dynamic process will have significant compounding effects on the trajectory of the wound track. The typically supine CT scanning position of the victim also means that stab wounds to the dependent part of the body (for example, the back of the chest or abdomen or buttocks) may be significantly obscured by compression of the wound in the skin surface and the wound track and adjacent soft tissue haematoma.

While skin surface wounds of stab injuries can only be crudely assessed on cross-sectional radiological imaging such as CT imaging compared to the direct inspection and evaluation by the forensic physician or forensic pathologist, grouping and multiplicity and orientation can be identified on acute CT imaging and is frequently important in determining specific causation – vertical chest stab wounds tend to be seen in homicides and, in one study, tended to be vertically orientated, whereas self-inflicted wounds were more likely to be horizontal (Karlsson 1998; Karger et al. 2000; Scolan et al. 2004; Bohnert et al. 2006; Shkrum and Ramsay 2007). Grouping may also be evident on emergency CT scanning of sharp force injuries of a self-inflicted causation – hesitation (or tentative) wounds are frequently discernible on such imaging and will be identifiable on forensic radiological evaluation. Hesitation injuries are characterised by superficial incised or stab wounds typically nearer the deeper life-threatening or fatal injury. The superficial incised wounds tend to run parallel to each other (Figure 7.6a–c), and small puncture wounds may be grouped together near a deeper life-threatening or fatal stab wound (Karlsson 1998; Levy et al. 2010).

One of the limitations of clinical acute CT imaging performed on victims of severe stab injuries is that wounds of the limbs, particularly the distal limbs, are rarely deemed clinically life threatening and are, therefore, not imaged, and therefore, the presence of defensive or fending injuries may not be captured in the acute setting.

Chop injury

Chop injuries are distinctive from a forensic perspective because they exhibit a combination of features of sharp and blunt force trauma. Typically, they are sharp force injuries caused by heavy implements with a cutting edge such as machetes or axes or meat cleavers (Mason and Purdue 2000; Di Maio and Di Maio 2001; Wittschieber et al. 2016). The rotating blades or propellers of helicopters or boats and ships may also produce chop injuries. Typically, the chop injury exhibits a deep gaping incised wound with injury to the underlying bony structures. Abrasions may be present at the edge or edges of the incised wound, sometimes attributed to the bluntness or thickness of the blade of the implement. Crushing of the adjacent tissue, a feature which is frequently encountered in lacerations caused by blunt trauma as well as bridging soft tissue in the deep aspect of the wound, may also be encountered in chop injuries. The common sites of chop injuries on the body of the victim are the head and the neck and, in the author's experience, the upper and lower limbs (the former probably related to defensive wounds).

In the surviving victims of chop injuries, radiological imaging plays a key part in wound assessment, in terms of both diagnostic characterisation and treatment and also forensic evaluation of the injuries. This may take the form of plain X-ray assessment where bone injury has been sustained, but more frequently, owing to the severity of the soft tissue injury associated with such wounds, cross-sectional imaging, typically in the form of CT scanning, is frequently performed and provides evaluation of both the extent of the soft tissue injury (including potential vascular injury) for treatment and also secondarily in the forensic evaluation of these injuries and their threat to life (Figures 7.8 and 7.9a,b). Volume-rendered (3D) CT reconstructions can be used to determine the number of potential impacts and to infer the angles of impact and therefore the direction from which the impact forces originated (Bauer et al. 2004; Levy et al. 2010; Grassberger et al. 2011; Aromatario et al. 2016) (Figure 7.8).

Injury patterns and causation

There are soft tissue subtleties evident on CT scanning performed on surviving victims of trauma undergoing CT scanning as part of the medical and surgical management of their injuries which have been overlooked for a long time by clinical radiologists and continue to be, largely because of the lack of relevance of some of these soft injuries to the treatment of the victim. However, in the author's experience, such CT scans, often performed within a short time interval between the index trauma and scanning, provide a wealth of forensic information which, if sought and correctly evaluated, can be of significant value to any overall forensic investigation not only in terms of supporting and offering possible mechanisms of injury and degree of force associated with such injuries but also supporting the veracity of proposed mechanisms of injury or throwing doubt on the proposed mechanisms of injury (Schuh et al. 2013). This is

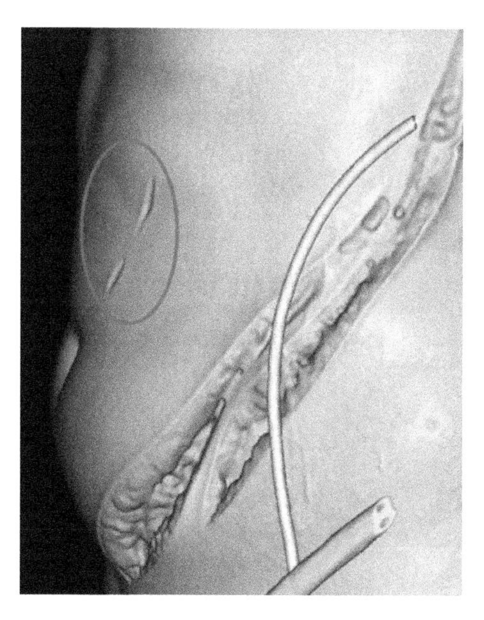

Figure 7.8 Volume-rendered reconstructed image (3D reconstruction) from the CT scan performed on a 19-year-old male victim of machete attack. Deep extensive left flank soft tissue wounds consistent with machete 'chop' injury. Smaller incised/slash wounds are present above the umbilicus on the left (annotated).

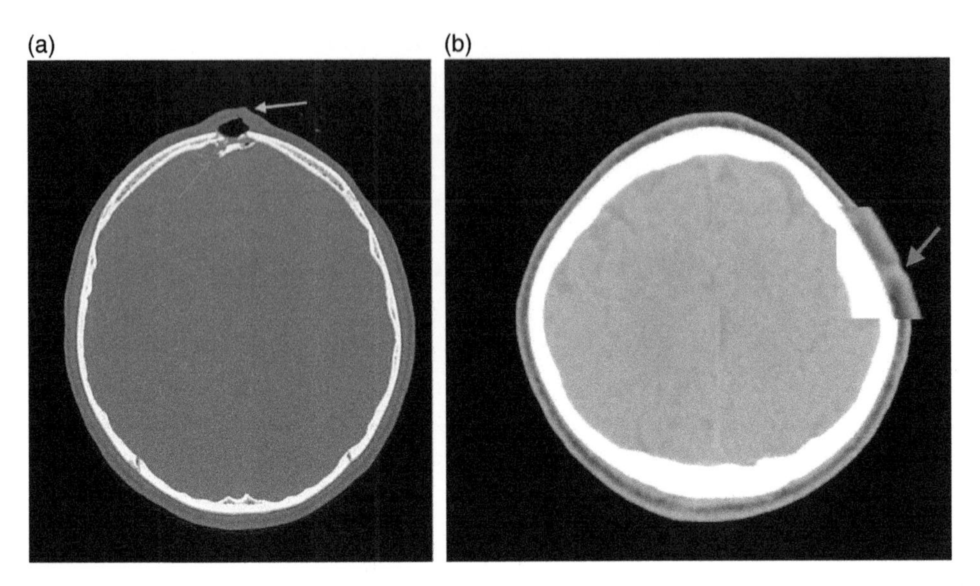

Figure 7.9 (a) Axial CT image (on bone windows) of admission CT cranial scan of a 27-year-old male admitted to hospital following a chop injury to the head with an axe – typical depressed acute fracture configuration in keeping with the presenting surface of an axe (red arrow) to the impacted skull consistent with chop injury mechanism and overlying scalp soft tissue wound which has been sutured (green arrow). (b) Axial CT image (with magnified inset) of the same scan of the same victim showing typical forensic radiological appearances of a separate old healed laceration of the left side of the frontal scalp (arrow) – note the retracted 'scarred' skin surface of the wound.

particularly relevant in trauma involving vulnerable victims such as the infant in cases of non-accidental injury, the elderly or infirm in cases of elder abuse, and the adult in cases of intimate-partner violence. Indeed, the radiologist is opportunely placed to raise concerns about the number and nature of separate imaging encounters by a particular patient or individual (i.e. X-rays and CT scans performed on an individual) in a similar vein to the emergency department consultant observing excessive separate attendances of a child or adult for distinct episodes of trauma (Hunsaker and Hunsaker 2006; Murphy et al. 2013; Bhole et al. 2014; Rosen et al. 2016; Wong et al. 2017; Lee et al. 2018, 2019; Flores and Narayan 2019; George et al. 2019; Alessandrino et al. 2020; Matoori et al. 2020). Indeed, in the period during which this chapter was being written, the world was struck by the COVID19 virus pandemic, and measures taken to reduce the spread of the COVID19 virus across many countries also saw a significant rise in the incidence of intimate-partner violence (Gosangi et al. 2020). The author is one of a number of radiologists encountering elder abuse and intimate-partner violence (in addition to paediatric abusive head trauma) on a frequent basis and is keen that radiologists are aware not only of the imaging presentations of such abuse and violence but also the central and pivotal role of the radiologist in alerting safe-guarding agencies of possible victims attending Radiology Imaging departments.

The punch injury to the face is frequently demonstrated on cross-sectional imaging such as CT scanning or even MRI scanning, and the former is frequently performed as part of treatment assessment with the victim seeking medical attention because of pain or typical 'raccoon eyes' frequently encountered where the punch injury to the face has been associated with fracture of the nose or anterior skull base fracture extension. Some of the hallmarks evident on radiological imaging can be supportive of the mechanism of injury of a punch to the face or eye, for example. The stamping injury as well as potentially causing bone fractures can also produce a much more geographical pattern of cutaneous and subcutaneous soft tissue injury on CT scanning, and while not as specific as direct visual evaluation of an area of soft tissue injury, the radiological appearances can be suggestive and supportive of such a mechanism of injury and direct considerations of such in the forensic investigation.

Non-fatal strangulation is increasingly recognised as a significant factor in intimate-partner violence, and this is further explored in Chapter 3.

Defence injuries resulting from sharp force provide specific evidence of injury inflicted by a third-party as well as indicating that the victim was conscious at some point during some or all of the attack (Di Maio and Di Maio 2001; Bohnert et al. 2006; Schmidt and Pollak 2006; Hugar et al. 2012). Defence injuries can be described as passive – occurring when the victim raises the hands or the arms for protection – or active – occurring when the victim attempts to seize the weapon or the hand of the attacker which is holding the weapon. Active defence injuries are frequently seen in the region of the thumb, index finger, and the palmar surface of the hand, particularly related to the thenar eminence and the web space between the thumb and the index finger (Shkrum and Ramsay 2007). The instinctive reflex response of the victim in the majority of such attacks is to protect the head and face, raising the hands and arms to do so. As a result, (passive) defence injuries are most frequently encountered on the fingers, thumbs, hands, and arms and, secondarily, the axilla exposed by such a defensive posture. Perforating (through-and-through) stab injuries to the upper limb held in front of the body may also be associated with second re-entry stab wounds to another part of the body. With defence wounds related to sharp force trauma, the typical sites are infrequently subject to cross-sectional imaging such as CT scanning unless there is concern for life-threatening vascular injury or in the case of the more severe chop-type injury to the upper limb. In the case of defence injuries resulting from blunt force injury assault, similar regions of the upper limbs of the victim may be injured, but particularly, the extensor surfaces of the forearms and upper arms as well as the side and the back of the trunk and the legs – this is particularly the case in conscious individuals who have fallen to the ground and the subject of blunt force impacts from blunt weapons or kicks or stamps – frequently, the foetal position is adopted protecting the front of the body and the head and the face but exposing the side and the back of the trunk and legs. Such areas may be subject to radiological imaging to diagnose fracture in the form of plain X-ray of the upper limbs (demonstrating the classical 'nightstick' fracture of the shaft of the ulna and overlying soft tissue swelling) or CT scanning of the chest, abdomen, and pelvis when cutaneous, and in particular, subcutaneous soft tissue injury may be evident if specifically sought by the radiologist: such CT-discernible soft tissue injury will be indeterminate/non-specific in causation without provision of supporting description of the specific circumstances of the incident from police or crime scene investigators or witness statements, for example.

Gunshot injuries

Non-fatal gunshot injuries will frequently undergo emergency CT scanning on presentation to the emergency department of a hospital or trauma centre. There is an absolute contraindication of MR imaging of firearms injuries in the acute setting because of the specific metallic nature and composition of the vast majority of gunshot injuries which may move within the body of the victim when subject to the strong magnetic fields associated with MRI scanners. However, on evaluation of acute CT imaging performed in the treatment of victims of firearms injuries, but for the purpose of assisting the forensic investigation, the forensic radiologist must be familiar with the basics of ballistics including the characteristics of entry and exit bullet-perforating wounds on CT imaging and factors which may affect the appearances of entry and exit wounds radiologically as well as the radiological appearances of tissues affected along the perforating bullet wound track such as bones (and bevelling patterns) which are also discernible on CT imaging and greatly assist in trajectory assessment and the potential life-threatening nature of non-fatal firearms injuries. In addition, the forensic radiologist should also be aware of the concept of re-entry re-exit wounds in relation to firearms injuries and atypical appearances where bullets may have ricocheted, struck an intermediate target, fragmented before entry or be associated with secondary injuries caused by displaced, expressed, or in-driven bone fragments or external foreign body fragments such as glass, for example as well as the radiological appearances of very close-range and contact (and frequently but not always fatal) 'execution-style' bullet injuries compared to gunshots that are not. As with macroscopic/gross assessment of bullet wounds (but not shotgun wounds), entry wounds tend to be round small neat holes which can look deceptively innocuous (McClay 2009; Levy and Harcke 2011; Payne-James and Jones 2019) and can be very difficult to locate, requiring detailed evaluation of the acute CT scans. Radiologically, there may be more localised subcutaneous soft tissue emphysema associated with entry wounds. Exit wounds tend to be larger and exhibit more complex surface features radiologically, and there may be less subcutaneous soft tissue emphysema than expected presumably because of bleeding (Figure 7.10a–d).

Where retained metallic fragments can be excluded radiologically in the cold assessment of non-fatal perforating gunshot injuries, because of its inherent greater soft tissue contrast, MRI scan can be very useful (compared to CT scan) in confirming (or excluding) the presence of old perforating bullet wound tracks and the historical non-acute nature of such perforating bullet wounds where this becomes a requirement forensically in corroborating the account of a victim in criminal investigations and also in war-crime investigations and asylum-seeking investigations (Figure 7.11): the author has used MRI scanning for these purposes in aiding such investigations.

Ligature soft tissue injuries

Both non-fatal ligature strangulation and near-hanging with ligature/noose can produce soft tissue injuries evident on acute CT scanning of victims as well as on acute and subacute MRI scanning, which demonstrate not only the ligature marking or furrowing on the skin surface but also subcutaneous/subdermal soft tissue contusion in the superficial fascia and deep cervical fascia including the deep cervical fascia surrounding the carotid sheaths as well as the strap muscles of the neck where

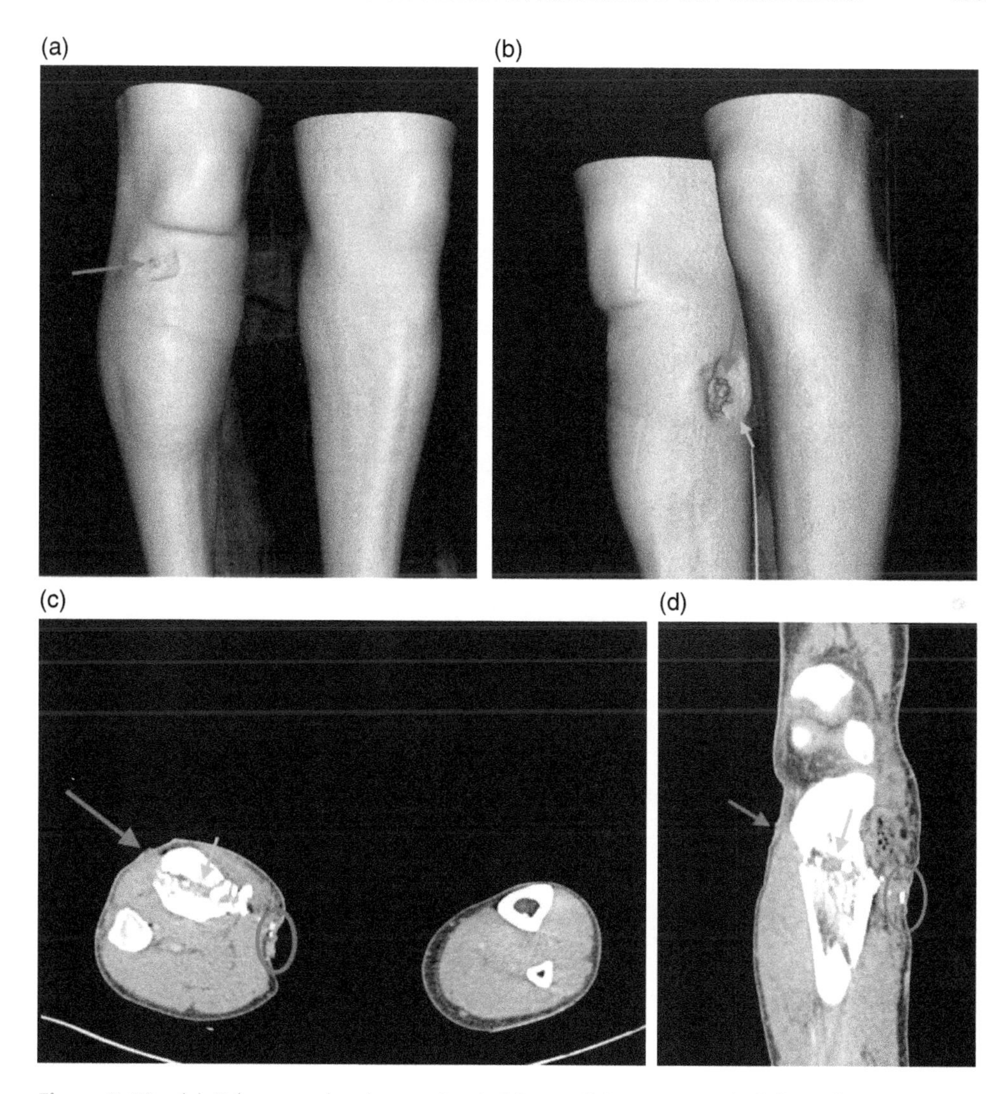

Figure 7.10 (a) Volume-rendered reconstructed image (3D reconstruction) from the CT scan performed on a 30-year-old male admitted to the emergency department with a perforating bullet wound to the right lower leg as part of an alleged gang-related 'punishment' shooting. The entry wound is arrowed. (b) Volume-rendered reconstructed image (3D reconstruction) from the CT scan performed on a 30-year-old male admitted to the emergency department with a perforating bullet wound to the right lower leg as part of an alleged gang-related 'punishment' shooting. The exit wound is arrowed (red arrow) with a small pressure dressing applied beneath a tourniquet (note the indentations caused by tourniquet (green arrows) – the tourniquet itself has been segmented out of the postprocessed 3D reconstruction). (c) Angled axial (multi-planar-reconstructed) CT image of the same scan of the same victim to show the entire perforating bullet wound track through the right lower leg with the entry wound (red arrow), bullet track through the right proximal tibia (green arrow), and the exit wound (red oval). The right-to-left (i.e. lateral to medial) and anterior-to-posterior trajectory through the limb is evident. (d) Angled coronal (multi-planar-reconstructed) CT image of the same scan of the same victim to show the entire perforating bullet wound track through the right lower leg with the entry wound (red arrow), bullet track through the right proximal tibia (green arrow), and the exit wound (red oval). The downward and right-to-left (i.e. lateral to medial) trajectory through the limb is evident.

mechanical compression associated with these mechanisms of injury has been sustained and is sufficient. The extent of such soft tissue changes may not be as evident on gross examination in non-fatal cases but may be discernible on both CT scanning (Figure 7.12a,b) and MRI scanning (Yen et al. 2004; Levy and Harcke 2011).

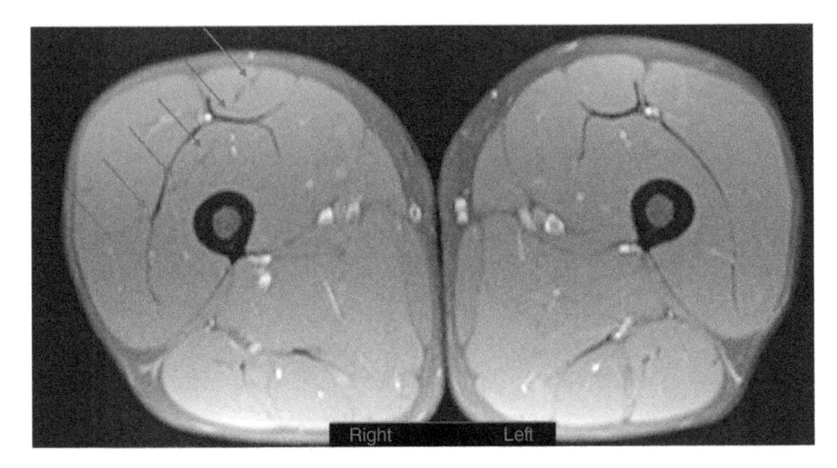

Figure 7.11 Axial MRI scan image of the right (and left) thigh of a 28-year-old male arrested and deported to the United Kingdom for prosecution for a shooting homicide two years prior. The criminal investigation required clarification of the defendant's claim that he had been shot in the right thigh during the homicide. After confirming the absence of any retained metal fragments in the right thigh on plain X-ray, MRI was performed, confirming a well-healed perforating bullet wound track (red arrows) of slender scar tissue extending between the healed entry and exit wound scars on the skin of the thigh. The MRI appearances would be consistent with a healed wound track of more than one to two years of age.

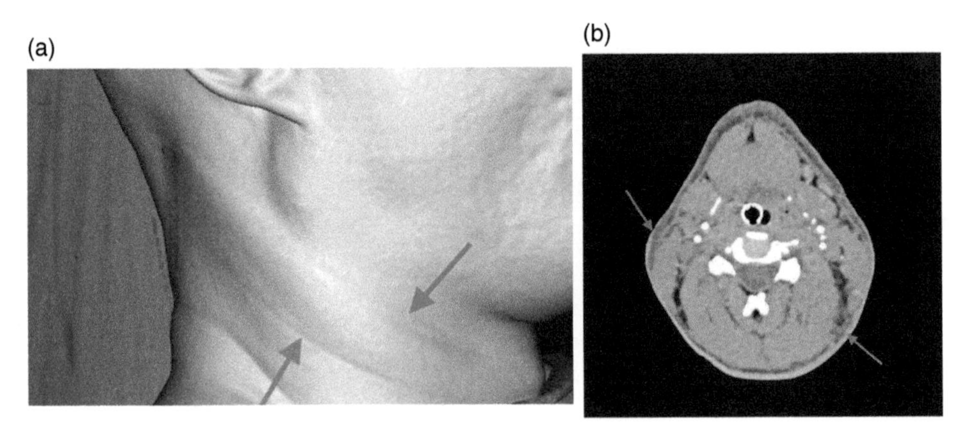

Figure 7.12 (a) Volume-rendered reconstructed image (3D reconstruction) from the CT scan performed on a 37-year-old male admitted to the emergency department following near-hanging, showing a part of the right side of the neck – the ligature was removed at the scene, but the ligature marks/impressions are discernible (arrows). (b) Axial CT image from the source dataset of the same victim – the appearances of subcutaneous soft tissue contusional changes caused by the ligature are demonstrated (arrows).

Conclusion

Cross-sectional imaging, performed most frequently with the imaging modality of CT scanning of the victim attending the hospital emergency department, contains a wealth of information of significant utility to the forensic investigation. While such emergency imaging is performed to facilitate the rapid diagnosis and treatment of injuries sustained by the victim, the existence of such imaging should always be borne in mind by the criminal investigation team and allied professionals associated with the forensic investigation pathway. Clinical radiologists evaluating such imaging for treatment purposes may not have the experience to identify forensically relevant findings on such radiological imaging, particularly given the fact that identification of such findings may not be relevant to the treatment of the victim at the time of presentation to the emergency department. While there are limitations in specificity and sensitivity of soft tissue injuries discernible on the most frequently used imaging modalities, when evaluated and interpreted appropriately in a forensic context, the findings can be of significant value to the forensic investigation.

References

Alessandrino, F., Keraliya, A., Lebovic, J. et al. (2020). Intimate partner violence: a primer for radiologists to make the "Invisible" visible. *Radiographics* 40 (7): 2080–2097. https://doi.org/10.1148/rg.2020200010.

Aromatario, M., Cappelletti, S., Bottoni, E. et al. (2016). Weapon identification using antemortem CT with 3D reconstruction, is it always possible?--A report in a case of facial blunt and sharp injuries using an ashtray. *Leg Med (Tokyo)* 18: 1–6. https://doi.org/10.1016/j.legal med.2015.11.003.

Bauer, M., Polzin, S., and Patzelt, D. (2004). The use of clinical CCT images in the forensic examination of closed head injuries. *J Clin Forensic Med* 11 (2): 65–70. https://doi.org/10.1016/j.jcfm.2003.10.003.

Bhole, S., Bhole, A., and Harmath, C. (2014). The black and white truth about domestic violence. *Emerg Radiol* 21 (4): 407–412. https://doi.org/10.1007/s10140-014-1225-1.

Bohnert, M., Hüttemann, H., and Schmidt, U. (2006). Homicides by sharp force. In: *Forensic Pathology Reviews. Forensic Pathology Reviews, vol 4* (ed. M. Tsokos). Humana Press. https://doi.org/10.1007/978-1-59259-921-9_3.

Borowska-Solonynko, A. and Solonynko, B. (2015). The use of 3D computed tomography reconstruction in medico-legal testimony regarding injuries in living victims – risks and benefits. *J Forensic Leg Med* 30: 9–13. https://doi.org/10.1016/j.jflm.2014.12.006.

Di Maio, V.J.M. and Di Maio, D.J. (2001). *Forensic Pathology*, 2e. Boca Raton: CRC Press.

Flores, E.J. and Narayan, A.K. (2019). The role of radiology in intimate partner violence. *Radiology* 291 (1): 70–71. https://doi.org/10.1148/radiol.2019190057.

George, E., Phillips, C.H., Shah, N. et al. (2019). Radiologic findings in intimate partner violence. *Radiology* 291 (1): 62–69. https://doi.org/10.1148/radiol.2019180801.

Gosangi, B., Park, H., Thomas, R. et al. (2020). Exacerbation of physical intimate partner violence during COVID-19 lockdown. *Radiology* 13: 202866. https://doi.org/10.1148/radiol.2020202866.

Grassberger, M., Gehl, A., Puschel, K., and Turk, E.E. (2011). 3D reconstruction of emergency cranial computed tomography scans as a tool in clinical forensic radiology after survived blunt head trauma – report of two cases. *Forensic Sci Int* 207: e19–e23.

Hugar, B.S., Harish, S., Girish Chandra, Y.P. et al. (2012). Study of defence injuries in homicidal deaths – an autopsy study. *J Forensic Leg Med* 19 (4): 207–210. https://doi.org/10.1016/j.jflm.2011.12.022.

Humphrey, C., Kumaratilake, J., and Henneberg, M. (2017). Characteristics of bone injuries resulting from knife wounds incised with different forces. *J Forensic Sci* 62 (6): 1445–1451. https://doi.org/10.1111/1556-4029.13467.

Hunsaker, D.M. and Hunsaker, J.C. (2006). Elder abuse. In: *Forensic Pathology Reviews. Forensic Pathology Reviews, vol 4* (ed. M. Tsokos). Humana Press. https://doi.org/10.1007/978-1-59259-921-9_2.

Karger, B., Niemeyer, J., and Brinkmann, B. (2000). Suicides by sharp force: typical and atypical features. *Int J Legal Med* 113 (5): 259–262. https://doi.org/10.1007/s004149900093.

Karlsson, T. (1998). Homicidal and suicidal sharp force fatalities in Stockholm, Sweden. Orientation of entrance wounds in stabs gives information in the classification. *Forensic Sci Int* 93 (1): 21–32. https://doi.org/10.1016/s0379-0738(98)00025-5.

Ko, P. and Dang, C. (2010). Blunt force trauma. In: *Manual of Forensic Emergency Medicine—A Guide for Clinicians* (ed. R.J. Riviello), 65–76. London: Jones and Bartlett Publishers.

Lee, M., Rosen, T., Murphy, K., and Sagar, P. (2018). A role for imaging in the detection of physical elder abuse. *J Am Coll Radiol* 15 (11): 1648–1650. https://doi.org/10.1016/j.jacr.2018.06.008.

Lee, M., Rosen, T., Murphy, K., and Sagar, P. (2019). A new role for imaging in the diagnosis of physical elder abuse: results of a qualitative study with radiologists and frontline providers. *J Elder Abuse Negl* 31 (2): 163–180. https://doi.org/10.1080/08946566.2019.1573160.

Levy, A.D., Harcke, H.T., and Mallak, C.T. (2010). Postmortem imaging: MDCT features of postmortem change and decomposition. *Am J Forensic Med Pathol.* 31 (1): 12–17. https://doi.org/10.1097/PAF.0b013e3181c65e1a.

Levy, A.D. and Harcke, H.T. (2011). *Essentials of Forensic Imaging: A Text-Atlas*. CRC Press Taylor & Francis Group.

Malli, N., Ehammer, T., Yen, K., and Scheurer, E. (2013). Detection and characterization of traumatic scalp injuries for forensic evaluation using computed tomography. *Int J Legal Med* 127 (1): 195–200. https://doi.org/10.1007/s00414-012-0690-x.

Mason, J.K. and Purdue, B.N. (2000). *The Pathology of Trauma*, 3e. Oxford University Press.

Matoori, S., Khurana, B., Balcom, M.C. et al. (2020). Intimate partner violence crisis in the COVID-19 pandemic: how can radiologists make a difference? *Eur Radiol* 30 (12): 6933–6936. https://doi.org/10.1007/s00330-020-07043-w.

McClay, W.D.S. (2009). *Clinical Forensic Medicine*, 3e. Cambridge University Press.

Murphy, K., Waa, S., Jaffer, H. et al. (2013). A literature review of findings in physical elder abuse. *Can Assoc Radiol J* 64 (1): 10–14. https://doi.org/10.1016/j.carj.2012.12.001.

Payne-James, J. and Jones, R. (2019). *Simpson's Forensic Medicine*, 14e. Hoboken: CRC Press.

Pooley, R.A. (2005). AAPM/RSNA physics tutorial for residents: fundamental physics of MR imaging. *Radiographics* 25 (4): 1087–1099. https://doi.org/10.1148/rg.254055027.

Prokop, M. (2003). General principles of MDCT. *Eur J Radiol* 45 (Suppl 1): S4–S10. https://doi.org/10.1016/s0720-048x(02)00358-3.

Rosen, T., Bloemen, E.M., Harpe, J. et al. (2016). Radiologists' training, experience, and attitudes about elder abuse detection. *AJR Am J Roentgenol* 207 (6): 1210–1214. https://doi.org/10.2214/AJR.16.16078.

Rydberg, J., Buckwalter, K.A., Caldemeyer, K.S. et al. (2000). Multisection CT: scanning techniques and clinical applications. *Radiographics* 20 (6): 1787–1806. https://doi.org/10.1148/radiographics.20.6.g00nv071787.

Schuh, P., Scheurer, E., Fritz, K. et al. (2013). Can clinical CT data improve forensic reconstruction? *Int J Legal Med* 127 (3): 631–638. https://doi.org/10.1007/s00414-013-0830-y.

Schmidt, U. and Pollak, S. (2006). Sharp force injuries in clinical forensic medicine--findings in victims and perpetrators. *Forensic Sci Int* 159 (2–3): 113–118. https://doi.org/10.1016/j.for sci int.2005.07.003.

Scolan, V., Telmon, N., Blanc, A. et al. (2004). Homicide-suicide by stabbing study over 10 years in the toulouse region. *Am J Forensic Med Pathol* 25 (1): 33–36. https://doi.org/10.1097/01.paf.0000113816.09035.01.

Shkrum, M.J. and Ramsay, D.A. (2007). *Forensic Pathology of Trauma: Common Problems for the Pathologist*, 1e. Humana Press.

Tarani, S., Kamakshi, S.S., Naik, V., and Sodhi, A. (2017). Forensic radiology: an emerging science. *J Adv Clin Res Insights* 4: 59–63.

Wittschieber, D., Beck, L., Vieth, V., and Hahnemann, M.L. (2016). The role of 3DCT for the evaluation of chop injuries in clinical forensic medicine. *Forensic Sci Int* 266: e59–e63. https://doi.org/10.1016/j.forsciint.2016.05.025.

Wong, N.Z., Rosen, T., Sanchez, A.M. et al. (2017). Imaging findings in elder abuse: a role for radiologists in detection. *Can Assoc Radiol J* 68 (1): 16–20. https://doi.org/10.1016/j.carj.2016.06.001.

Woźniak, K., Moskała, A., and Rzepecka-Woźniak, E. (2015). Imaging for homicide investigations. *Radiol Med* 120 (9): 846–855. https://doi.org/10.1007/s11547-015-0529-x.

Yen, K., Vock, P., Tiefenthaler, B. et al. (2004). Virtopsy: forensic traumatology of the subcutaneous fatty tissue; multislice computed tomography (MSCT) and magnetic resonance imaging (MRI) as diagnostic tools. *J Forensic Sci* 49 (4): 799–806.

Yen, K., Ranner, G., and Dirnhofer, R. (2009). Clinical forensic imaging. In: *The Virtopsy Approach: 3D Optical and Radiological Scanning and Reconstruction in Forensic Medicine* (ed. M. Thali, R. Dirnhofer and P. Vock), 363–378. Boca Raton: CRC Press.

8 Abusive head trauma in children – a clinical diagnostic dilemma

John A.M. Gall[1,2,3]
[1] *Department of Paediatrics, The University of Melbourne, Melbourne, Australia*
[2] *Victorian Forensic Paediatric Medical Service, The Royal Children's Hospital and Monash Medial Centre, Melbourne, Australia*
[3] *Era Health, Melbourne, Australia*

Controversy in paediatric forensic medicine is not new, particularly as the stakes are high, and reputations have been damaged. Allegations of abuse may lead to significant family disruption, separation of children from their parents, and parents/caregivers being charged with an offence and being sent to prison or, in some cases where the child may have died, sentenced to death. In the late 1980s, erroneous injury interpretation in relation to anal reflex dilatation led to the disruption of a number of families, wide publicity, and a subsequent enquiry headed by Lord Justice Butler-Sloss (BMJ 1988; Butler-Sloss 1988). A similar but more geographically extensive controversy has developed during the past 20 years in relation to what is now referred to as abusive head trauma (AHT). Although well established, the diagnosis of AHT is not incontrovertible. There is substantial evidence (e.g. observational studies and confessions) for a diagnosis of AHT, but the diagnosis is based on a hypothesis rather than actual experimental evidence, evidence that can never ethically be obtained. This chapter considers this controversy and discusses AHT from the basis of our current approach to and diagnosis of this condition and the challenges it may present in court.

Definitions

Before discussing this matter further, some definitions. For the purposes of this chapter, the term 'parent' or 'parents' will be used to encompass guardians, other family members, foster parents, and other caregivers. The term 'forensic physician' encompasses a number of specialty fields including forensic physicians, forensic paediatricians, paediatricians working in the area of child abuse, and forensic pathologists who also engage in the area of clinical child abuse.

It is also important to define the area of AHT being discussed. AHT has been defined by the US Centers for Disease Control and Prevention as an injury to the skull or intracranial contents of an infant or young child (<5 years of age) due to inflicted blunt impact and/or violent shaking (Parkes et al. 2012). It replaces the

Current Practice in Forensic Medicine, Volume 3, First Edition. Edited by John A.M. Gall and J. Jason Payne-James.

older terminology of Shaken Baby Syndrome as the condition may be due to not only shaking but also impact injuries. It may occur in children aged up to 5 years old but more commonly occurs in children less than two years of age in which it is one of the leading causes of death (Choudhary et al. 2018). AHT in a child, who has not been involved in a significant accident, often presents with multiple other injuries, which may include bruises, abrasions, lacerations, fractures (especially the posterior ribs and metaphyseal fractures involving the long bones), and internal organ damage. Making an assessment of possible abuse in cases of AHT where there are also external and/or other internal injuries is relatively straightforward and generally not in dispute. And this is a point that needs to be considered throughout this chapter; features of AHT in the presence of other injuries are not generally in dispute.

For the purposes of this discussion, however, the aspect of AHT in discussion is being restricted to those non-fatal cases that present without any evidence of physical injury but with a subdural haemorrhage and one or both of retinal haemorrhages and encephalopathy. A child presenting with either only a subdural haemorrhage or a subdural haemorrhage and one or a combination of retinal haemorrhages and encephalopathy is not uncommon. It is, however, a vastly more challenging medical dilemma for the forensic physician and is the subject of this chapter.

To place things into perspective, subdural haemorrhage is the most common finding in AHT and is present in about 90–95% of fatal cases and in over 50% of living cases (Case 2014). Retinal haemorrhages are similarly common findings being found in as many as 85% of cases.

A brief history

It is not the purpose of this chapter to discuss in depth all aspects of AHT. A brief overview of the history of the condition does provide some perspective, however, with respect to the period over which AHT, under varying terminology, has been described and studied, particularly where there is an absence of external injuries on the head and injuries elsewhere involving other parts of the body.

Al-Holou et al. (2009) and Narang (2011) have described the history of this condition and make a salient point that our current knowledge of AHT is the result of decades of meticulous documentation and research around the world.

Life as a child was difficult in previous centuries, and child abuse is a century-old problem, but it was not until the mid-1900s that it was recognised as a problem requiring the attention of medical practitioners (Evans 2004). Al-Holou et al. (2009) inform us that it was the French forensic physician, Augusta Ambrose Tardieu, who initially described child abuse in the mid-1800s. In an 1860 publication, *Etude Medico-Legale sur les Services et Mauvais Traitements Exerces sur des Enfants*, Tardieu detailed 32 cases of child abuse in which there were bruises, skeletal fractures, and subdural haemorrhages. He described cases of infanticide that did not display external signs of injuries but in which were haemorrhages in the brain and collections of blood over the brain. Unfortunately, Tardieu's work was essentially ignored.

With advances in medical knowledge, the causation of subdural haemorrhages during the late 1800s and early 1900s was thought to be inflammatory and infectious

(Al-Holou et al. 2009), a theory presented by one of the foremost German leaders in medicine and pathology of that time, Rudolf Virchow. His theory was referred to as 'pachymeningitis haemorrhagica interna'. This theory remained intact until the early 1900s. Sir Wilfred Trotter, a prominent British neurosurgeon, who had a particular interest in subdural haemorrhages (Rosen 2006), considered Virchow's theory to be misleading and reported trauma to be the real cause, a view increasingly supported by other neurosurgeons of the time (Weigel et al. 2004; Al-Holou et al. 2009). Trotter was of the view that the haemorrhages resulted from the tearing of veins that coursed between the brain and the dural sinus. A re-examination of the pathophysiology of subdural haemorrhages began to question Virchow's theory that they were caused primarily by inflammation and infection.

During this period, other authors identified an association between the development of subdural haematoma, and in some cases retinal haemorrhages, and trauma (Greeley 2015). In 1946, Dr John Caffey, a paediatric radiologist, described an association between long bone fractures in children and chronic subdural haemorrhages although he acknowledged that the causal mechanism remained obscure (Caffey 1946). In 1953, a British neurosurgeon, Norman Guthkelch, published a report on 18 infants with a subdural haemorrhage in which a small percentage had retinal haemorrhages. Birth trauma was considered the cause of the subdural haemorrhage in 75% of these cases (Guthkelch 1953). In 1962, Henry Kempe and colleagues use the term, The Battered-Child Syndrome, to describe a constellation of injuries including subdural haemorrhages that resulted from trauma (Kempe et al. 1962, 1985). Studies showed that subdural haematoma was a far more common complication of head injury in young children and particularly those under two years of age and more so for those aged under six months (Guthkelch 1971). Guthkelch (1971, p. 430) stated that the most common cause of infantile subdural haemorrhage was rupture of '. . .*the delicate bridging veins which run from the cerebral cortex to the sinuses, the mode of injury being either a single acceleration or deceleration due to a heavy moving object striking the head or the rapidly moving head being brought up against a stationary mass; multiple applications of force would increase the total strain on the bridging veins and result in an increased incidence of rupture*.' He was also of the opinion that . . .'*the relatively large head and puny neck muscles of the infant must render it particularly vulnerable to whiplash injury*. . ..' Further that '*subdural haematoma is one of the commonest features of the battered child syndrome, yet by no means all patients so affected have external marks of injury on the head*.'

In 1974, Caffey coined the term, whiplash shaken infant syndrome, to describe those children suffering from subdural haemorrhage, retinal haemorrhage, and neurologic injury, attributing the findings to shaking (Caffey 1974). Referring to earlier works of Ommaya and Yarnell (1969) and Guthkelch (1971), he believed that the pathogenesis of the subdural haematomas arose from the effect of rotational acceleration and deceleration forces of the whiplash. This mechanism explained why there were frequently external marks of injury. He also believed that this similarly explained the aetiology of the retinal haemorrhages found in these children.

Despite the efforts of Tardieu in the 1800s, it was not for about another century, and the efforts of Caffey, Guthkelch, and other colleagues of that time, that the medical profession began to consider the concept of child abuse and accept that parents injured their young children either intentionally or (arguably) innocently during periods of stress or anger. Since the mid-1900s, a very large amount of research has been undertaken and published in relation to this subject. Despite some contrary views, and the reported presence of idiopathic chronic subdural haematomas in newborn babies (Cheung et al. 2004; Hadzikaric et al. 2006), the astute analysis by Tardieu, Caffey, Guthkelch, and their colleagues effectively remains intact in that in the absence of a medical condition, a subdural haemorrhage in a young child, and more particularly in an infant, is the result of trauma, and trauma most likely inflicted by a parent.

Current hypothesis on the development of subdural haemorrhage, retinal haemorrhage, and hypoxic–ischaemic encephalopathy in AHT

For the purposes of clarity in relation to the current working hypothesis for AHT:

- The mechanism of subdural haemorrhage results from a tearing of the bridging veins that ascend from the surface of the cerebrum to the dural sinuses. This tearing results from movement (acceleration–deceleration) of the brain with its attached arachnoid within the skull with its attached dura (Guthkelch 1971).
- Retinal haemorrhages are similarly the result of this movement developing at the areas of greatest stress at the retinal periphery and posterior pole.
- Hypoxic–ischaemic encephalopathy is multi-factorial and includes axonal injury, apnoea (arising from shaking with or without impact), hypotension and acidosis, and the development of cytotoxic oedema and hypoxic ischaemia.

The presentation and diagnosis of AHT

The presentation of young children and infants for investigation of potential AHT in the absence of injuries elsewhere on the head and body is variable. Some, particularly with only a subdural haemorrhage, may be identified following investigation for an entirely unrelated medical condition. Others will present with medical symptoms of variable severity from the relatively minor to the potentially life threatening or may be found dead. The symptoms may be non-specific but also may be more specific and include some form of autonomic dysfunction such as a bradycardia, pallor, vomiting or apnoea, seizures, signs of increased intracranial pressure (e.g. bulging fontanelle and increasing head size), and some form of neuromotor disorder.

The assessment, investigation, and diagnosis of AHT is an extraordinarily complex, challenging, and time-consuming procedure, particularly in the absence of other external and internal injuries. It requires the involvement of a multi-disciplinary team. The diagnosis is essentially one of exclusion that is based on all information acquired through the collection of an exhaustive clinical history, third-party information, physical examination, laboratory and imaging data, and where all other plausible aetiologies have been ruled out.

Following investigation, the findings that indicate a more definitive diagnosis of AHT include:

1. subdural haemorrhage, particularly where the haemorrhage is bilateral, forms a thin film (particularly over the cerebral convexities and frontal regions), and is parafalcine in location (Gennarelli et al. 1982; Piteau et al. 2012; Bradford et al. 2013; Hedlund 2015);

2. retinal haemorrhages, especially where these are bilateral, multiple, multi-layered, extend to the ora serrata (retinal edge), and occur with retinoschisis with or without retinal folds (Vinchon et al. 2005; Binenbaum et al. 2009; Bhardwaj et al. 2010; Levin 2010; Morad et al. 2010; Maguire et al. 2013; RCPCH 2013); and, in more severe cases,

3. hypoxic–ischaemic encephalopathy (RCPCH 2019; Oates et al. 2021).

The presence of all three of these conditions, sometimes referred to the 'triad', is not necessarily diagnostic of AHT. So too, the presence of only one or two of these findings in a patient does not necessarily exclude a diagnosis of AHT. There are many causes for each of these conditions. Importantly, unlike diagnoses of diabetes or heart attacks, there is no one constellation of signs or symptoms or a diagnostic tool that identifies AHT. It should be noted that the term 'triad', although fashionable many years ago, tends now not to be used medically but is still raised during legal argument (Choudhary et al. 2018).

The development of a controversy

Very few areas of medicine come under the degree of scrutiny as does forensic medicine and those medical disciplines contributing to the forensic evidence. The expert evidence and opinions provided by forensic practitioners may be critical in enabling either a judge or jury to reach a verdict. Not surprisingly, given the potential stakes involved, the evidence and opinions provided may be subjected to significant scrutiny with any potential flaw pursued; not an unreasonable process. AHT is an emotive diagnosis that warrants scrutiny and is the only area of child abuse that has courted ongoing controversy for well over a two decades. The diagnosis has been contested by a small medical minority and an increasing number of legal practitioners despite the expansion of knowledge in this area and an acceptance by the medical majority including many of the national and international professional organisations directly involved with this area of medicine (Narang et al. 2016; Choudhary et al. 2018; Holmgren 2013). Just because a medical majority agrees with a particular concept, however, does not necessarily establish it as fact. This is in keeping with the view of Guthkelch (2012; p. 207–208) who was of the view that it '*is not what the majority of doctors (or lawyers) think but rather what is supported by reliable scientific evidence, the evidence should be reviewed by individuals who have no personal stake in the matter, and have a firm grounding in scientific principles, including the difference between hypothesis and evidence.*' What is required is reasoned debate and reliable supporting scientific evidence. Unfortunately, reasoned discussion and debate have been replaced by a hardening of opinions and the development of animosity on both sides.

Contrary to some views (Innocence Network 2019; Brook 2019), the diagnosis of AHT is not based on a dearth of evidence and in the absence of validation but on an extensive body of peer-reviewed research (Dias 2011; Narang 2011; Narang et al. 2013; Vinchon 2017). It is not a diagnosis that has been rapidly accepted as fact and has been open to very extensive critique over the years. It is also a finding that occurs in the presence of injuries elsewhere on the body resulting from abuse. Unfortunately, however, despite the extensive amount of research undertaken over the past 60 years, our understanding of the pathogenesis has not progressed and remains hypothetical, a situation that gives rise to ongoing debate regarding the very existence of AHT.

In 2017, the Swedish Agency for Health Technology Assessment published a systematic review for 'shaken baby syndrome' in which they claim that there is '*insufficient scientific evidence on which to assess the diagnostic accuracy of the triad in identifying traumatic shaking (very low-quality evidence). It was also demonstrated that there is limited scientific evidence that the triad and its components can be associated with traumatic shaking (low-quality evidence)*' (Lynoe et al. 2017). This systematic review (referred to as the SBU Report) was itself subjected to careful assessment with the finding being that the methodology utilised was flawed (Saunders et al. 2017; Debelle et al. 2018; Strouse 2018). They argue that any systematic review must be focused, answerable, and clinically relevant. The SBU Report undertook to determine whether the *healthcare principle that the triad is attributable exclusively to traumatic shaking*. This is a principle or concept of AHT that does not exist, and thus the question posed is irrelevant. The authors of the SBU Report, in other articles (Rosén et al. 2017; Lynoe et al. 2018), appeared to be unfamiliar with the progress in the diagnosis of AHT and that the triad was no longer considered a reliable diagnostic tool, the diagnostic basis, as always, being multi-faceted. Beyond the flawed primary question, Debelle et al. (2018) further argue that the search strategy used, the lack of standardised definitions for terms, inadequate inclusion and exclusion criteria, critical appraisal tools, and the synthesis of included studies were all flawed. Levin (2017) similarly criticises their search criteria in relation to retinal haemorrhages, suggesting that it failed both clinically and scientifically. Demands were unsuccessfully made for this review to be withdrawn (Debelle et al. 2018). The SBU Report has been defended by its authors (Rosén et al. 2017; Lynoe et al. 2017). With only a working hypothesis and in the presence of a highly critical medical publication, albeit based on a flawed methodology, it is not surprising that the diagnosis of AHT continues to court controversy, particularly within the legal profession, as to its existence.

Over the years, despite the pathogenesis of AHT remaining an hypothesis, our understanding of the condition has improved, and the list of conditions that mimic AHT have expanded. Alternative hypotheses have been presented and disproven. At the current time, the main points of contention relate to the scientific basis of AHT. Both sides of the argument acknowledge that there is a circularity of argument, a reliance on confession evidence, and an absence of a biomechanical model, a situation likely to remain for some time, and in the case of the biomedical model, perhaps never to be found.

Clinical medicine and the medical diagnosis

There is a significant difference between the two oldest professions, that of the law and medicine, and an understanding of each profession's processes may sometimes be clouded during the pursuit of a goal. Importantly, medicine is not a science but a combination of science and an art. This is a fact that too frequently is overlooked. Laboratory medicine is more scientific in nature, but clinical medicine, that area of medicine that deals with direct patient interaction, is not exclusively scientific. If it were, the need for medical practitioners would no longer exist as their role could easily be undertaken by computers and algorithms.

Arriving at a medical diagnosis is a complex process. Pappworth (1984, p. 27) summed it up by stating that it '*is a creative art and belongs to the realm of discovery. It is detective work controlled by a system of logical analysis, and it should always be an intellectual exercise which may be based entirely on personally observed evidence. Medicine is both an art and a science, the art necessitating a skilful, judicious and timely application of any relevant science*'. Further, and quoting the epidemiologist, AR Feinstein, '*medicine is the most scientific art and the most humanistic science. The art and the science are symbiotic, intermingled and inseparable*'.

The assessment and diagnosis of AHT is a complex and lengthy process that collects and considers information from a number of sources in relation to the history and the events surrounding it, involves an examination of the patient and the undertaking of radiological and pathological investigations, relies upon advice from other medical specialties, and considers all other medical conditions that may mimic the pathology identified. Once all this information is collected and collated, and the list of potential differential diagnoses considered and subsequently excluded, the diagnosis of AHT is subsequently reached. It involves lengthy medical detective work, that is as described by Pappworth, controlled by a system of logical analysis requiring a skillful, judicious, and timely application of the relevant science. It is not a diagnosis made in isolation and is one, because of the experience required, that should be only undertaken by senior forensic physicians in consultation with a multi-disciplinary team.

Alternative hypotheses

Alternative hypotheses regarding the pathogenesis of subdural haemorrhage, retinal haemorrhages, and encephalopathy appear periodically within the medical and scientific literature and subsequently within the courtroom (Jenny 2014). In the past, the two gained some degree of notoriety. These were the 'unified hypothesis' (Geddes et al. 2003) and the 'dural immature vascular plexus theory' (Squier and Mack 2009; Mack et al. 2009). Geddes and co-workers hypothesised that the subdural haematomas identified in some cases of infant head injury were not due to traumatic rupture of bridging veins but due to a combination of severe hypoxia, brain swelling, and raised central venous pressure, resulting in a leakage from the intracranial veins into the subdural space. This same hypoxia and brain swelling was hypothesised to account for the observed retinal haemorrhages. Under cross-examination, Dr Geddes accepted that the hypothesis was meant to stimulate debate only and was not fact (Supreme Court of

Judicature Court of Appeal (Criminal Division) Neutral Citation Number 2005a; Richards et al. 2006). A Court of Appeal subsequently determined that the unified hypothesis could no longer be regarded as a credible alternative cause for, what was then referred to, as the 'triad of injuries' (Supreme Court of Judicature Court of Appeal (Criminal Division) Neutral Citation Number 2005b; Richards et al. 2006). Squirer and Mack suggested a dural origin for the film subdural haemorrhages associated with AHT. They commented that the dura contains an inner vascular plexus that is larger in the infant than in the adult, and that although subdural haemorrhage is frequently traumatic, the inner dural plexus is a likely source of bleeding in non-traumatic circumstances. As with the unified hypothesis of Geddes, hypoxia was considered the principal factor causing these immature vessels to leak with the subsequent development of a subdural haemorrhage. Both hypotheses led to controversy within the medical literature and the courts. Neither theory has been supported by subsequent investigation.

It has also been raised in court that subdural haematomas may result from venous sinus thrombosis (Jenny 2014). A study by Choudhary et al. (2015) indicated that about 70% of children with AHT have some abnormality (including thrombosis) of the intracranial venous system. They attributed these abnormalities to be secondary to trauma. This is contrary to the hypothesis presented in some courts that the cortical sinus and venous thrombosis is the primary event with subsequent subdural and retinal haemorrhages. They point out that there is no evidentiary support in the literature for a relationship between cortical sinus and venous thrombosis and either subdural or retinal haemorrhages.

Immunisation has also been raised as a cause of subdural haemorrhage and brain swelling. An extensive review of the epidemiological, clinical, and biological evidence regarding the adverse effects of immunisation did identify some cases of head injury, including a few cases of subdural haemorrhage, but these were the result of falls occurring because of a vasovagal event (Stratton et al. 2012).

Choking and dysphagia (acute/apparent life-threatening event) have been raised as a potential cause of subdural haemorrhages, retinal haemorrhages, and hypoxic–ischaemic encephalopathy (Talbert 2005; Barnes et al. 2010), the hypothesis being that choking causes a paroxysmal cough injury, resulting in excessive intraluminal pressure causing vascular rupture and subsequent haemorrhage. In addition, it has been further hypothesised that the initial choking causes some degree of hypoxia which contributes to oozing from the meningeal vessels, fitting with the 'dural immature vascular plexus theory' of Squier and Mack as briefly discussed above. Hansen et al. (2017) undertook a retrospective case-controlled study over a five-year period of 170 children aged less than two years with subdural haemorrhages, both associated with and not associated with apparent life-threatening events. They were unable to confirm that these events caused features of AHT, a view also expressed by Edwards (2015) in his assessment of the argument for an association.

Short-distance falls

Short-distance falls are generally defined as falls of less than 1.5 m in height (Bilo et al. 2010). In 2001, Plunkett (2001) published findings of a review of the United States Consumer Product Safety Commission database for head injury associated with the use of playground equipment during a period of just over 10 years. The study revealed 18 fall-related head injury fatalities, with the youngest child being 12 months old and the eldest 13 years, and the fall height being from 0.6 to 3 m. Retinal haemorrhages were identified in four of six children who underwent funduscopic examination, leading the author to conclude that an infant or child may suffer fatal head injury from a fall of less than 3 m and also sustain retinal haemorrhages. It is important to note, however, that in this study no children were less than one year of age and that the fall heights were in cases greater than the accepted 1.5 m for short falls. This finding by Plunkett is also at odds, with the literature showing an absence of intracranial injuries in short-distance falls even in children younger than 12 months of age (Helfer et al. 1977; Nimityongskul and Anderson 1987; Lyons and Oates 1993; Tarantina, et al. 1999; Narang et al. 2013). Based on the literature, short-distance falls do not result in serious or life-threatening injuries such as intracranial haemorrhage and other cerebral pathology in the absence of complicating factors involved during the fall (i.e. complex falls such as falls downs stairs, from a baby walker, and falls onto an object). Falls from the arms of carers, particularly at heights closer to the 1.5 m, do have attendant risks and may lead to intracranial haemorrhages (Bechtel et al. 2004; Luder 2005).

The circular argument

It has been argued that the diagnosis of AHT is based on circular reasoning and not scientific criteria (Piteau et al. 2012; Lynoe et al. 2017a; Brook 2019). Further, that it is a form of scientific misconduct to undertake research ignoring the issue of circular reasoning. Circular reasoning or circular logic is where the reasoner begins with what they are trying to end with. In the case of AHT, the premise is that when the triad is present, then the child has been shaken violently so that when the triad is identified in the absence of other explanations, then the infant has been shaken violently. Lynoe and colleagues criticise child-protection teams for making a decision which they claim is not independent of the general assumption (i.e. that always when the triad is present, then the child has been violently shaken) and instead have developed criteria on which their decisions are made, criteria which have consequences for the scientific classification of study cases and controls. Further, that these criteria are not based solely on medical observations but on other factors such as the supposedly trustworthiness of the parent. Levin (2017) suggested that Lynoe and colleagues have ignored the full breadth of scientific evidence in relation to AHT and misrepresented the knowledge base in their systematic review. As the triad is not a basis for a diagnosis of AHT in isolation, the review and arguments in support of the SBU Repot findings as presented by Lynoe and colleagues are essentially irrelevant.

The logical fallacy of circular reasoning and the associated theoretical biases alleged to be associated with research into the diagnosis of AHT has been questioned by Narang (2011). Specifically, he notes that the early physicians who identified the association between subdural haemorrhage, retinal haemorrhages, and abuse did so before the diagnosis was established and were, therefore, not influenced by biases. Further, that these physicians, who were separated, some by time in history, and others by significant geographical distances, and who worked in unrelated medical fields, managed to reach similar conclusions; conclusions that cannot be accounted for by chance. He also importantly noted that there have been no large research trials demonstrating a lack of association between subdural haemorrhages and retinal haemorrhages with AHT, a situation still applicable today.

Confession evidence

Accompanying the argument of circularity is the issue of confession evidence. As it is unethical to undertake experiments of this nature on young children and infants, and in the absence of independent witnesses, much of the research linking shaking and/or impact to the development of subdural haemorrhages, retinal haemorrhages, and hypoxic–ischaemic encephalopathy have relied upon confessions (Vinchon et al. 2010; Narang 2011; Findley et al. 2012). It has been argued that confession evidence is unreliable, and thus, using this as a basis for scientific investigation is engaging a circular argument which is itself unscientific (Lynoe et al. 2017; Brook 2019). False confessions do occur, have been well documented, but may be minimised utilising appropriate interrogation techniques including ensuring that the interviewee is medically and physically fit for interview (Gall and Freckelton 1999; Drizin and Leo 2004; Leo 2009; Stark and Rix 2020). Young people (aged under 18 years) and those people with medical, mental health, and intellectual disabilities are more susceptible to providing false confessions (West and Meterko 2016). False confessions are estimated to occur in as many as up to 25% of cases (Drizin and Leo 2004; West and Meterko 2016), and unsubstantiated claims of 25% (Innocence Network 2019). But the real rates are difficult to ascertain and in some studies rely, like AHT confession evidence, on information from the accused. A circular argument? Even if up to 25% of confessions in AHT accusations were false, that leaves the remaining 75% being correct. This, therefore, must mean that at least 75% of the confessions made to the investigators of AHT are correct and that the trauma that they state that they committed on the infant/child may only be viewed as being correct. Dias (2011) and Edwards et al. (2020) noted the consistent volunteered mechanism of injury by the perpetrators in cases of AHT, and this is the shaking of the infant with or without impact of the head. Shaking has been a constant and universal feature of AHT, and it is implausible that if the confessions were false, they would provide the same mechanism of injury. Thus, the research undertaken that is based on confession evidence, although containing inherent weaknesses from a scientific perspective, is justifiable in the absence of direct experimentation on humans and supports the existence of AHT as a valid diagnosis.

The missing biomechanical model

The contentious matters of a circular argument and confession evidence would become less relevant if a suitable biomechanical model could be established. To date, this has remained elusive and is utilized as an argument against the diagnosis of AHT in court.

Biomechanical models (including animal experimentation and biofidelic and computer models) alone are not predictive of outcomes in humans, but they may be informative particularly as direct experimentation on infants is not feasible (for a detailed discussion on the mechanism and modelling of AHT, see Case, 2014). Although unable to fully replicate an infant's/child's head and brain, the findings from some, but not all, of these experiments have been supportive of the AHT hypothesis and include:

- experimental whiplash injury in rhesus monkeys (rotational displacement of the head on the neck alone) may result in subdural haemorrhages and cerebral contusions (Ommaya et al. 1968);
- in porcine brain, there are age-dependent material properties of the cerebrum that affect the mechanical response of the brain to inertial loading, with adult brain being more tolerant (Thibault and Margulies 2002);
- repetitive axial rotation forces to piglet head resulted in a greater degree of injury (i.e. axonal injury) than a single rotational force (Raghupathi et al. 2004); and
- using a lamb model (that have a large brain and week neck muscles as in the human infant), shaking was shown to result in axonal injury and death without the need for head impact (Finnie et al. 2012; Anderson et al. 2014).

The clinician's approach to a diagnosis of AHT

Most clinicians will only be involved with live infants or children. The presentation may be a coincidental finding of a subdural haemorrhage or retinal haemorrhages without any obvious symptoms. Alternatively, the symptoms may range from minor to the very significant where the child is effectively moribund. Those children who arrive in a deceased state or die upon admission transfer to the paediatric pathologist.

The general diagnostic process has been discussed above. The process involves the undertaking of a holistic assessment of the patient, which includes obtaining information from around that patient. Details regarding the approach to the medical assessment, diagnosis, and management of AHT are described elsewhere (see Hymel and Deye 2011; Duhaime and Christian 2019; O'Meara et al. 2020). In brief, the assessment will involve the following:

- *History* – This involves the taking of a full and independent history of the present illness, details of the child's past medical history, the family medical history, a social history, a review of the child's body systems, and a review of birth, inpatient, and outpatient records. It also involves obtaining information from third persons including child-protection workers, police, social workers, and others who may have been involved with the child.

- *Examination* – A full and careful external examination of the child followed by an examination of the various body systems. Measuring of the child's height, weight, and head circumference is part of this examination.

- *Investigations* – The extent of investigations undertaken will depend on the history of the presenting illness and the findings on examination. Both pathology and radiological examinations will be required, and a suggested baseline series of investigations are provided in Table 8.1. The appropriate radiological investigations and experienced interpretation are vitally important (for further discussion, see Rao and Smith 2016; Choudhary et al. 2018).

- *Exclusion of 'mimics'* – Depending on the case presentation, specific examinations and investigations may be required to exclude the mimics of AHT (for reviews, see Sirotnak 2006; Greeley 2011; Narang et al. 2013), keeping in mind that mimics of AHT and AHT may coexist. A non-exhaustive list of mimics is given in Table 8.2.

- *Medical management* – The infant/child should be under the care of a medical or surgical unit independent of the forensic investigation. This unit, rather than the forensic physician, is responsible for the day-to-day care of the patient. Separating the responsibilities enables the treating unit to engage in a more trusting and therapeutic relationship with both the patient and the parents.

Table 8.1 Suggested preliminary/baseline investigations and referrals for the assessment of a child presenting with a possible head injury but in the absence of apparent injuries elsewhere on or within the body.

Pathology	Radiology/imaging	Referral
• Full blood examination • Clotting profile/platelet function including von Willebrand factor, Factor VIII, IX, XI, XIII • Serum biochemistry including liver, kidney, and pancreatic function • Urine metabolic screen • Ward urine test • Urine drug screen	• MRI and/or CT brain Plus, in children aged <2 years of age: • MRI neck • Skeletal survey	• Ophthalmology • Neurology/ neurosurgery

- *Referral* – Referral to other medical specialties may be necessary in the overall assessment of the case.

- *Ophthalmology* – Referral to an ophthalmologist at the earliest opportunity is essential for the documentation of retinal haemorrhages, which should also be photographed. It is important to exclude other causes of retinal haemorrhages (Table 8.3) (for a review and listing of other causes, see Levin 2011).

- *Communication* – The forensic assessment of AHT involves input from a multi-disciplinary team where information is shared and obtained. This team may involve the treating unit, other specialist medical units, allied health, child protection, and police. Communication with the parents is also essential even if one of those parents (or both) may be the alleged assailant. It needs to be remembered that it is not the physician's role to determine guilt, that is the role of the court.

- *Diagnosis* – The diagnosis is essentially one of exclusion and is only arrived at after all avenues of investigation have been completed and the findings considered in relation to the body of scientific evidence available.

- *Report* – As with most things of a forensic nature, a full and detailed medico-legal report is required providing full details of all information relied upon in arriving at the diagnosis.

Terminology

There is ongoing controversy regarding the terminology of AHT. Guthkelch (2012) and Byard (2014) very correctly commented that the current terminology and past terminology (e.g. AHT, non-accidental head injury, and shaken baby syndrome) are terms that imply both mechanism (trauma) and intent (abusive). As indicated throughout this chapter, the diagnosis of AHT is a medical conclusion based on information obtained through information gathering, an exhaustive clinical history, a physical examination, laboratory and imaging data, and the exclusion of plausible causes. It is not a legal investigation of the intent of the alleged perpetrator (Janson 2020). That is a matter for the judicial process. It is perhaps time that the diagnostic terminology for AHT is revisited and a more general diagnostic term agreed upon. A diagnostic term that avoided all references to either the mechanism or intent may allow for a more objective discussion of possible scenarios that have resulted in the injuries identified. AHT is not an appropriate term for a diagnosis.

Table 8.2 Mimics of AHT-related subdural haemorrhages (non-exhaustive list).

Trauma	Haematological/ haemostasis related/ vascular	Metabolic	Genetic	Infections	Tumours	Other
• Birth-related	• Haemophilia A and B	• Glutaric aciduria Type 1	• Osteogenesis imperfecta	• HSV encephalitis	• Leukemia	• Scurvy
• Accidental trauma	• Haemorrhagic disease of the newborn	• Homocystinuria	• Ehlers–Danlos Syndrome	• Bacterial meningitis	• Solid CNS tumors	• Congenital malformations
• Mild trauma associated with Benign enlargement of subarachnoid space of infancy	• Coagulopathies (e.g. factor XIII deficiency; factor VIII deficiency, Von Willebrand Disease; vitamin K deficiency)		• Menkes kinky hair disease	• Malaria		• Toxins
	• Arteriovenous malformations		• Hereditary haemorrhagic telangiectasia			• Latrogenic causes
	• Disseminated intravascular coagulation		• Marfan syndrome			

Table 8.3 Mimics of AHT-related retinal haemorrhages (non-exhaustive list).

Trauma	Haematological/ haemostasis related/vascular	Metabolic	Genetic	Infections	Tumours	Other
• Birth-related • Accidental trauma	• Haemophilia • Coagulopathies • Anemia • Vasculitis	• Glutaric aciduria Type 1	• Osteogenesis imperfecta • Ehlers–Danlos syndrome Type II	• HSV encephalitis • Bacterial meningitis • Malaria	• Lymphoblastic leukemia	• Carbon monoxide poisoning • Papilloedema/ Increased intracranial pressure • Hypoxia • Hypotension/ hypotension • Hyponatremia/ hypernatremia • Endocarditis

Conclusion

AHT is not a recent concept but has been documented within the medical literature for over 160 years. The concept of child abuse did not gain attention with the medical profession until the mid-1900s, and since this time, an enormous amount of research has been undertaken. AHT can occur in the presence of not only one or more of subdural haemorrhages, retinal haemorrhages, and hypoxic–ischaemic encephalopathy (the subject of this chapter and a cluster of findings sometimes referred to as the 'triad') but also with the presence of both external and internal injuries about the head and body generally. The pathogenesis of AHT was once considered to be solely the result of shaking of an infant/child but has subsequently been modified to include impact, this mechanism resulting in tearing of the bridging veins to cause haemorrhages and, in some cases, axonal injury and the development of hypoxic–ischaemic encephalopathy. In the absence of finding a suitable biomechanical model, and the absence of direct human experimentation for obvious ethical reasons, the diagnosis of AHT has remained an hypothesis which has subsequently led to challenges to its validity in the courts over the past few decades. It is acknowledged on both sides of the argument that there is some circularity of argument, a reliance on confession evidence, and an absence of a biomechanical model for AHT. Despite this, there is an enormous amount of research in support of a diagnosis of AHT, and it is a diagnosis that has general acceptance by the medical community. Arriving at a diagnosis of AHT involves an extensive multi-faceted medical investigation, a multi-factorial puzzle. The diagnosis is one of exclusion. Those currently disputing the diagnosis of AHT have displayed an absence of understanding of the medical diagnostic approach and the extensive body of supportive research. Some erroneously believe that the 'triad' is a diagnostic tool that forensic practitioners rely upon. Others selectively misuse, either intentionally or unintentionally, the scientific literature. AHT remains a valid diagnosis.

References

Al-Holou, W.N., O'Hara, E.A., CohenGadol, A.A., and Maher, C.O. (2009). Nonaccidental head injury in children. Historical vignette. *J Neurosurg Pediatr* 3: 474–483. https://doi.org/10.3171/2009.1.PEDS08365.

Anderson, R.W., Sandoz, B., Dutschke, J.K. et al. (2014). Biomechanical studies in an ovine model of non-accidental head injury. *J Biomech* 47: 2578–2583.

Barnes, P.D., Galaznik, J., Gardner, H. et al. (2010). Infant acute life-threatening event--dysphagic choking versus nonaccidental injury. *Semin Pediatr Neurol* 17: 7–11. https://doi.org/10.1016/j.spen.2010.01.005.

Bechtel, K., Stoessel, K., Leventhal, J.M. et al. (2004). Characteristics that distinguish accidental from abusive injury in hospitalized young children with head trauma. *Pediatrics* 114: 165–168.

Bhardwaj, G., Chowdhury, V., Jacobs, M.B. et al. (2010). A systematic review of the diagnostic accuracy of ocular signs in pediatric abusive head trauma. *Ophthalmology* 117: 983–992. https://doi.org/10.1016/j.ophtha.2009.09.040.

Bilo, R.A.C., Robben, S.G.F., and van Rijn, R.R. (2010). *Forensic Aspects of Pediatric Fractures. Differentiating Accidental Trauma from Child Abuse*. Heidelberg: Springer.

Binenbaum, G., Mirza-George, N., Christian, C.W. et al. (2009). Odds of abuse associated with retinal hemorrhages in children suspected of child abuse. *J AAPOS* 13: 268–272.

BMJ (1988). Summary of the Cleveland inquiry. *BMJ* 297: 190–191.

Bradford, R., Choudhary, A.K., and Dias, M.S. (2013). Serial neuroimaging in infants with abusive head trauma: timing abusive injuries. *J Neurosurg Pediatr* 12: 110–119.

Brook, C. (2019). Is there an evidentiary basis for shaken baby syndrome? The conviction of Joby Rowe. *Aust J Forens Sci* 53: 1. https://doi.org/10.1080/00450618.2019.1626483.

Butler-Sloss, E. (1988). *Report of Inquiry Into Child Abuse in Cleveland in 1987*. London: HMSO.

Byard, R.W. (2014). "Shaken baby syndrome" and forensic pathology: an uneasy interface. *Forensic Sci Med Pathol* 10: 239–241.

Caffey, J. (1946). Multiple fractures in the long bones of infants suffering from chronic subdural hematoma. *Am J Roentgenol Radium Ther* 56: 163–173.

Caffey, J. (1974). The whiplash shaken infant syndrome: manual shaking by the extremities with whiplash -induced intracranial and intraocular buildings, with residual permanent brain damage and mental retardation. *Pediatrics* 54: 396–403.

Case, M.B.S. (2014). Fatal pediatric craniocerebral and spinal trauma. In: *Forensic Pathology of Infancy and Childhood* (ed. K.A. Collins and R.W. Byard), 391–434. New York: Springer.

Cheung, P.Y., Obaid, L., and Rajani, H. (2004). Spontaneous subdural haemorrhage in newborn babies. *Lancet* 363: 2001–2002.

Choudhary, A.K., Bradford, R., Dias, M.S. et al. (2015). Venous injury in abusive head trauma. *Pediatr Radiol* 45: 1803–1813.

Choudhary, A.K., Servaes, S., Slovis, L. et al. (2018). Consensus statement on abusive head trauma in infants and young children. *Pediatr Radiol* 48: 1048–1065. https://doi.org/10.1007/s00247-018-4149-1.

Debelle, G.D., Maguire, S., Watts, P. et al. (2018). Abusive head trauma and the triad: a critique on behalf of RCPCH of 'Traumatic shaking: the role of the triad in medical investigations of suspected traumatic shaking'. *Arch Dis Child* 103: 606–610. https://doi.org/10.1136/archdischild-2017-313855.

Dias, M.S. (2011). The case for shaking. In: *Child Abuse and Neglect: Diagnosis, Treatment and Evidence* (ed. C. Jenny), 364–372. St. Louis: Elsevier Saunders.

Drizin, S.A. and Leo, R.A. (2004). The problem of false confessions in the post-DNA world. *North Carolina Law Rev* 82: 891–1004.

Duhaime, A.-C. and Christian, C.W. (2019). Abusive head trauma: evidence, obfuscation, and informed management. *J Neurosurg Pediatr* 24: 481–488. https://doi.org/10.1016/j.jbiomech.2014.06.002.

Edwards, G.A. (2015). Mimics of child abuse: can chocking explain abusive head trauma? *J Forensic Leg Med* 35: 33–37.

Edwards, G.A., Maguire, S.A., Gaither, J.R., and Leventhal, J.M. (2020). What do confessions reveal about abusive head trauma? A systematic review. *Child Abuse Rev* 29: 253–268. https://doi.org/10.1002/car.2627.

Evans, H.H. (2004). The medical discovery of shaken baby syndrome and child physical abuse. *Pediatr Rehabil* 7: 161–163. https://doi.org/10.1080/13638490410001715340.

Findley, K.A., Barnes, P.D., Moran, D.A. et al. (2012). Shaken baby syndrome, abusive head trauma, and actual innocence: getting it right. *Hous J Health Pol'y* 12: 209–312.

Finnie, J.W., Blumbergs, P.C., Manavis, J. et al. (2012). Neuropathological changes in a lamb model of a non-accidental head injury (the shaken baby syndrome). *J Clin Neurosci* 19: 1159–1164.

Gall, J.A. and Freckelton, I. (1999). Fitness for interview: current trends, views and an approach to the assessment procedure. *J Clin Forensic Med* 6: 213–223. https://doi.org/10.1016/s1353-1131(99)90000-7.

Geddes, J.F., Tasker, R.C., Hackshaw, A.K. et al. (2003). Dural haemorrhage in non-traumatic infant deaths: does it explain the bleeding in 'shaken baby syndrome'? *Neuropathol Appl Neurobiol* 29: 14–22.

Gennarelli, T.A., Spielman, G.H., Langfit, T.W. et al. (1982). Influence of the type of intracranial lesion on outcome from severe head injury: a multicentre study using a new classification system. *J Neurosurg* 56: 26–31.

Greeley, D.S. (2011). Conditions confused with head trauma. In: *Child Abuse and Neglect. Diagnosis, Treatment, and Evidence* (ed. C. Jenny), 441–450. St Louis, Elsevier Saunders.

Greeley, C.S. (2015). Abusive head trauma: a review of the evidence base. *AJR* 204: 967–973. https://doi.org/10.2214/AJR.14.14191.

Guthkelch, A.N. (1953). Subdural effusions in infancy: 24 cases. *BMJ* 1: 233–239.

Guthkelch, A.N. (1971). Infant subdural haematoma and its relationship to whiplash injuries. *BMJ* 2: 430–431.

Guthkelch, A.N. (2012). Problems of infant retino-dural haemorrhage with minimal external injury. *Hous J Health Pol'y* 12: 201–208.

Hadzikaric, N., Al-Habib, H., and Al-Ahmad, I. (2006). Idiopathic chronic subdural hematoma in the newborn. *Childs Nerv Syst* 22: 740–742.

Hansen, J.B., Frazier, T., Moffatt, M. et al. (2017). Evaluation of the hypothesis that Choking/ALTE may mimic Abusive Head Trauma. *Acad Pediatr* 17: 362–367.

Hedlund, G. (2015). Abusive head trauma: extra-axial hemorrhage and nonhemic collections. In: *Diagnostic Imaging of Child Abuse* (ed. P.K. Kleinman), 394–452. Cambridge: Cambridge University Press.

Helfer, R.E., Slovis, T.L., and Black, M. (1977). Injuries resulting when small children fall out of bed. *Pediatrics* 60: 533–535.

Holmgren, B.K. (2013). Ethical issues in forensic testimony involving abusive head trauma. *Acad Forensic Pathol* 3: 317–328.

Hymel, K.P. and Deye, K.P. (2011). Abusive head trauma. In: *Child Abuse and Neglect. Diagnosis, Treatment, and Evidence* (ed. C. Jenny), 349–358. St Louis: Elsevier Saunders.

Innocence Network (2019). Statement of the innocence network on shaken baby syndrome/abusive head Trauma. http://ncip.org/wp-content/uploads/2019/12/STATEMENT-OF-THE-INNOCENCE-NETWORK-ON-SHAKEN-BABY-SYNDROME-2.pdf (accessed 30 December 2021).

Janson, S. (2020). Towards a deeper understanding of abusive head trauma. *Acta Paediatr* 109: 1290–1291. https://doi.org/10.1111/apa.15164.

Jenny, C. (2014). Theories of causation in abusive head trauma: what the science tells us. *Pediatr Radiol* 44 (Suppl 4): S543–S547.

Kempe, C.H., Silverman, F.N., Steele, B.F. et al. (1962). The battered-child syndrome. *JAMA* 181: 17–24. https://doi.org/10.1001/jama.1962.03050270019004.

Kempe, C., Silverman, F., Steele, B. et al. (1985). The Battered-Child Syndrome. *Child Abuse Neglect* 9: 143–154.

Leo, R.A. (2009). False confessions: causes, consequences, and implications. *J Am Acad Psychiatry Law* 37: 332–343.

Levin, A.V. (2010). Retinal hemorrhage in abusive head trauma. *Pediatrics* 126: 961–970.

Levin, A.V. (2011). Eye injuries in child abuse. In: *Child Abuse and Neglect. Diagnosis, Treatment, and Evidence* (ed. C. Jenny), 402–412. St Louis: Elsevier Saunders.

Levin, A.V. (2017). The SBU report: a different view. *Acta Paediatr* 106: 1037–1039.

Luder, G.T. (2005). Retinal haemorrhages in accidental and nonaccidental injury. *Pediatrics* 115: 192.

Lynoe, N., Elinder, G., Hallberg, B. et al. (2017). Insufficient evidence for 'shaken baby syndrome' – a systematic review. *Acta Paediatr* 106: 1021–1027.

Lynoe, N., Elinder, G., Hallberg, B. et al. (2017a). Is accepting circular reasoning in shaken baby studies bad science or misconduct? *Acta Paediatr* 106: 1445–1446. https://doi.org/10.1111/apa.1394.

Lynoe, N., Rosen, M., Elinder, G. et al. (2018). Pouring out the dirty bathwater without throwing out either the baby or its parents. *Pediatr Radiol* 35: 1036–1040.

Lyons, T.J. and Oates, R.K. (1993). Falling out of bed: a relatively benign occurrence. *Pediatrics* 92: 125–127.

Mack, J., Squier, W., and Eastman, J.T. (2009). Anatomy and development of the meninges: implications for subdural collections and CSF circulation. *Pediatr Radiol* 39: 200–210.

Maguire, S.A., Watts, P.O., Shaw, A.D. et al. (2013). Retinal haemorrhages and related findings in abusive and non-abusive head trauma: a systematic review. *Eye* 27: 28–36.

Morad, Y., Wygnansky-Jaffe, T., and Levin, A.V. (2010). Retinal haemorrhage in abusive head trauma. *Clin Exp Ophthalmol* 38: 514–520.

Narang, S. (2011). A Daubert analysis of abusive head trauma/shaken baby syndrome. *Hous J Health Pol'y* 11: 505–633.

Narang, S.K., Melville, J., Greeley, C. et al. (2013). A daubert analysis of abusive head Trauma/shaken baby syndrome — Part II: an examination of the differential diagnosis. *SSRN Electron J* 98. https://doi.org/10.2139/ssrn.2288126.

Narang, S.K., Estrada, C., Greenberg, S., and Lindberg, D. (2016). Acceptance of shaken baby syndrome and abusive head trauma as medical diagnoses. *J Pediatr* 177: 273–278. https://doi.org/10.1016/j.jpeds.2016.06.036.

Nimityongskul, P. and Anderson, L.D. (1987). The likelihood of injuries when children fall out of bed. *J Pediatr Orthop* 7: 184–186.

O'Meara, A.M.I., Sequeira, J., and Ferguson, N.M. (2020). Advances in future directions of diagnosis and management of paediatric abusive head trauma: a review of the literature. *Front Neurol* 11: 118. https://doi.org/10.3389/fneur.2020.00118.

Oates, A.J., Sidpra, J., and Mankad, K. (2021). Parenchymal brain injuries in abusive head trauma. *Pediatr Radiol* 52: 898–910. https://doi.org/10.1007/s00247-021-04981-5.

Ommaya, A.K. and Yarnell, P. (1969). Subdural haematoma after whiplash injury. *Lancet* 2: 237–239.

Ommaya, A.K., Faas, F., and Yarnell, P. (1968). Whiplash injury and brain damage: an experimental study. *JAMA* 204: 285–289.

Pappworth, M.H. (1984). *A Primer of Medicine*, 5e, 27. Bodmi: Butterworths.

Parkes, S.E., Annest, J.L., Hill, H.A. et al. (2012). *Paediatric Abusive Head Trauma: Recommended Definitions for Public Health Surveillance and Research*. Atlanta, GA: Centres for Disease Control and Prevention.

Piteau, S.P., Ward, M.G.K., Barrowman, N.J. et al. (2012). Clinical and radiographic characteristics associated with abusive and nonabusive head trauma: a systematic review. *Pediatrics* 130: 315–323.

Plunkett, J. (2001). Fatal pediatric head injuries caused by short-distance falls. *Am J Forensic Med Pathol* 22: 1–12.

Raghupathi, R., Mehr, M.F., Helfaer, M.A. et al. (2004). Traumatic axonal injury is exacerbated following repetitive closed head injury in the neonatal pig. *J Neurotrauma* 21: 307–316.

Rao, P. and Smith, J.A.S. (2016). The utility of radiological investigation of suspected abusive head trauma in children. In: *Current Practice in Forensic Medicine*, vol. 2 (ed. J.A.M. Gall and J.J. Payne-James), 207–241. Chichester: Wiley.

RCPCH (2013). Abusive head Trauma and the eye in infancy. 2013-SCI-292-ABUSIVE-HEAD-TRAUMA-AND-THE-EYE-FINAL-at-June-2013.pdf (rcophth.ac.uk) (accessed 30 December 2021).

RCPCH (2019). Child protection evidence. Systematic review on head and spinal injuries. https://www.rcpch.ac.uk/sites/default/files/2021-02/Child%20Protection%20Evidence%20-%20Head%20and%20spinal%20injuries.pdf (accessed 30 December 2021).

Richards, P.G., Bertocci, G.E., Bonshek, R.E. et al. (2006). Shaken baby syndrome. *Arch Dis Child* 91: 205–206. https://doi.org/10.1136/adc.2005.090761.

Rosen, I.B. (2006). Wilfred Trotter: surgeon, philosopher. *Can J Surg* 49: 278–280.

Rosén, M., Lynøe, N., Elinder, G. et al. (2017). Shaken baby syndrome and the risk of losing scientific scrutiny. *Acta Paediatr* 106: 1905–1908. https://doi.org/10.1111/apa.14056.

Saunders, D., Raissaki, M., Servaes, S. et al. (2017). Throwing the baby out with the bathwater – response to the Swedish Agency for Health Technology Assessment and Assessment of Social Services (SBU) report on traumatic shaking. *Pediatr Radiol* 47: 1386–1389.

Sirotnak, A.P. (2006). Medical disorders that mimic abusive head trauma. In: *Abusive Head Trauma in Infants and Children. A Medical, Legal, and Forensic Reference* (ed. L. Frazier, K. Rauth-Farley, R. Alexander and R. Parrish), 191–226. St Louis: G.W. Medical Publishing.

Squier, W. and Mack, J. (2009). The neuropathology of infant subdural haemorrhage. *Forensic Sci Int* 187: 6–13. https://doi.org/10.1016/j.forsciint.2009.02.005.

Stark, M.M. and Rix, K.J.B. (2020). Fitness to be interviewed and fitness to be charged. In: *Clinical Forensic Medicine. A Physician's Guide*, 4e (ed. M.M. Stark), 393–420. Cham: Springer.

Stratton, K., Ford, A., Rusch, E. et al. (2012). *Committee to Review Adverse Effects of Vaccines, Board on Population Health and Public Health Practice, Institute of Medicine. Adverse Effects of Vaccines: Evidence and Causality*. Washington, DC: National Academies Press.

Strouse, P.J. (2018). Shaken baby syndrome is real. *Pediatr Radiol* 48: 1043–1047.

Supreme Court of Judicature Court of Appeal (Criminal Division) Neutral Citation Number (2005a). EWCA Crim 1980. Case Nos: 200403277, 200406902, 200405573, 200302848, Approved Judgement. Paragraph 58.

Supreme Court of Judicature Court of Appeal (Criminal Division) Neutral Citation Number (2005b). EWCA Crim 1980. Case Nos: 200403277, 200406902, 200405573, 200302848, Approved Judgement. Paragraphs 68, 69.

Talbert, D.G. (2005). Paroxysmal cough injury, vascular rupture and 'shaken baby syndrome'. *Med Hypotheses* 64: 8–13. https://doi.org/10.1016/j.mehy.2004.07.017.

Tarantina, C.A., Dowd, M.D., and Murdock, T.C. (1999). Short vertical falls in infants. *Pediatr Emerg Care* 15: 5–8.

Thibault, K.L. and Margulies, S.S. (2002). Age-dependent material properties of the porcine cerebrum: effect on pediatric initial head injury criteria. *J Biomech* 31: 1119–1126.

Vinchon, M. (2017). Shaken baby syndrome: what certainty do we have? *Childs Nerv Syst* 33: 1721–1733.

Vinchon, M., Defoort-Dhellemmes, S., Desurmont, M. et al. (2005). Accidental and nonaccidental head injuries in infants: a prospective study. *J Neurosurg* 102 (4 suppl): 380–384.

Vinchon, M., de Foort-Dhellemmes, S., Desurmont, M. et al. (2010). Confessed abuse versus witnessed accidents in infants: comparison of clinical, radiological, and ophthalmological data in corroborated cases. *Childs Nerv Syst* 26: 637–645. https://doi.org/10.1007/s00381-009-1048-7.

Weigel, R. and Department of Neurosurgery, University Hospital Mannheim, Mannheim, Germany Correspondenceralf.weigel@nch.ma.uni-heidelberg.de Department of Neurosurgery, University Hospital Mannheim, Mannheim, GermanyDepartment of Neurosurgery, University Hospital Mannheim, Mannheim, Germany (2004). Concepts of neurosurgical management of chronic subdural haematoma: historical perspectives. *Br J Neurosurg* 18: 8–18. https://doi.org/10.1080/02688690410001660418.

West, E. and Meterko, V. (2016). Innocence project: DNA exonerations, 1989-2014; Review of data and findings from the first 25 years. *Albany Law Rev* 79: 717–795.

9 The ageing population: needs and problems of the older person in prison

Caroline Watson
Royal College of General Practitioners, UK, Clinical Champion Healthcare in Secure Environments, Chair Secure Environments' Group.

Overview

A review of the incidence and nature of health issues for the older person in custody and the steps taken to address their specific vulnerabilities.

Introduction

The global prison population is increasing, with older people forming the fastest growing subset. All-age figures rose by 24% between 2000 and 2018, with high-income countries reporting the greatest proportional increases of older people; in Japan, over 60s made up 7% of the detained population in 2008 and 19% by 2016 (PRI Global Prison Trends Report 2019); in England and Wales, people over 50 comprised 7% in 2002 and 16.7% of the total prison population in 2020 (MOJ 2020).

The International Committee of the Red Cross has attributed the rise in older people in prison to a number of factors, including lengthy pre-trial periods, criminal justice policy changes leading to longer sentences, stipulations attached to early release, and convictions of older people for historic offences (ICRC 2020).

In England and Wales, 45% of men over 50 in prison are serving sentences for sexual offences, 23% for violence against the person, and 9% for drug offences (House of Commons Justice Committee 2020). Those imprisoned for sexual offences tend to be older at the time of conviction and receive very long sentences. They usually serve the whole of their sentence and, not infrequently, die in prison. Parole is only rarely granted, and compassionate release applications fail due to a complexity of factors, including a failure to acknowledge the crime committed; rehabilitation may be impaired, in part, due to relational repercussions; and societal intolerance of sexual offences. Older women are most likely to be in prison for violence against the person, for drug offences, or for acquisitive crime (theft, robbery, or burglary); only approximately 4% are in prison for sexual offences (CPA 2016).

The physical and mental health of people in prison, and particularly older people, tends to be poorer, and the need for health and social care greater, than the general

Current Practice in Forensic Medicine, Volume 3, First Edition. Edited by John A.M. Gall and J. Jason Payne-James.

population (House of Commons HSC Committee 2018). This is thought to be due to a range of factors impacting people prior to prison and possibly due to 'accelerated ageing' by up to 10 years while in prison (HMIP 2004).

Adverse socioeconomic, psychological, and biological factors that contribute to poor health are frequently present from before birth or early years and accumulate through childhood. They include parental smoking, drug use, poor nutrition and living conditions, exposure to domestic violence, and direct abuse. The adverse effects of these factors on health may be compounded further in adulthood by poverty, unemployment, poor diet, smoking, alcohol, substance misuse, and other lifestyle choices. Despite high levels of need, there is also frequently a reluctance of individuals to engage with community health and care agencies.

Against this backdrop of disparity of risk and poor health prior to prison, adverse environmental and social factors such as overcrowding, poor access to sanitation, limited opportunity for exercise, poor dietary choices, and psychological stresses of prison life can accelerate the progression of chronic health conditions and the process of ageing in the prison population. It is therefore unsurprising that older people frequently have complex medical and social care needs, which are distinct from the younger prison population.

This chapter looks at the incidence and nature of health issues for the older person in custody, to highlight barriers to providing care, and to identify steps that can be taken to address the vulnerabilities of this growing population in prisons.

Health and social care needs of older people in prison

Older people in prison are not a homogeneous group, but they tend to have common features and often have complex medical and social care needs, including physical and mental illness, sensory impairment, low mobility and difficulties managing personal care. A co-ordinated approach by prison and health and social care staff is required to meet these needs. Hayes and Burns (2012) found the needs of men aged 50–59 to be similar to those aged 60–69 and distinct from the younger population; over 50% had three or more moderate or severe health conditions. Over 10 years earlier, Fazel and Hope (2001) found 85% of those over 60 had one or more major illness recorded, and a study of over 170 000 people in prison in Texas (Baillargeon et al. 2000) showed that 61% of the over-50 cohort had two or more chronic medical conditions. Figures from the UK Ministry of Justice estimate that more than half of the older people in prison have a disability, with 28% having a physical disability (MOJ 2014).

Case A (see box) gives an example of the type of background and history of an 83 year old.

Complex co-morbidities

Diseases that are responsible for morbidity and mortality in the general population affect older people in prison disproportionately, and health outcomes are worse than for people of the same age in the wider community. Physical inactivity, obesity, and alcohol dependence are among the most important disease risk factors for the over 50s, and smoking, substance misuse, and poor diet for the under 45s in prison.

Case Study A

Mr A had been in prison for many years for the murder of a close relative and had been transferred to a secure psychiatric hospital to receive treatment for schizophrenia. He was 83 when he was transferred back to prison for the repair of an abdominal aortic aneurysm, detected through a screening programme. He had three myocardial infarctions over 15 years and had a pacemaker in situ. Having smoked heavily for 72 years, he had also developed chronic obstructive pulmonary disease (COPD) and had been treated for pneumonia on several occasions.

When Mr A first arrived, he was able to look after himself and lived in a cell on the ground floor of one of the main wings. He was able to walk short distances with a frame but used a wheelchair outside his cell to collect his meals and medication and frequently spent the night sitting in the chair, resulting in recurrent cellulitis and leg oedema. Within a month of arriving, he had decided against the elective surgical aneurysm repair.

Six months into his stay, Mr A was brought to the prison inpatient unit, after a cell inspection revealed evidence of urinary incontinence and soiled underwear, and no sheets on his bed. Once there, a social care assessment was arranged, and adaptations were made to his new cell; it had originally been designed as a dormitory and was large enough to accommodate a hoist, a supportive chair, and a hospital bed. Nursing staff provided 24-hour care but struggled to persuade Mr A to sleep in his bed at night or to use the armchair during the day. He continued to have problems with leg oedema and cellulitis and began to develop marks on his skin as a result of pressure from the standard-sized wheelchair which was becoming too small for him, as he gained weight due to inactivity, unhealthy choices from the menu, and regular orders of high-fat and sugary snacks and carbonated drinks bought from the canteen. A tailored powered wheelchair was provided, and Mr A enjoyed the independence he gained from this. His hygiene habits did not improve while in the healthcare unit, and he would regularly spit and throw used cups and rubbish onto the floor, where he would also urinate if he became angry.

One evening Mr A was found confused and drowsy, lying on the floor, having fallen out of his wheelchair. His breathing was rapid, and his oxygen saturations were low. He was transferred to the local hospital as an emergency where he was found to have diverticulitis and a bowel perforation. He required a laparotomy and a stoma and was admitted to the intensive treatment unit (ITU). He developed a post-operative bowel infection and dehiscence of his wound but eventually recovered sufficiently to return to the prison. He was treated with opiate medication post-operatively which he did not want to stop, even when his wound was fully healed.

During the months that followed, Mr A remained in the prison inpatient unit and became regularly frustrated as he waited for his parole hearing, hoping to transfer to a lower security prison. He called out repeatedly during the day when nursing staff left his room, and intermittently, he would pull off his stoma bag, emptying its contents onto the floor. He gradually became more frail and agreed that he did not wish to be resuscitated in the event of a cardiac arrest but would like to be admitted to hospital for intravenous fluids and antibiotics if he were to become acutely unwell. He was admitted to hospital again with low saturations and breathlessness, and after two weeks on the ward, he died from a myocardial infarction, with COPD – chronic obstructive pulmonary disease, pulmonary embolus, and aortic aneurysm.

In UK prisons, the most common diseases in older people are cardiovascular, musculoskeletal, and respiratory diseases. Fazel and Hope (2001) reported prevalence figures of 34% cardiovascular, 24% musculoskeletal, and 15% respiratory disease. Comparable data from over 10 years later (Hayes and Burns 2012) identified cardiovascular and musculoskeletal disease in more than half of over 60s in prison, and respiratory disease in almost one-third (Sturup-Toft and O'Moore 2018). Global non-communicable disease prevalence figures for older people in prison were explored by Munday and colleagues in a systematic review and meta-analysis (Munday and Leaman 2019). Heterogeneity across studies was high, but pooled prevalence data showed cardiovascular disease 38% (hypertension 39%, ischemic heart disease 21%, and stroke 6%), arthritis 34%, diabetes 14%, cancer 8%, COPD 8%, and asthma 7%.

Mr A is an example of someone who grew old in prison; his long history of smoking, compounded by his poor diet, physical inactivity, and obesity, resulted in progressive cardiovascular and respiratory diseases and poor health in his later years.

Age-related syndromes

Besides discrete chronic medical diseases, older people in prison frequently live with sensory, functional, and cognitive impairments and with the burden of multiple symptoms, including persistent pain, breathlessness, dizziness, constipation, incontinence, and agitation or anxiety. These age-related syndromes often have several different causes and affect the quality of life of older people in prisons as much as discrete medical diseases.

Falls and functional impairment

The incidence of falls in people over 65 is about 30% and increases with age. A number of factors heighten the risk of falling: poor vision; altered gait, impaired balance, or loss of muscle strength; foot problems or badly fitting shoes; multiple co-prescribed medicines; cognitive impairment; long periods of inactivity; and physical demands exceeding functional capability (WHO 2004). A study of older women in prison in the United States found half of them fell, during a one year time period (ranging from 33% to 63%). The incidence varied depending on pre-existing functional impairment, measured in terms of activities of daily living, including prison-specific tasks such as climbing on to a top bunk, getting to meals, queueing for medication, and hearing orders given by discipline staff (Williams et al. 2006).

Incontinence

In 2014, a study from the United States found 40% of people in prison aged 60 or over reported some level of incontinence (WHO 2014). This figure may well have been an underestimate of the true prevalence, since it is known that embarrassment can result in reluctance to report symptoms, which in turn leads to underdiagnosis, lack of treatment, and a reduced quality of life. Incontinence can be particularly distressing in the prison context due to several unique factors, including restricted access to showering or laundry facilities, close communal living, and poor access to toilet facilities during out-of-cell activities such as education, guided groups, work, and 'association' (time for socialising). The potential for hygiene problems, with accompanying malodour and possible bullying, can lead to a reluctance to attend structured rehabilitative activities and to isolation and depression.

Sensory impairment

Sensory impairment (visual and hearing loss) can reduce quality of life by contributing to communication difficulties, social isolation, poor balance, falls, and disability. Prison-specific challenges include preparation for trial and attendance in court; the obligation to follow signs and verbal orders to avoid breaking prison rules; difficulties of identifying people and holding conversations in crowded areas; and the potential for angering other residents by watching television on high volume, resulting in bullying. Munday et al. identified visual impairment in 65% of older people in prison and hearing loss in 28%.

Symptom burden

Older people may experience multiple symptoms due to a number of different causes, which frequently interact with one another. Persistent pain (Croft and Mayhew 2015), breathlessness, dizziness, and constipation are commonly reported. Case Study B (see box) explores some of these issues.

Mental health and substance misuse

Mood disorders, alcohol, and substance use disorders are common among older people in prison. Hayes and Burns (2012) found that 61% of older people had a 'mental disorder' and that those aged 50–59 were most frequently affected and at highest risk of suicide. Another study in 2013 found over half of older male prison residents to have depression – 31% mild and 23% severe (Senior et al. 2013). Mr B had pre-existing anxiety and depression and evidence of obsessional thoughts which were exacerbated by coming into prison. Another problem that may be seen in older people, such as Mr A, who have spent many years in prison, is that of 'institutionalisation', described by Crane as a 'biopsychosocial state' brought on by imprisonment and characterised by anxiety, depression, hypervigilance, and a disabling combination of social withdrawal and/or aggression (Crane 2019).

The next case study (Case Study C) illustrates the complex and sometimes tragic consequences that can result from alcohol and substance misuse in older people in contact with the criminal justice system.

Problematic use of drugs and alcohol may begin in an attempt to relieve anxiety, to mask voices or thoughts accompanying severe mental illness, or to counteract the distress of trauma experienced either as a child or as an adult, for example violence in the context of crime or war. It is often associated with a pattern of 'revolving door' imprisonment, with deteriorating physical and mental health and with a downward spiralling of social circumstances, not infrequently culminating in homelessness and exclusion from supportive relationships.

Case Study B

Mr B was very distressed when he first arrived at the prison. At the age of 75, he had received a two-year custodial sentence that he had not been expecting. He had a diagnosis of anxiety and depression and described obsessional thoughts that were intrusive. When his case had first come to trial, Mr B had taken a serious overdose of beta-blockers but had suffered no recognised immediate or ongoing adverse effects from this. He also had hypertension and stage 3 chronic kidney disease and was prescribed four different classes of medication for his mental and physical health. In the first few weeks of being in prison, the emergency nurse was repeatedly called to see Mr B with episodes of dizziness and collapse. His examination findings were normal each time, and ECG and blood tests were reassuring. He was admitted to hospital on two occasions, but no cause for the collapses was identified. After nurses began to find Mr B positioned on the floor with his eyes closed and a pillow under his head, they began to explore whether anxiety and difficulties with adjustment were underlying his collapses. With supportive input from staff and other residents, Mr B's collapses stopped.

In addition to physical symptoms, distressing emotions may be suffered in the presence of cognitive impairment, or related to adjustment to prison life as an older person, and the stress of receiving a long custodial sentence. Mr B's case illustrates the difficulties of adjustment and the ways in which it may present, with attempts at self-harm or suicide, an exacerbation of pre-existing mental health issues, or physical symptoms.

Case Study C

Mr C had almost become 'part of the furniture'. In his 60s, he was regularly in and out of the local prison for breach of his licence conditions, returning usually within a week of being released, often for theft of alcohol. This had been the pattern for years. He admitted that, during the winter months particularly, having three meals a day and a roof over his head was a more comfortable existence than living on the streets in the biting cold, being repeatedly moved on by police, scraping together enough money to buy alcohol, and under constant threat of being attacked by passers-by.

He suffered with atrial fibrillation, hypertension, and cirrhosis. He usually stayed on the main wings but occasionally spent the early days of his recall period in the inpatient unit for assisted alcohol withdrawal treatment, if he had remained out of prison for long enough to be at risk of withdrawal. During one inpatient stay, Mr C developed progressive weakness in his face and arm. He was treated at the nearest stroke unit and returned to prison for his rehabilitation. Staff regularly commented that prison was a safer place for Mr C and would be worried about his welfare, if he had not returned to custody within a few weeks of his release. One time, a resettlement placement was arranged for him at a charitable community, but he was rearrested before completing the train journey there. So, when he had not been seen for over three months, and a missing person's enquiry had drawn a blank, there was anxiety among healthcare and prison staff. Mr C was eventually located after 18 months. He had died while sleeping among undergrowth many months before being found. There was a genuine sense of grief among staff.

Cognitive impairment

Dementia is defined as a decline in two or more areas of cognitive functioning resulting in impaired performance. It is recognised by the WHO to be a worldwide public health priority (WHO 2020), and it is a growing problem in the ageing prison population, affecting around 8% of people. Risk factors for being detained in custody include post-traumatic stress disorder, traumatic brain injury, low educational achievement, alcohol, and substance misuse, which overlap with risk factors for dementia. The prevalence of dementia is estimated to be twice as high in 60–69 year olds in prison as in the community and four times as high in the over 70s (Forsyth and Heathcote 2020); however, cognitive decline may be initially masked by the repetitive nature of prison regimes and by removal of domestic tasks, for example cooking, managing money, and laundry. Where prison staff lack training in identifying and handling dementia, early signs may also be missed, and misunderstanding may arise; agitation, disinhibition, aggression, or failure to follow instructions may be believed to be acts of deliberate disruption. As dementia progresses and capacity declines, care needs will escalate, and the older person may no longer be able to understand the reason behind their imprisonment or demonstrate progress in their rehabilitation. An example is seen in Case Study D.

Case Study D

Mr D had 65 convictions for 111 offences and was recalled to prison aged 67, just prior to the COVID-19 pandemic, early in 2020. He had been in the armed forces and had been drinking and smoking heavily since he left. He had been diagnosed with schizophrenia and type 2 diabetes. He had been rough sleeping for a number of years and had sustained an ankle fracture which he had walked on. He had been found to be confused on a number of occasions, and it had been presumed that his confusion was linked to his drinking.

Shortly after he arrived, Mr D was assessed by a psychologist and found to have a reduced cognitive score. He was referred to the memory clinic and had blood tests arranged to assess him for possible causes of cognitive decline. He was located on the main wings and collected his meals every day. He kept his medicine in his possession and did not come out of his cell otherwise. He made no phone calls and had no showers. There were no group activities to attend, and only very few residents secured jobs during the lockdown period. Officers did not raise concerns about Mr D. His initial appointment with the memory clinic was a telephone consultation, but Mr D declined to speak. The consultant then visited the prison, but Mr D would not leave his cell to come to the healthcare department. One of the nurses went to see Mr D in his cell and found him sitting on his bed, moving tea bags and sugar sachets back and forth between two plastic bags repeatedly and without any clear intent. He had very poor hygiene and appeared to be confused. He spoke in non-coherent sentences, and some words were neologisms. He was unable to articulate where he was or the month. His cell mate spoke no English and could give no further information about Mr D's behaviour. It was decided that Mr D needed to be admitted to the inpatient unit for a period of assessment and for assistance with activities of daily living. He was found to have seven months of medication stored in his property. His cognitive decline had been unnoticed for seven months due to the restricted regime and lack of demands made on him in prison compared to the community.

Polypharmacy

As the range of medication available to manage long-term conditions widens and evidence-based national guidelines evolve, the older person with multiple co-morbidities may end up being prescribed a large number of different medicines to manage multiple individual conditions. Chronic non-cancer pain, mental health disorders, and substance misuse may also contribute to polypharmacy.

When more than four different medicines are prescribed concurrently, regardless of the type, an older person is at increased risk of falling, fear of falling, and cognitive impairment. Specific drugs or classes of drugs can also have detrimental effects, and it is important that prescribers are aware of this. For example, drugs with anticholinergic properties can cause urinary retention, constipation, impaired cognition and confusion, falls, and potentially even increase the risk of mortality, particularly if multiple drugs with anti-cholinergic properties are co-prescribed.

It is also important that prescribers are aware of medicines that cause withdrawal or discontinuation symptoms when stopped abruptly, for example opioids and antidepressants. Tapering regimes are essential in order to avoid precipitating unpleasant and debilitating symptoms, including pain, anxiety, agitation, and dizziness, which might lead to a fall.

Palliative care

The standardised mortality rate of people in prison in England is 50% higher than the general population, and the average age of death is 56 years compared to 81 years in the general population (House of Commons HSC Committee 2018). Natural-cause deaths are the leading cause of mortality in prisons in the United Kingdom (MOJ 2018), reflecting the ageing population and the complex health needs of people in contact with the criminal justice system.

An application for early release from prison may be considered if a person is approaching the final months of life; however, many older people will not meet the eligibility criteria for release due to the nature of their offence. Others who have served long sentences or whose offence has caused a breakdown of relationship with family or friends may prefer to remain in prison for their death.

A dignified death is an integral part of any humane society. Compassionate high-quality palliative and end-of-life care underpins this; however, provision in prisons is highly variable. Some establishments deliver coordinated multi-professional holistic care and others have poor interdisciplinary communication, haphazard recognition of patient need, suboptimal care, coordination, and delayed provision, potentially resulting in repeated acute admissions to hospital which may be distressing for the patient and costly for the establishment.

Resettlement

Safe and effective resettlement of older people leaving prison requires specific preparation and planning. For the patient with complex medical and social needs, it is essential to transfer information and to set up appointments with community services (e.g. substance misuse and primary care) ahead of release. Referrals should be made to hospital specialists in the locality of release to facilitate continuity of care, and a tailored package of social care will need to be prepared, including arranging any adaptations to accommodation before the person leaves the prison. Adequate quantities of medication and other necessary medical supplies (e.g. incontinence pads and walking frame) should be available to cover the initial weeks following release, until community provision is effectively established. Additionally, the older person will require support with housing, benefits, and pension applications together with assistance in becoming familiar with information technology, in order to successfully encounter the demands that will be made of them in the community, e.g. paying bills and food shopping.

The impact of COVID-19 on older people in prison

The COVID-19 pandemic has exacerbated health and social inequalities globally, and the virus has had a disproportionately negative impact on older people, people with multiple health conditions, particular ethnic groups, those from a background of social deprivation, and those in residential settings – all characteristics common to the older person in prison.

At the start of the pandemic, the rates of infection and death from COVID-19 were anticipated to be high in prisons, and countries across the world adopted a range of prison-specific infection prevention and control (IPC) measures, beyond those introduced for the general population. In the United Kingdom, modelling predicted more than 77 000 COVID-19 cases and 2000 deaths in prisons (Braithwaite et al. 2021), and so, people were required to remain in their cells for 23 hours a day, with very limited opportunities for exercise, work, or education. Group rehabilitation activities and family visits were suspended. In addition to social distancing measures and hygiene, compartmentalisation of the population was introduced; people newly arriving in prison, those at most clinical risk, and those symptomatic or testing positive for COVID-19 were required to isolate from the main population. Mass testing was also introduced. Although fewer deaths attributable to COVID-19 occurred in UK prisons than predicted, the stringent IPC measures and mass testing did not prevent large outbreaks; more than 4000 new cases across 85% of prison and youth custody facilities were seen in January 2021 alone (MOJ 2021).

COVID-19 has had an impact on the mental and physical health of people in prison, reaching beyond directly attributable morbidity and mortality. Some residents have experienced relief from the pressure of usual prison interactions, but many people have experienced worsening anxiety, depression, and increasing rumination due to months of being locked up 22–23 hours each day, isolated, and prevented from seeing family members. Distress and obsessive compulsive disorder (OCD) – obsessive compulsive disorder, symptoms have been seen increasingly in people already on mental health caseloads, and lack of purposeful activity and stimulation has led to deteriorating cognitive abilities in older people.

Detrimental physical effects of being locked down and inactive have included reduced mobility, strength, and balance, increasing the likelihood of falls among older people. Lack of exercise, together with increased junk food consumption to modulate emotions, has led to reports from across the UK prison estate of weight gain and increasing problems with obesity. Some immediately attributable effects have included increasing insulin requirements in patients with diabetes; however, there is likely to be a wider evolving picture of morbidity.

The true extent of the impact of COVID-19 on older people who have been in prison during the pandemic is likely to unfold over the coming months and years. It will, therefore, be important to take a proactive approach to identifying and addressing health needs in older people in prison, starting with screening for emerging physical and mental health conditions and for cognitive decline.

Key steps in addressing the needs of the older person in prison

Achieving continuity of health information

Continuity of information through access to patient health records is one of the most important factors influencing the safety of the older person transferring between community, police custody, court, and prison. Regularly, however, people arrive in prison with multiple medical problems, no medication, and little or no health information available. They may have neglected their health or may have outstanding hospital appointments. Without rapid access to essential health-related information, continuity of care is compromised, and the older person's health may rapidly deteriorate due to disruption to medication, investigations, or planned specialist treatment.

Full access to the patient electronic health record, achieved through connectivity between community and prison health information systems, facilitates uncompromised continuity of care. Where this cannot be achieved and a custodial sentence is anticipated, preparation of summarised salient health information for the patient to take into prison, by the community practice, will assist the prison healthcare team in providing timely care. Information should include hospital letters, investigation results, and current problems together with an up-to-date list of medication providing clinical indications for each drug. Where a custodial sentence has not been expected and the older person has multiple health and care needs, it would be helpful for the court to consider whether it would be safe to delay incarceration until appropriate information has been gathered and preparations made in the receiving prison. Where this is not considered safe or practicable, the prison healthcare team should obtain patient consent and request urgent health information transfer by secure email from the community care provider.

Screening and assessing for age-related syndromes

Ad hoc reactive management and lack of oversight of a person's vulnerabilities can result in gaps in essential care, accelerated decline in health, escalation of frailty, unplanned admissions, and premature mortality.

Systematic screening of the older person on arrival in prison should effectively identify:

- medical conditions that require urgent assessment and treatment
- sensory, functional, or cognitive impairments
- disabilities
- problems that may not be volunteered, e.g. incontinence, falls, and alcohol or drug dependence
- social care needs
- any areas of immediate risk requiring safeguarding actions.

Within the first week, mental health screening (NICE 2017) and a more substantial secondary screening (NICE 2016) should be carried out to identify any issues not picked up on entry, any that have arisen since arrival, and any age-related problems (falls and functional impairment, incontinence, sensory impairment, symptom burden, and dementia). Health promotion advice, vaccinations, and screening tests should be offered for, e.g. blood-borne viruses, and arrangements made for, e.g. vascular, bowel cancer, and abdominal aortic aneurysm screening.

It is important to reassess older people in prison at regular intervals to identify emerging needs and reduce the risk of acute events, e.g. falls, stroke which may trigger admission and increase morbidity, disability, and dependence. Some prisons train residents to carry out aspects of screening and preventive care such as taking blood pressure on the wings. Where residents are employed, there should be good governance structures, so that both the older person and the employed resident are protected by adequate training and effective supervision.

Communication, co-ordination, and care pathways

Effective care provision for the older person in prison with complex needs requires clear timely communication between different healthcare professionals and teams. Good partnership working between prison, health, and social care providers is also necessary, and it can be beneficial to set out a memorandum of understanding to ensure all parties are aware of their roles and responsibilities.

Clearly defined pathways and systems are also important to facilitate effective timely care and provide a framework for accountability. For example, medicines optimisation (NICE 2015) involves utilising systems to achieve timely medicines reconciliation, to determine ongoing clinical suitability and acceptability of medicines to the patient, to assess patient suitability for keeping medicines in possession, to monitor for concordance and side effects, and to set up repeat medication re-ordering. Without these systems in place, there is considerable risk to the older person, including dose omission, drug side effects, drug–drug interactions, toxicity, and worsening health. A medication review soon after the arrival of the older person is advisable to rationalise prescribing by assessment of the ongoing clinical indication and acceptability of each medicine, and identification of potential drug–drug interactions and problems with concordance, which may contribute to deterioration of health. Decision aids, e.g. the STOPP/START tool (NHS England 2017), can assist with the process. It should be remembered that drugs associated with withdrawal or discontinuation symptoms should be tapered rather than stopped abruptly, if no longer required.

A clear protocol is needed to ensure rearrangement of outstanding hospital appointments when the older person first comes into prison from the community or from another prison, in order to avoid disruption and delays to investigations and treatment, which may result in avoidable or accelerated morbidity and even premature death. Additionally, pathways should be set up to refer them to the appropriate healthcare professional or team within the prison, where screening has identified a new problem that needs investigation and treatment, or a long-term medical condition that requires monitoring, e.g. new leg swelling or long-standing COPD.

For the older patient with multiple morbidities and frailty, e.g. heart and lung problems, diabetes, leg ulcers, poor hearing and eyesight, memory loss, and reduced mobility, the traditional model of medical care, which involves following national or local disease-specific evidence-based guidance for individual conditions, may result in a large number of appointments for monitoring, investigations, and the burden of multiple medications. A multi-morbidity care model (NICE 2016) is an alternative approach, involving coordinated interdisciplinary care which aims to reduce polypharmacy, the number of appointments, and episodes of unplanned care.

A team-based approach is used in which health professionals (e.g. general practitioner (family practitioner, family physician), nurse, pharmacist, occupational therapist, and physiotherapist) and social care providers work together to assess the older person, share information with one another, and provide education to enable the patient to weigh up the risks and benefits of each proposed intervention and reach a shared decision on prioritising needs. A care plan is drawn up which may include a self-management plan, supporting the patient to retain as much independence as possible. The care plan should be available to all professionals responsible for providing care and to the patient.

Where it is not suitable or desired by the older person approaching the end of their life to be released back into the community, it is important to provide good-quality palliative and end-of-life care, with support from community specialists so that the holistic needs and wishes of the individual are anticipated, planned for, and met, within a framework of effective governance. In 2018, The Dying Well in Custody Charter was launched in the United Kingdom, setting out ambitions and standards for palliative and end-of-life care in prisons (Ambitions for Palliative and End of Life Care Partnership 2018).

Access to specialist services

It is important that regular health and social care needs assessments are carried out in order to plan and commission appropriate services to meet the needs of the ageing population in prisons (PHE 2017), including access to specialist services.

When attending specialist appointments in the community, patients have to be accompanied by prison security staff. These escorts can be difficult to facilitate at times, due to competing pressures on staff to maintain the prison regime, which can result in cancellation of appointments and delays in access to care.

A variety of solutions are employed at different establishments to resolve problems in access to specialist care. Some specialist services can be effectively delivered on site, either regularly or for individual patient referrals, depending on the demographics of the population. These may include, for example psychiatrist, optician, podiatrist, physiotherapist, audiologist, and occupational therapist.

Increasingly, and particularly since the onset of the COVID-19 pandemic, hospitals have begun to offer greater numbers of remote consultations to patients in the community, utilising telephone or video calls. In prisons, remote consultations are in their infancy. Co-ordination within the prison is needed to bring the older person from the wing to a dedicated clinic room in the healthcare centre in time for the appointment, and a member of the healthcare team is required to be present. This is costly in terms of staffing but increases the likelihood of effective engagement of the older person, particularly where there may be confusion or memory problems. It also bypasses the difficulties of arranging escorting officers.

In prisons with large cohorts of older people serving long sentences, it is important to ensure access to memory clinics and to establish good links with the local hospice team, giving consideration to engaging a visiting palliative care consultant.

Provision of suitable accommodation

Providing health and social care that can be accessed by an ageing population may require adaptations to the physical surroundings. Most prisons were designed to hold young, fit people, with buildings often spread over a substantial footprint, requiring residents to cover large distances in order to access programmes, dining areas, the gymnasium, and their accommodation. Surfaces may be uneven, lighting and signage poor, and steps steep, with some areas having no access to lifts. Communal areas may have no seating, and accommodation is often cramped. Narrow doorways, bunk beds, and little in-cell floor space frequently provide difficulties for the older person who requires a walking aid, commode, supportive chair, or any in-cell care assistance. An unsuitable physical environment can even potentially create care requirements for the older person that had not been present in the community and can contribute to social isolation, loneliness, and lack of access to healthcare, programmes, or exercise.

Adaptations to the physical surroundings may include installation of lifts and stairlifts to facilitate access to care and washing facilities, changes to lighting and signage, modifications to accommodation, such as widening doorways to accommodate wheelchairs and frames, installation of grab rails and toilet seat raisers, lowering beds, and providing additional emergency alarms.

For the more frail person requiring in-cell assistance and larger equipment, e.g. a hoist, wheelchair, and commode, a larger room is necessary. Some prisons require the older person to be transferred to the inpatient unit, to a specific wing, or to another prison better able to meet their needs, which may trigger isolation and loneliness due to loss of contact with peers. Other prisons have designed accessible cells with in-cell showers and hospital beds within each house block, to enable the older frail person to remain with peers and younger residents while receiving the care they need.

Peer mentor ('buddy') schemes, voluntary sector organisations, and resettlement planning

The social care requirements of the older person should be identified on admission to prison. Where these cannot be fully met by commissioned providers, some prisons will transfer the resident to the inpatient unit if there is one, or require healthcare staff to assist with care on the wings. This can be demanding in terms of time and detract from the needs of other patients with medical care requirements.

Some establishments have set up effective peer mentor or 'buddy' schemes in which selected residents are trained to assist older, frail, or disabled peers by, e.g. collecting meals, pushing wheelchairs, and providing companionship. These peer supporters are not involved in providing intimate care. Adequate governance of peer mentor or 'buddy' schemes is essential, with provision of training and ongoing supervision of peer carers, to avoid putting both the patient and carer at risk.

Voluntary sector organisations can also provide valuable support for the older person by partnering with prisons in a variety of ways. Examples include offering support for older women to keep up family relationships while in custody; running day-care programmes to provide age-appropriate activities, e.g. seated exercise classes; facilitating groups for residents with cognitive impairment; and holding events to support residents receiving palliative care and their families and friends.

Effective resettlement planning requires good communication and coordination between the prison, probation, and healthcare teams well in advance of the expected release date, in order to secure housing that is appropriate to the needs of the older person, transfer information to community health teams and social care providers in the locality of release, and set up any necessary appointments immediately following release.

Resettlement planning is, however, often focused on the needs of the younger adult and does not prepare the older person for the unique challenges they may face, particularly if they have become 'institutionalised' after a long sentence. They may find themselves ineligible for suitable housing if their offence was of a sexual nature and, in a worst-case scenario, may end up being discharged into the community with little notice, no housing and without timely transfer of essential health information, adequate provision of medicines, or community appointments arranged. The resulting disruption to care can cause rapid deterioration of the older person leaving prison.

RECONNECT is a recent innovative scheme in the United Kingdom, commissioned to assist the vulnerable person leaving prison, by making contact with community services, facilitating pre-release registration, and arranging through-the-gate support. Voluntary sector organisations can also assist with the preparation for release by teaching, e.g. cooking, finance, and IT skills and providing support with housing, pensions, and benefits applications.

Where next?

Anticipating need, future planning, financial provision, and staff training

Future planning is needed to avoid the prison system being overwhelmed by the escalating needs of its ageing population. Strategic decisions should be informed by health and social care needs assessments, and financial resourcing should be released to meet the needs identified.

The increasing demand for specialist and primary care should be recognised, with staffing, medicines, and equipment budget implications and rising hospital escorts and hospital bed watch provision. Alternative models for care should be considered, including local multi-morbidity team working, with named individuals providing oversight to ensure implementation of co-ordinated care; access to specialists on site or via video consultations; and commissioning of regional dedicated prison older person (geriatric) medicine teams. Social care can only be delivered within suitable accommodation, and therefore, future estate development programmes need to incorporate adaptations to pre-existing facilities and development of purpose-built accommodation to provide a suitable environment for the older person, which meets their changing needs, often typified by escalating frailty and declining security risk.

Prisons need to recognise the importance of adjusting the regime for older people and the need to train both prison and healthcare staff in the recognition and management of the differing needs of the older person. Good governance is essential and at an operational level, co-ordinated multi-professional partnerships which require effective communication and agreement on delineation of responsibilities. Above all, the needs of the older person should remain at the heart of the custodial estate.

References

Ambitions for Palliative and End of Life Care Partnership (2018). Dying well in custody charter. A national framework for local action. http://endoflifecareambitions.org.uk/wp-content/uploads/2018/06/Dying-Well-in-Custody-Self-Assessment-Tool-June-2018.pdf (accessed 31 October 2020).

Baillargeon, J., Black, S.A., John Pulvino, P.A., and Dunn, K. (2000). The disease profile of Texas prison inmates. *Annals of Epidemiology* 10 (2): 74–80. https://www.ncjrs.gov/pdffiles1/nij/grants/194052.pdf (accessed 3 November 2020).

Braithwaite, I., Edge, C., Lewer, D., and Hard, J. (2021). High COVID-19 death rates in prisons in England and Wales, and the need for early vaccination. *The Lancet Respiratory Medicine* 9: 569–570. https://doi.org/10.1016/S2213-2600(21)00137-5 (accessed 28 March 2021).

Centre for Policy on Ageing - Rapid review (2016). Diversity in older age – older offenders. https://www.ageuk.org.uk/globalassets/age-uk/documents/reports-and-publications/reports-and-briefings/equality-and-human-rights/rb_may16_cpa_rapid_review_diversity_in_older_age_older_offenders.pdf (accessed 3 October 2020).

Crane, J. (2019). Becoming institutionalized: incarceration and 'slow death'. Social Science Research Council. https://items.ssrc.org/insights/becoming-institutionalized-incarceration-and-slow-death/ (accessed 1 November 2020).

Croft, M. and Mayhew, R. (2015). Prevalence of chronic non-cancer pain in a UK prison environment. *British Journal of Pain* 9 (2): 96–108. https://doi.org/10.1177/2049463714540895 https://www.ncbi.nlm.nih.gov/pmc/articles/PMC4616963/ (accessed 1 November 2020).

Fazel, S. and Hope, T. (2001). Health of elderly male prisoners: worse than the general population, worse than younger prisoners. *Age and Ageing* 30 (5): 403–407. https://doi.org/10.1093/ageing/30.5.403. https://pubmed.ncbi.nlm.nih.gov/11709379/ (accessed 1 November 2020).

Forsyth, K. and Heathcote, L. (2020). Dementia and mild cognitive impairment in prisoners aged over 50 years in England and Wales: a mixed-methods study. *Health Services and Delivery Research* 8 (27): 1–116. https://doi.org/10.3310/hsdr08270 https://pubmed.ncbi.nlm.nih.gov/32609458/ (accessed 31 October 2020).

Hayes, A. and Burns, A. (2012). The Health and social care needs of older male prisoners. *International Journal of Geriatric Psychiatry* 27 (11): 1155–1162. https://doi.org/10.1002/gps.3761 https://pubmed.ncbi.nlm.nih.gov/22392606/ (accessed 1 November 2020).

HM Inspectorate of Prisons (2004) 'No Problems – Old and Quiet': Older Prisoners in England and Wales. A Thematic Review by HM Chief Inspector of Prisons. Her Majesty's Inspectorate of Prisons. Older Prisoners-2004.pdf (justiceinspectorates.gov.uk).

House of Commons Health and Social Care Committee (2018). Prison health. Twelfth report of sessions 2017-19. https://publications.parliament.uk/pa/cm201719/cmselect/cmhealth/963/963.pdf (accessed 3 October 2020).

House of Commons Justice Committee (2020). Ageing prison population. https://publications.parliament.uk/pa/cm5801/cmselect/cmjust/304/30403.htm (accessed 3 October 2020).

International Committee of the Red Cross (2020). Ageing and detention. https://www.icrc.org/en/publication/4332-ageing-and-detention (accessed 31 October 2020).

Ministry of Justice (2014). Analytical summary: the needs and characteristics of older prisoners: results from the Surveying Prisoner Crime Reduction (SPCR) survey. https://assets.publishing.service.gov.uk/government/uploads/system/uploads/attachment_data/file/368177/needs-older-prisoners-spcr-survey.pdf (accessed 1 November 2020).

Ministry of Justice (2018). (Safety in custody statistics, England and Wales: deaths in prison custody to June 2018 assaults and self-harm to March 2018. https://assets.publishing.service.gov.uk/government/uploads/system/uploads/attachment_data/file/729496/safety-in-custody-bulletin-2018-Q1.pdf (accessed 23 October 2020).

Ministry of Justice (2020). Prison population statistics (England and Wales).

Ministry of Justice (2021). HM prison and probation service COVID-19 official statistics. https://assets.publishing.service.gov.uk/government/uploads/system/uploads/attachment_data/file/960361/HMPPS_COVID19_JAN21_Pub_Doc.pdf (accessed 29 March 2021).

Munday, D. and Leaman, J. (2019). The prevalence of non-communicable disease in older people in prison: a systematic review and meta-analysis. *Age and Ageing* 48 (2): 204–212. https://doi.org/10.1093/ageing/afy186. https://pubmed.ncbi.nlm.nih.gov/30590404/ (accessed 3 November 2020).

NHS England (2017). Toolkit for general practice in supporting older people living with frailty. Appendix 5 STOPP, Screening Tool of Older Persons' potentially inappropriate Prescriptions; START, Screening Tool to Alert to Right Treatment. https://www.england.nhs.uk/wp-content/uploads/2017/03/toolkit-general-practice-frailty-1.pdf (accessed 23 October 2020).

NICE guideline (NG5) (2015). Medicines optimisation: the safe and effective use of medicines to enable the best possible outcomes. https://www.nice.org.uk/guidance/ng5 (accessed 3 October 2020).

NICE guideline (NG56) (2016). Multimorbidity: clinical assessment and management. https://www.nice.org.uk/guidance/ng56 (accessed 14 November 2020).

NICE guideline (NG57) (2016). Physical health of people in prison. https://www.nice.org.uk/guidance/ng57 (accessed 12 November 2020).

NICE guideline (NG66) (2017). Mental health of adults in contact with the criminal justice system. https://www.nice.org.uk/guidance/ng66 (accessed 12 November 2020).

Penal Reform International and Thailand Institute of Justice (2019). Global prison trends 2019. https://cdn.penalreform.org/wp-content/uploads/2019/05/PRI-Global-prison-trends-report-2019_WEB.pdf (accessed 31 October 2020).

Public Health England (2017). Health and social care needs assessments of the older prison population. A guidance document. https://assets.publishing.service.gov.uk/government/uploads/system/uploads/attachment_data/file/662677/Health_and_social_care_needs_assessments_of_the_older_prison_population.pdf (accessed 31 October 2020).

Senior, J., Forsyth, K., Walsh, E., et al. (2013). Health and social care services for older male adults in prison: the identification of current service provision and piloting of an assessment and care planning model, NHS National Institute for Health Research. https://pubmed.ncbi.nlm.nih.gov/25642504/ (accessed 8 November 2020).

Sturup-Toft, S. and O'Moore, E. (2018). Looking behind the bars: emerging health issues for people in prison. *British Medical Bulletin* 125 (1): 15–23. https://doi.org/10.1093/bmb/ldx052 https://academic.oup.com/bmb/article/125/1/15/4831244 (accessed 27 October 2020).

Williams B, Lindquist, K., Sudore, R. et al. (2006). Being old and doing time: functional impairment and adverse experience of geriatric female prisoners. *Journal of American Geriatric Society* 54 (4): 702–707. https://doi.org/10.1111/j.1532-5415.2006.00662.x. https://pubmed.ncbi.nlm.nih.gov/16686886/ (accessed 7 November 2020).

World Health Organisation (2004). What are the main risk factors for falls amongst older people and what are the most effective interventions to prevent these falls? https://www.euro.who.int/__data/assets/pdf_file/0018/74700/E82552.pdf (accessed 3 November 2020).

World Health Organisation (2014). Prisons and health. https://www.euro.who.int/__data/assets/pdf_file/0007/249208/Prisons-and-Health,-19-The-older-prisoner-and-complex-chronic-medical-care.pdf (accessed 31 October 2020).

World Health Organisation (2020). Dementia. https://www.who.int/news-room/fact-sheets/detail/dementia (accessed 31 October 2020).

10 Fitness to plead and stand trial – from the Ecclesfield Cotton Mill dam to Capitol Hill

Nicholas Hallett[1] *and Keith J.B. Rix*[2]
[1] *Essex Partnership University NHS Foundation Trust, Brockfield House, Essex, SS11 7FE, UK*
[2] *School of Medicine, University of Chester, Chester, CH1 4BJ, UK*

Introduction

The case of Esther Dyson

One Saturday night in September 1830, at about 8 p.m., William Graham was returning home by a footpath from Wortley to Ecclesfield, then a village on the road from Leeds to Sheffield in the north of England. On the footpath he met a young woman who was carrying something in her apron. This was about 600 yards from the cotton mill dam. Shortly after this he met Henry Woodhouse who asked him if it had been 'the deaf and dumb girl' who had passed by. He confirmed that it was indeed her. Fanny Guest had also seen the dumb girl go past with something in her apron.

Esther Dyson

The 'deaf and dumb girl' was Esther Dyson. She was about 26 years old, and according to the local press, she was of exceedingly good appearance, tall, slender, with light hair and complexion, and of a rather pleasing and pensive cast of feature. She lived in Ecclesfield with her older brother, William, who was also 'deaf and dumb'. They both worked at the thread or cotton mill (later the Hallamshire Paper Mill) (Figure 10.1)) and had done so for 11 years.

Suspicious behaviour

Esther Dyson's behaviour was suspicious. Three or four months previously, her next-door neighbour, Ellen Greaves, wife of William Greaves, a file cutter, had challenged her about being 'in the family way'. She denied it. A month previously she had challenged her again. This time Esther Dyson reacted angrily, and using signs, she

Current Practice in Forensic Medicine, Volume 3, First Edition. Edited by John A.M. Gall and J. Jason Payne-James.

Figure 10.1 Scene of the crime and Esther Dyson's place of work – Ecclesfield Cotton Mill. *Source:* Reproduced with the permission of Frank Cooper as the custodian of the Cyril Slinn collection of Ecclesfield photographs.

said that some stuff that she had applied to her throat, inwardly and outwardly, had made her body swell. The neighbour made signs to the effect that Esther Dyson should make some clothes for her child, but she responded dismissively and made signs to the effect that she was not 'with child'.

On the Friday, the day before she was seen on the footpath with something in her apron, her neighbour had seen her at about 2 p.m. at the door of her house, and she had appeared 'quite big in the family way'. She next saw her on the Saturday morning at about 9 a.m. washing her floor. She appeared pale, languid, and weak. She had a flannel tied round her neck. Her neighbour asked her how she was, and she motioned to say that she had thrown up a large substance, and it had settled her body. The following day, Sunday, the neighbour visited, and her brother motioned to the effect that she was in bed very sick. Later in the afternoon, the neighbour saw Esther Dyson. She appeared poorly and weak. She suggested that her brother should make her some tea, and she stayed until she had taken it. The neighbour was in no doubt from her appearance that she had been delivered of a child.

The crime scene investigation

Later on Sunday, James Henderson, the overlooker from the cotton mill, went to the Dysons' house. Her brother allowed him to see his room and open his boxes. There was nothing untoward. However, there was blood on the floor of Esther's bedchamber, some attempt had been made to wipe up the blood, there was blood sprinkled on the wall, and there was also blood on the bottom of the window. Mr. Henderson found two blood-stained aprons and a skirt. Convinced that something was wrong, he sent for the vestry clerk, and together, they searched Esther Dyson's box and found several articles from which it was evident that they belonged to a person who had been delivered of a child.

James Machin was alerted, and with W. Shaw, the parish constable, he commenced a search of the cotton mill dam and pulled from the dam the headless body of a full-grown baby girl wrapped in a piece of green cloth. The head was found, also wrapped in a piece of green cloth. The green cloth was part of the sofa cover from the Dysons' house.

Forensic examination

Ann Briggs, who had been a midwife for 20 years, was shown the body and the head. She was of the opinion that the head had been cut off with a blunt instrument.

In the meantime, Esther Dyson had been taken to the Ecclesfield Workhouse (Figure 10.2). The midwife took the body parts to the Workhouse and laid them down beside her. Esther Dyson communicated to her brother that she had not thrown the baby into the dam but merely laid it there 'pretty and nice'. The midwife asked her to explain how she was delivered of the child, and she put to her the fact that the child's head was cut off. On this and subsequent occasions, Esther insisted that the head had just come off.

Figure 10.2 Remanded in custody – Ecclesfield Workhouse and in the background St. Mary's Church where Esther Dyson was baptised. *Source:* From the Peter Higginbotham Collection and is reproduced with the permission of the Mary Evans Picture Library.

The governess of the Workhouse, Sarah Ingham, examined Esther Dyson's breasts and found 'a deal of milk' in them. Esther Dyson repeated her story about how the head came off. Miss Ingham showed her a knife and put it to her by signs that she had cut off the head, but Esther threw herself to one side and shunned the idea.

Mr. William Jackson, lecturer on anatomy, was summoned. Mr. Jackson was a member of the Royal College of Surgeons and later one of its first fellows. He went on to lecture in midwifery, and he may also have become lecturer in forensic medicine at the Sheffield Medical School. Mr. Jackson found that Esther Dyson had every appearance of having been recently delivered. He was decidedly of the opinion that the head had not been torn or screwed off. He was also in no doubt, from his examination of the baby, that it had been born alive. Mr. Joseph Campbell, a surgeon, also examined the mother and child and corroborated Mr. Jackson's opinion.

The inquest

The following Thursday, at *The Black Bull Inn* in Ecclesfield (Figure 10.3), Mr. B. Badge, coroner for that district of Yorkshire, conducted the inquest. The body of the child was on full view. The various witnesses gave evidence, the coroner summed up, the jury – 'a respectable body of men' – retired, and a few minutes later, they returned with a verdict of wilful murder. The coroner then issued a warrant for Esther Dyson's committal to York Castle for trial at the following Lent Assizes and on a charge of the wilful murder of her female bastard child. The local Sheffield newspaper

Figure 10.3 The inquest – The Black Bull Inn. *Source:* Reproduced with the permission of the Sheffield City Library and Archives.

commented not only on her good looks but also on the observation of the coroner that she was very shrewd and cunning.

Remand in custody

It is likely that Esther Dyson walked from Ecclesfield to York where she was remanded in custody in the Women's Prison, the cells of which now form part of the York Castle Museum.

Inquiry as to fitness to plead

At the 1831 York Spring Assizes, held in the eighteenth century court house which is still in use today (Figure 10.4), Esther Dyson was indicted for the wilful murder of her bastard child by cutting off its head (*R v Dyson* (1831) 7 Car & P 303–305). As she had been born 'deaf and dumb', the case excited the greatest interest, and the public galleries were crowded. She was dressed in a coloured silk bonnet, a light calico printed dress, and a red cloak. She had the appearance of 'a respectable female in the lower walks of life'.

She stood mute. A jury was impanelled to try whether she was mute of malice or mute by the visitation of God. The trial judge, Mr Justice Parke, then examined on oath

Figure 10.4 The trial – York Spring Assizes. *Source:* Stephen Richards / CC BY 2.0.

James Henderson, the overlooker from the mill, who had known her for ten years. He swore that she could be made to understand some things by signs and could give her answers by signs. He was then sworn to 'well and truly interpret' for her. He explained to her by signs what she was charged with, and she made signs that obviously amounted to a denial. Mr Justice Parke then directed that a plea of not guilty should be recorded. Mr. Henderson was then called upon to explain to Esther that she was to be tried by a jury and that she might object to such as she pleased. He and another witness stated that it was impossible to make her understand a matter of that nature, although 'upon common subjects of daily occurrence, which she had been in the habit of seeing, she was reasonably intelligent'. One of the witnesses had instructed her in the 'dumb alphabet', but she was not so advanced as to make words. This witness swore that, although she was incapable of understanding the nature of the proceedings against her and incapable of making her defence, he also had no doubt that with time and pains she might be taught to do so by the means used by instructors of the deaf and dumb.

Directions to the jury

Mr Justice Parke directed the jury to be impanelled and sworn to try whether she was sane or not. He charged the jury in the words of Lord Hale and quoting from his *Pleas of the Crown* (Box 10.1). The jury was told that, if they were satisfied that she had not then, from the defect of her faculties, intelligence enough to understand the nature of the proceedings against her, they ought to find her not sane. The same witnesses were then sworn and examined as to her capacity, at the time, to understand the mode of her trial or to conduct her defence. However, there is no indication that the jury was informed of the coroner's observation as to how very shrewd and cunning she was. The jury returned a verdict that the prisoner was not sane. Her incapacity to understand the mode of her trial or to conduct her defence was proved. She was ordered to be kept in strict custody until His Majesty's pleasure was known.

Box 10.1 Sir Matthew Hale on fitness to plead and stand trial

If a man in his sound memory commits a capital offence, and before his arraignment he becomes absolutely mad, he ought not by law to be arraigned during his phrensy, but be remitted to prison until that incapacity be removed. The reason is, because he cannot advisedly plead to the indictment. And if such person after his plea, and before his trial, becomes of non-sane memory, he shall not be tried; or if after his trial he becomes of non-sane memory, he shall not receive judgment; or if after judgment he become of non-sane memory, his execution shall be spared; for, were he of sound memory, he might allege somewhat in stay of judgment or execution. But because there may be great fraud in this matter, yet if the crime be notorious, as treason or murder, the Judge before such respite of trial or judgment, may do well to impannel a jury to inquire *ex officio* touching such insanity, and whether it be real or counterfeit.

Source: Modified from Hale (1736).

His Majesty's pleasure

Esther Dyson was committed to the Wakefield Asylum, otherwise known as the West Riding Pauper Lunatic Asylum (and later known as Stanley Royd Hospital) (Figure 10.5). She died in the Lunatic Asylum, Stanley, on 18 March 1869, about a month short of her 60th birthday. Her 'occupation' was stated as 'Single from near Sheffield'. Her cause of death was 'Chronic disorganisation of the Brain' and 'Bronchitis'.

Dyson's legacy

When Pritchard, who was also 'deaf and dumb', was indicted for the capital offence of bestiality a few years later (*R v Pritchard* (1836) 7 Car & P 303–305), the trial judge referred to the procedure followed in *Dyson*. However, the adoption by Lord Chief Justice Parker, in *R v Podola* [1960] 1 QB 325, of the direction to the jury in *Pritchard* and his statement that it had become 'firmly embodied in our law' has confirmed the status of *Pritchard* as the leading case in England and Wales. So, the criteria for fitness to plead and stand trial (FTP) are known as the Pritchard test (Box 10.2), albeit that it might be more accurate to refer to it as the Dyson test.

Figure 10.5 Detained according to His Majesty's Pleasure – The Wakefield Asylum. *Source:* Reproduced with the permission of the West Yorkshire Archive Service (C85/1361).

Box 10.2 The *Pritchard* test

There are three points to be inquired into. First, whether the prisoner is mute of malice or not; secondly, whether he can plead to the indictment or not; thirdly whether he is of sufficient intellect to comprehend the course of proceedings on the trial, so as to make a proper defence – to know that he might challenge [any jurors] to whom he may object – and to comprehend the details of the evidence, which in a case of this nature must constitute a minute investigation . . . if you think that there is no certain mode of communicating the details of the trial to the prisoner, so that he can clearly understand them, and be able properly to make his defence to the charge; you ought to find that he is not of sane mind. It is not enough that he may have a general capacity of communicating on ordinary matters.

In this chapter, we describe:

- the application of the *Pritchard* test in England and Wales;
- the relevance of physical illness or disability to FTP in England and Wales;
- the related provisions in some other common law jurisdictions; and
- a practical approach to assessment.

The application of the Pritchard test in England and Wales

Several cases have shaped the application of the *Pritchard* test in England and Wales in the twenty-first century. *R v John (M)* [2003] EWCA Crim 3452 (Box 10.3) has operationalised the *Pritchard* test into six capabilities. *Marcantonio v R* [2016] EWCA Crim 14 in which *R v John (M)* [2003] EWCA Crim 3452 was quoted at length has clarified the unitary nature of the test despite sympathy with the view that fitness to plead could in the future be distinguished from fitness to stand trial. *Marcantonio* also clarified the case-specific nature of the test:

> '(T)he court is required to undertake an assessment of the accused's capabilities in the context of the particular proceedings. An assessment of whether an accused has the capacity to participate effectively [. . .] should require the court to have regard to what that legal process will involve and what demands it will make on the accused. It should be addressed [. . .] in the context of the particular case [. . .] (considering), for example, the nature and complexity of the issues arising in the particular proceedings, (their) likely duration [. . .] and the number of parties'.

The case-specific nature of the test was also explained in *JD v R* [2013] EWCA Crim 465; '(t)he evidence was not complicated. It was before the court in a readily comprehensible form [. . .] The Appellant was able to convey his defence to his legal

Box 10.3 The *Pritchard* test as operationalised in *R v John (M)* [2003] EWCA Crim 3452 (with expansions *in italic* based on *R v Whitefield* Leeds Crown Court, unreported, 1995 (Rix 1996)) (Rix and Nathan 2021)

Are any of the following beyond the accused's capabilities?

i. Understanding the *nature and effect of the* charges
ii. Deciding whether to plead guilty or not
iii. Exercising his right to challenge jurors
iv. Instructing solicitors and counsel *so as to prepare and make a proper defence in this case including understanding the details of the evidence which can reasonably be expected to be given in his case and advising his solicitor and counsel in relation to that evidence. This applies to his ability to instruct his legal advisers before and/or during his trial*
v. Following the course of the proceedings
vi. Giving evidence in his own defence

team. He understood sufficiently that his potential involvement went beyond his own actions'.

Similarly, whether an intermediary is required is also case specific and must take into account not only the defendant's vulnerabilities but also the circumstances of the particular trial as described in *R v Thomas* [2020] EWCA Crim 117; '(w)hat is required by way of assistance for a defendant may vary greatly between, for instance, a simple shoplifting case and a multi-handed trial involving cut-throat defences and complex bad character applications. Context is critical'.

Box 10.3 shows how the *Pritchard* test is now applied in England and Wales.

FTP is governed by the Criminal Procedure (Insanity) Act 1964 and the Criminal Procedure (Insanity and Unfitness to Plead) Act 1991. The burden of proof is on the defence to prove unfitness on the balance of probabilities although if raised by the prosecution it must be proved beyond reasonable doubt. To fail at one or more of the six capabilities renders a defendant unfit. FTP may be relevant at any point during the court process (*R v Orr* [2016] EWCA Crim 889).

Although not explicitly mentioned in *R v John (M)*, the ability to understand the nature and effect of the charges is likely to include the nature and seriousness of the offence, an understanding of the evidence on which the prosecution relies, and any penalty which may be imposed. Similarly, although not explicitly mentioned, an ability to decide whether to plead guilty or not guilty is likely to include the ability to understand the charges and to understand the consequences of such a plea but does not include the need to act in one's best interests (*R v Robertson* [1968] 1 WLR 1767).

The ability to instruct solicitors and counsel involves the accused being able to convey intelligibly to his legal team the case that he wishes them to advance on his behalf. He must be able (i) to understand the lawyers' questions, (ii) to apply his mind to answering them, and (iii) to convey intelligibly to the lawyers the answers that he wishes to give. It is not necessary that his instructions should be plausible or believable or reliable, nor is it necessary that he should be able to see that they are implausible, unbelievable, or unreliable.

Being able to follow the course of proceedings means (i) to understand what is being said by the witness and by counsel in their speeches to the jury and (ii) to communicate intelligibly to his lawyers any comment which he may wish to make on anything that is said by the witnesses or counsel. He does not have to understand, or be capable of understanding, every point of law or evidential detail, only to have 'a grasp of the essential issues' (*JD v R*). Even having delusional beliefs that the jury are possessed by evil does not prevent the accused being able to follow proceedings (*R v Moyle* [2008] EWCA Crim 3059).

The ability to give evidence in his own defence means (i) to understand the questions he is asked in the witness box, (ii) to apply his mind to answering them, and (iii) to convey intelligibly to the jury the answers that he wishes to give. This includes being cross-examined (*R v Orr* [2016] EWCA Crim 889).

In addition to the six abilities operationalised in *R v John (M)*, the importance of the right to a fair trial under Article 6 of the European Convention on Human Rights has led to the introduction of the concept of 'effective participation' by the European Court of Human Rights in *SC v United Kingdom* [2004] 6 WLUK 252:

> 'In the case of a child, it was essential that the proceedings take full account of his age, level of maturity and intellectual and emotional capacities, and that steps (are) taken to promote his ability to understand and participate, including conducting the hearing in such a way as to reduce as far as possible his feelings of intimidation and inhibition'.

Effective participation was considered further in *JD v R* where *SC v United Kingdom* was quoted as describing the need for:

> (1) A broad understanding of the nature of the trial process and of what is at stake for him or her, including the significance of any penalty which may be imposed. (2) It means that he or she, if necessary with the assistance of, for example, an interpreter, lawyer, social worker or friend, should be able to understand the general thrust of what is said in court. (3) The defendant should be able to follow what is said by the prosecution witnesses and, (4) if represented, to explain to his own lawyers his version of events, point out any statements with which he disagrees and make them aware of any facts which should be put forward in his defence.

While FTP is often considered to focus on understanding and communication, effective participation is more focused on decision-making ability (Owusu-Bempah 2018). Effective participation is also more important in the magistrates' court or youth court as there are no statutory provisions for FTP outside the Crown Court. Although FTP should be synonymous with effective participation these are often considered to be different concepts. The English Law Commission (2016) has proposed a number of areas for reform including introducing the concept of capacity to participate effectively.

Physical illness or disability and fitness to plead and fitness to stand trial in England and Wales

Where an accused is physically too ill to be tried or for their trial to continue, there is a provision for the court to order the proceedings to be stayed on the grounds that it would be an abuse of process to continue with the prosecution. This usually means that the case is stopped permanently, so it may not be appropriate where the physical illness, albeit serious or severe, is not likely to be, or become, chronic. This process is governed by the Prosecution of Offences Act 1985. The basis for this provision is concern that trying the accused in the particular circumstances of the case would '(offend) the court's sense of justice and propriety' (*R v Horseferry Road Magistrates' Court, ex p Bennett* [1994] 1 AC 42) or undermine public confidence in the criminal justice system and bring it into disrepute (*R v Latif* [1996] 1WLR 104).

As this is an exceptional course of action to take (*R v Maxwell* [2010] UKSC 48), there must be 'a firm factual basis' for doing so (*DPP v Gowing* [2013] EWHC 4614 (Admin)). The burden of proof is on the defence to establish abuse of process on the balance of probabilities. As a stay of the proceedings is effectively a permanent remedy, it is a remedy of last resort (*R v Crawley* [2014] EWCA Crim 1028). Thus, the appropriate course, where physical illness is likely to be short-lived, is to adjourn the proceedings until the accused sufficiently recovers.

Related provisions in some other common law jurisdictions

Australia

The statutory provisions for FTP that have been enacted in most Australian jurisdictions rely for their minimum requirements on the judgment in *R v Presser* [1958] VR 45. The accused has:

> 'to be able to understand what he is charged with. He needs to be able to plead to the charge and to exercise his right of challenge. He needs to understand generally the nature of the proceedings, namely that it is an inquiry as to whether he did what he is charged with. He needs to be able to follow the course of the proceedings so as to understand what is going on in court in a general sense, though he need not, of course, understand the purpose of all the court formalities. He needs to be able to understand, I think, the substantial effect of any evidence that may be given against him; and he needs to be able to make his defence or answer the charge. Where he has counsel he needs to be able to do this through his counsel by giving any necessary instructions and by letting his counsel know what his version of the facts is and, if necessary, telling the court what it is. He need not, of course, be conversant with court procedure and he need not have the mental capacity to make an able defence; but he must, I think, have sufficient capacity to be able to decide which defence he will rely upon and to make his defence and his version of the facts known to the court and to his counsel, if any'.

This is based on *Pritchard*. Although the judgment refers to the effect of a 'mental defect', it has been accepted that physical as well as mental impairment can result in unfitness (*R v Abdulla* [2005] SASC 399) such as resulting from head injuries (*R v Bradley (No 2)* (1986) 85 FLR).

Freckelton (2018) has observed that it is essentially a capacity for comprehension and communication test although he points out that the consideration of rationality is found in South Australia where the Criminal Law Consolidation Act 1935 (SA) provides that:

> 'A person is unfit to stand trial on a charge of an offence if the person's mental processes are so disordered or impaired that the person is (a) unable to understand, or to respond *rationally* to, the charge or the allegations on which the charge is based; or (b) unable to exercise (or to give *rational* instructions about the exercise of) procedural rights (such as, for example, the right to challenge jurors); or (c) unable to understand the nature of the proceedings or to follow the evidence or the course of the proceedings'.

Rationality also figures in an important Western Australian judgement which refers to 'the ability to consider in an informed way and make rational choices about defences that are reasonably open on the evidence and the submissions that should be made by the defence in respect of the evidence presented by the prosecution' (*State of Western Australia v Tekle* [2017] WASC 170).

Canada

In 1992, FTP was put on a statutory basis in Canada when the mental disorder provisions of the *Criminal Code* were amended. Section 2 states that:

> 'unfit to stand trial' means unable on account of mental disorder to conduct a defence . . . or to instruct counsel to do so, and in particular, unable on account of mental disorder to

a. understand the nature or the object of the proceedings,

b. understand the possible consequences of the proceedings, or

c. communicate with counsel.

As Ferguson (2018) has observed, this is not a sufficiently comprehensive test, and its subsequent judicial interpretation has rendered it even more inadequate. This is because the Court of Appeal concluded that fitness requires only a 'limited cognitive capacity' and not an 'analytical or rational capacity' (*R v Taylor* (1992) 77 CCC (3d) 551 (Ont CA)), a judgment that received a boost from the Supreme Court in *R v Whittle* [1994] 2 SCR 914 (SCC)). As there are occasional liberal interpretations of *Taylor*, such as *R v Adam* 2013 ONSC 373, best practise for practitioners in Canada might be to set out the evidence for not only 'cognitive capacity' but also 'analytical and rational capacity'.

Box 10.4 Criteria for unfitness to be tried in the Republic of Ireland (Criminal Insanity Act 2006, s 4(2))

An accused person shall be deemed unfit to be tried if he or she is unable by reason of mental disorder to understand the nature or course of the proceedings so as to –
- plead to the charge,
- instruct a legal representative,
- in the case of an indictable offence which may be tried summarily, elect for a trial by jury,
- make a proper defence,
- in the case of a trial by jury, challenge a juror to whom it might be wished to object, or
- understand the evidence.

Ireland

The Criminal Law (Insanity) Act 2006, s 4, governs fitness to be tried in Ireland (Box 10.4).

Jersey

Until 2016 the test in Jersey was the capacity to participate effectively taking into account the accused's ability 'to make rational decisions in relation to his participation in the proceedings (including whether or not to plead guilty) which reflect true and informed choices on his part' (*Attorney General v O'Driscoll* [2003] JRC 117). However, this was considered not be any different in principle from that which has been held to apply in England (*Harding v Attorney General* [2010] JCA 091). In 2016 the *O'Driscoll* test was codified in the Mental Health (Jersey) Law 2016, s 57:

1. The court determining an issue as to the defendant's incapacity shall have regard (so far as each of the following factors is relevant in the particular case) to the ability of the defendant -

   a. to understand the nature of the proceedings so as to be able to instruct his or her lawyer and to make a proper defence;

   b. to understand the nature and substance of the evidence;

   c. to give evidence on his or her own behalf;

   d. to make rational decisions in relation to his or her participation in the proceedings (including entering any plea) which reflect true and informed choices on his or her part.

2. The issue as to the defendant's incapacity shall be determined on the balance of probabilities.

3. For the purpose of determining the issue of incapacity -

    a. the court must obtain, and have regard to, medical evidence on that issue ...

New Zealand

In New Zealand, FTP has been on a statutory basis, albeit based on *Pritchard*, since the 1950s. The present definition states that unfitness to stand trial (Criminal Procedure (Mentally Impaired Persons) Act 2003, s 4)

a. means a defendant who is unable, due to mental impairment, to conduct a defence or to instruct counsel to do so; and

b. includes a defendant who, due to mental impairment, is unable-

    i. to plead;
    ii. to adequately understand the nature or purpose or possible consequences of the proceedings;
    iii. to communicate adequately with counsel for the purposes of conducting a defence.

Brooksbanks (2018) describes this as 'a performance-based standard which tests an offender's ability to function meaningfully in the nominated domains'. He points out that 'includes' permits consideration of a 'more discriminating' list of incapacities (*P v Police* [2007] 2 NZLR 528) than the three traditional capacities. Commenting on the meaning of 'mental impairment', he has noted that someone with personality disorder may be mentally impaired although the question will always be whether it renders the person incapable of doing what is set out in s 4(b) of the Act.

Northern Ireland

The procedure for determining fitness to be tried in Northern Ireland is set out in Mental Capacity Act (Northern Ireland) 2016, s 204. There are no statutory criteria, but the Northern Ireland Law Commission (2013) has recommended that the 'Pritchard test' is updated to incorporate the language of capacity.

Scotland

There are statutory criteria for unfitness for trial in Scotland in the Criminal Procedure (Scotland) Act 1995, s 53F (Box 10.5). The test is whether the person is 'incapable, by reason of a mental or physical condition, of participating effectively in a trial'. Medical or psychological evidence is needed. The defence must prove unfitness on the balance of probabilities.

Box 10.5 Statutory criteria for unfitness for trial in Scotland (Criminal Procedure (Scotland) Act 1995, s 53F)

In determining unfitness for trial the court is to have regard to the ability of the person to:

i. understand the nature of the charge,
ii. understand the requirement to tender a plea to the charge and the effect of such a plea,
iii. understand the purpose of, and follow the course of, the trial,
iv. understand the evidence that may be given against the person,
v. instruct and otherwise communicate with the person's legal representative, and
vi. any other factor which the court considers relevant.

United States

What in the United States is called competence to stand trial has its origins in *Pritchard*, and its influence is evident in the two key US Supreme Court judgments. In *Dusky v United States* 362 US 402 (1960), the Court set down as the test in the federal courts for competence to stand trial 'whether the defendant has sufficient present ability to consult with his lawyer with a reasonable degree of rational understanding – and whether he has a rational as well as factual understandings of the proceedings against him'. In *Drope v Missouri* 420 US 162 (1975), the Court held that:

'It has long been accepted that a person whose mental condition is such that he lacks the capacity to understand the nature and object of the proceedings against him, to consult with counsel, and to assist in preparing his defense may not be subjected to a trial'.

Bonnie (2018) has broken down the 'foundational' requirement of competence to assist counsel (decisional competence) as entailing capacities

'(a) to understand the charges and the basic elements of the adversary system (understanding);

(b) to appreciate one's situation as a defendant in a criminal prosecution (appreciation); and

(c) to relate pertinent information to counsel concerning the facts of the case (reasoning)'.

Further, following the decision of the Court in *Godinez v Moran* 509 US 389 (1993), he suggests that decisional competence entails the following abilities:

'(a) capacity to understand information relevant to the specific decision at issue (understanding);

(b) capacity to appreciate the significance of the decision as applied to one's own situation (appreciation);

(c) capacity to think rationally (logically) about the alternative courses of action (reasoning); and

(d) capacity to express a choice among alternatives (choice)'.

A practical approach to assessment

Although the determination of FTP is a matter for the court, in forming their own view a medical or legal practitioner should carefully consider both the disabilities or impairments of the accused and how they might affect their participation in the actual proceedings having regard to the nature and complexity of the issues arising in the particular proceedings, their likely duration, and the number of parties.

How the accused responds to being asked to consent to the assessment may indicate whether they understand the adversarial criminal proceedings and the nature of the offence(s) although it may be worth exploring whether they appreciate the full implications of a conviction. Questions as to proposed plea may reveal an understanding of the available pleas and their effects. Asking why they are pleading not guilty and putting some of the evidence to the accused may give an indication of their ability to give instructions, understand the evidence, and give evidence in their own defence. The test of challenging a juror is satisfied if the accused is capable of understanding that he should tell his lawyers if he knows or recognises a juror and can tell them so.

It is important not to consider the case in isolation merely on the basis of a one-off assessment. How the accused is able to function in everyday life may provide evidence of their abilities to understand evidence, give instructions, and give evidence in court although the court procedure itself is likely to place additional demands on them. Evidence of the accused's everyday functioning may be revealed by witness statements, reports, and medical records. Their interview by the police may reveal their previous understanding of the allegations, their ability to understand evidence, and explanations that may suggest what their instructions to their legal representatives might be. How they reacted to searching and challenging questions by the police may assist as to their ability to give evidence under cross-examination.

Practitioners should consider the various capabilities required for FTP in their own jurisdiction and any adjustments that may be needed in court including rephrasing questions or using simple language; allowing an opportunity for the accused to communicate regularly with counsel; frequent breaks; the use of an intermediary; and the use of an interpreter where appropriate. All of these aim to improve effective participation and increase the likelihood that despite the accused's disabilities they are fit to plead.

If a practitioner is of the view that an accused is not fit to plead even with adjustments, they must explain how their specific disabilities or impairments lead to a lack of ability in a particular domain. It is not sufficient to assert that they are not fit to plead simply because of a mental or physical disorder. Although FTP will always remain subject to individual judgment, the use of standardised assessment instruments such as that developed by Brown (2018) can be useful aids in assessing FTP.

Conclusion

Although FTP has been fundamentally shaped by the *Pritchard* test, the way this has been applied varies across jurisdictions and has led to substantial criticism that FTP is no longer fit for the purpose. The shift in language away from mere understanding towards decision-making capacity indicates that reform is urgently needed as the English Law Commission has made clear. The concept of effective participation has gained momentum in championing the rights of the accused in a way to which FTP has given little attention in the past following recommendations from the Howard League Working Paper (2014). Although this chapter has not focused on the consequence of a finding of unfitness, the implications for the accused and the criminal justice system more broadly are significant. Nevertheless, despite the problems with the various differing FTP procedures, practitioners should make clear the basis on which the accused may be unfit in line with relevant statute and case law. Where possible they should recommend relevant support and assistance which the accused can receive in order to enable them to become fit to plead and to participate effectively with the criminal justice system.

Acknowledgements

The historical section of this chapter is based on newspaper reports of the Dyson case in the *York Courant* and the *Yorkshire Gazette*, the *Calendar of Felons* consulted in York City Library, and Dr. C.E.H. Orpen's *Anecdotes and Annals of the Deaf and Dumb*. It first appeared in Rix, K.J.B. Fitness to plead and stand trial – then, now and in the future, *The Expert and Dispute Resolver* 17(2), 16–20, 2012, and is reproduced with the permission of The Academy of Experts. KR has been greatly assisted by Dr. Roger Finlay of Barlow Moor Books, Manchester; Mr. Nigel Womersley a local historian in Ecclesfield; and the late Siân Busby, the author of *McNaughten: A Novel* (Short Books, 2009), who tracked down Esther Dyson to the Wakefield Asylum.

The photograph of the Hallamshire Paper Mill (Figure 10.1) is reproduced with the permission of Frank Cooper as the custodian of the Cyril Slinn collection of Ecclesfield photographs. The photograph of the Ecclesfield Workhouse (Figure 10.2) is from the Peter Higginbotham Collection and is reproduced with the permission of the Mary Evans Picture Library. Sheffield City Council owns the copyright of the photograph of The Black Bull (Figure 10.3), and it is reproduced with the permission of the Sheffield City Library and Archives (www.picturesheffield.com). The photograph of the former Assize Courts (Figure 10.4), The Castle, York, is reproduced under the Creative Common Licence with the permission of the copyright holder, Stephen Richards, and licenced for further reuse. The photograph of the Wakefield Asylum (Figure 10.5) is reproduced with the permission of the West Yorkshire Archive Service (C85/1361).

References

Bonnie, R.J. (2018). Fitness for criminal adjudication: the emerging significance of decisional competence in the United States. In: *Fitness to Plead: International and Comparative Perspectives* (ed. R. Mackay and W. Brookbanks). Oxford University Press.

Brown, P., Stahl, D., Appiah-Kusi, E. et al. (2018). Fitness to plead: development and validation of a standardised assessment instrument. *PLoS one 13* (4): e0194332.

Brookbanks, W. (2018). The development of unfitness to stand trial in New Zealand. In: *Fitness to Plead: International and Comparative Perspectives* (ed. R. Mackay and W. Brookbanks). Oxford University Press.

Ferguson, H. (2018). Unfit to stand trial: Canadian law and practice. In: *Fitness to Plead: International and Comparative Perspectives* (ed. R. Mackay and W. Brookbanks). Oxford University Press.

Freckelton, I. (2018). Fitness to stand trial under Australian law. In: *Fitness to Plead: International and Comparative Perspectives* (ed. R. Mackay and W. Brookbanks). Oxford University Press.

Hale, M. (1736). *The History of Pleas of the Crown*, vol. I, 34. London: Sollom Emlyn.

Kirby, A., Jacobson, J., and Hunter, G. (2014). Effective participation or passive acceptance: how can defendants participate more effectively in the court process? *Howard League Working Paper*.

Law Commission (2016). Unfitness to Plead. *Law Commission* No 364.

Northern Ireland Law Commission (2013). Unfitness to Plead. Report NILC 16.

Owusu-Bempah, A. (2018). The interpretation and application of the right to effective participation. *The International Journal of Evidence & Proof* 22 (4): 321–341.

Rix, K.J.B. (1996). Psychiatric reports for criminal proceedings in England and Wales. *Hospital Update* 22: 240–244.

Rix, K. and Nathan, R. (2021). Reports for criminal proceedings and in prison cases. In: *Rix's Expert Psychiatric Evidence*, 2nd edition), Chapter 7 (ed. K. Rix, L. Mynors-Wallis and C. Craven). Cambridge University Press.

Law reports

Australia

R v Abdulla [2005] SASC 399
R v Bradley (No 2) (1986) 85 FLR
State of Western Australia v Tekle [2017] WASC 170

England and Wales

DPP v Gowing [2013] EWHC 4614 (Admin)
JD v R [2013] EWCA Crim 465
Marcantonio v R [2016] EWCA Crim 14
R v Crawley [2014] EWCA Crim 1028
R v Dyson (1831) 7 Car & P 303-305.
R v Horseferry Road Magistrates' Court, ex p Bennett [1994] 1 AC 42
R v John (M) [2003] EWCA Crim 3452
R v Latif [1996] 1WLR 104
R v Maxwell [2010] UKSC 48
R v Moyle [2008] EWCA Crim 3059
R v Orr [2016] EWCA Crim 889

R v Podola [1960] 1 QB 325
R v Pritchard (1836) 7 Car & P 303-305
R v Robertson [1968] 1 WLR 1767
R v Thomas [2020] EWCA Crim 117
R v Whitefield Leeds Crown Court, unreported, 1995
SC v United Kingdom [2004] 6 WLUK 252

Canada

R v Adam 2013 ONSC 373
R v Taylor (1992) 77 CCC (3d) 551 (Ont CA)
R v Whittle [1994] 2 SCR 914 (SCC)

Jersey

Attorney General v O'Driscoll [2003 JLR 390]
Harding v Attorney General [2010] JCA 091

New Zealand

P v Police [2007] 2 NZLR 528

United States

Drope v Missouri 420 US 162 (1975)
Dusky v United States 362 US 402 (1960)
Godinez v Moran 509 US 389 (1993)

11 Quality standards for healthcare professionals working with victims of torture in detention

Juliet Cohen[1] *and Peter Green*[2,3,4]
[1] *Oxford, UK*
[2] *South West London Integrated Care Board, London, UK*
[3] *St George's University Hospital, London, UK*
[4] *Health and Community Services, The Government of Jersey, The Channel Isles*

Introduction

The Quality Standards for Healthcare Professionals Working with Victims of Torture in Detention were developed by a working party of the Faculty of Forensic and Legal Medicine of the Royal College of Physicians in the United Kingdom and published in 2019 (FFLM 2019). The standards are applicable to victims of torture and other ill-treatment in any place of detention and provide a process for the recognition and management of their healthcare needs. There is clinical overlap with other vulnerable groups, particularly those in detention who have suffered other forms of serious harm, so the principles set out in the quality standards apply also to them. Particular attention should be drawn to victims of slavery and trafficking in this regard.

The United Nations have set out general guidance on the minimum rules for the treatment of prisoners (UNODC 2015) – 'also referred to as the Mandela Rules' – and the Council of Europe Committee for the Prevention of Torture and Inhuman or Degrading Treatment or Punishment also sets out general standards for care of victims of torture in detention (Council of Europe 2010). The World Medical Association Tokyo Declaration (last revised 2016) also addresses this issue (WMA 1975). This document provides standards not only on ***what*** should be done, but ***why, how,* and *how we can establish that it has been done.***

Detention is acknowledged to be harmful to the health (Royal College of Psychiatrists 2016; Bosworth 2016). Healthcare professionals (HCPs) have an obligation to identify and report torture – Nelson Mandela Rule 34 (UNODC 2015). Torture victims have a right to rehabilitation as set out in the UN Committee Against Torture General Comment 3, paragraphs 11–15 (UNCAT 2012) and other documents (OHCHR 2005). Rehabilitation cannot be effectively undertaken whilst they are in detention.

Current Practice in Forensic Medicine, Volume 3, First Edition. Edited by John A.M. Gall and J. Jason Payne-James.

HCPs working in places of detention are well positioned to identify victims of torture, have a duty to report the fact of torture, and will continue to have the obligation to meet the healthcare needs of the victims, insofar as it is possible to do so, until such time as they are released. This is particularly important in those circumstances when the authorities may delay, or find reasons to refuse, their release. HCPs do not have the option of doing nothing in such cases.

Why were quality standards needed?

Victims of torture and ill-treatment may be found in all places of detention including psychiatric institutes, immigration removal centres (IRCs), prisons, and police stations. The effect of torture on the ability to trust others, and on mental health, regularly confers on victims of torture additional vulnerability. There should be a presumption of vulnerability in all victims of torture in the same way that there is an inherent vulnerability in all children.

Prior health conditions and the specific impact of detention processes are aggravating factors that can increase the harmful effects of torture. A review of the welfare of vulnerable persons in detention (Shaw 2016) has drawn attention to concerns about their complex healthcare needs.

Prevalence of torture

Estimates vary widely in different parts of the world. In the United Kingdom, healthcare professionals are most likely to come across victims of torture in patients who are seeking asylum. It is likely that over 30% of asylum seekers (Kalt et al. 2013) have been torture victims. A US meta-analysis suggests that 44% of refugees have suffered torture (Higson-Smith 2016).

Clinical consequences of prior torture

Detention of a victim of torture has an impact beyond that of the torture itself or of detention on a different individual. These different impacts will be considered below:

- the impacts of torture on health;
- the impact of detention on health; and
- the impact of detention on torture.

Methods of torture

The methods of physical torture are shown in Table 11.1. These methods have the potential for causing a wide variety of injury including superficial or deep wounds, fractures, joint injury, nerve damage, and injury to internal organs. Environmental conditions may cause sunburn, frostbite, dehydration, and malnutrition. While some

Table 11.1 Methods of torture.

Examples of physical torture	Examples of sexual torture	Examples of psychological torture
• Blunt force	• Forced nakedness	• Solitary confinement
• Sharp force	• Being forced to witness sexual violence	• Sensory deprivation
• Burns	• Sexual assault by touching or forced insertion of objects	• Manipulation of the environment – including exposure to extremes of heat, cold, light, or dark
• Electrical	• Oral, vaginal, or anal rape	• Mock execution
• Chemical	• Injury to the anogenital area	• Humiliation and threats
• Extremes of heat and cold	• Sexually transmitted infections	• Behavioural coercion
• Stress positions	• Forced transgression of sexual orientation and gender identity	• Forcing the victim to witness the torture of others
• Asphyxia		
• Drowning simulation such as water boarding		

victims have visible scars or marks as a result of torture, such treatment commonly leaves little or no external physical evidence.

Because of these factors, victims of torture and ill-treatment may often have chronic, generalised, or specific pain and untreated infections. Blunt force head impacts and associated traumatic brain injury may have adversely affected their cognitive function and behaviour regulation, and they may develop post-traumatic seizure activity.

Examples of sexual torture, which may often involve multiple assailants, are listed in Table 11.1. Exposure to such activities may have both physical and psychological effects on the individual. The very profound psychological effects include overwhelming feelings of shame, stigma, and fear. Difficulties in disclosing experiences of sexual violence, whatever the gender of the victim, may add to the impact of those experiences on the individual due to the consequent delay in accessing the treatment they need.

Table 11.1 identifies some psychological methods of torture. The effects of psychological torture are commonly assessed in diagnostic categories such as anxiety, depression, post-traumatic stress disorder (PTSD), and psychosis. The Istanbul Protocol, the UN Manual on the Effective Investigation and Documentation of Torture, and other cruel inhuman or degrading treatment or punishment (OHCHR 2004) make clear that not everyone who has been tortured necessarily develops a diagnosable mental illness. Behavioural change may result in and include social withdrawal, fearfulness, insomnia, hypervigilance, dissociation, paranoia, aggression, panic attacks, substance misuse, self-harming behaviour, and suicidality.

Victims of torture and ill-treatment may suffer flashbacks or intense intrusive memories of their experiences, and these may be triggered by sights, sounds, and smells. In some cases, psychological effects of torture show even years or decades later.

Detention in the United Kingdom and risks for patients' health

Detention is potentially very disruptive to healthcare, with pre-detention medical care being stopped abruptly, hospital appointments missed, and scheduled treatment cancelled.

Specific effects of detention on the physical health of patients, whether or not they are victims of torture and ill-treatment, include:

- limited access to movement and the ability to exercise outside;
- poor ventilation and close confinement, which increases the risk of rapid spread of infections; and
- changes in diet may lead to loss of appetite and poor nutrition.

Psychologically, detention can lead to:

- anxiety and depression;
- over the longer term, passivity, hopelessness, and despair;
- loss of self-esteem and mood changes are well recognised as consequences of immigration detention which serves administrative and not criminal justice purposes (Von Werthern et al. 2018; Cleveland et al. 2018); and
- extreme fear and anxiety may also trouble immigration detainees who can be held for an unknown period while also at risk of being removed to a country where they fear for their own safety.

The trauma informed care approach is being more widely adopted by those working with people in detention and with other vulnerable groups (Miller and Najavits 2012; Sweeney et al. 2016). This approach acknowledges that past experiences of trauma can have deep and long-lasting effects on mental health and behaviour. The approach requires the use of 'universal precautions' in vulnerable groups (i.e. assume that trauma has occurred and act accordingly). Key to this is to change the question when faced with, for example challenging behaviour, from questioning 'What is wrong with him/her?' to questioning 'What has happened to him/her?' A trauma-responsive service will educate staff about the effect of trauma on the brain and body, eliminate unnecessary triggers to re-traumatisation, and foster feelings of safety, collaboration, and empowerment in patients.

Research on the Shaw review (Shaw 2016) concluded that the predominant forms of mental disorder (in IRCs) were depression, anxiety, and PTSD. The key predictors of negative psychological outcomes of detention include:

- duration of detention;
- pre-existing trauma;
- pre-existing mental and physical health problems; and
- poor healthcare services in detention.

Asylum seekers with a history of torture were identified as particularly vulnerable to negative mental health outcomes (Royal College of Psychiatrists 2016).

Effects of detention on victims of torture

Specific experiences of detention may trigger powerful and traumatising memories of torture experiences. For example, the sounds of keys jangling, guards' footsteps, male voices shouting, and metal doors banging have all been identified as powerful triggers that recall both detention and torture in their country of origin (Medical Justice 2015). The faces of those who have tortured them may be seen in the faces of the detention personnel, guards, and other officers all around them. The recall of sexual violence may be powerfully invoked by something as simple as the smell of sweat. Freedom of movement and communication is severely limited, and personal control is once again usurped by all-powerful forces.

These effects not only exacerbate greatly any pre-existing mental health problems but also specifically elicit the symptoms due to their torture, thereby increasing the frequency and intensity of flashbacks, intrusive recall and nightmares, hypervigilance, irritability, avoidance symptoms, and withdrawal (Medical Justice 2015).

These can lead to behavioural problems (as noted above), aggression, emotional lability, and avoidance of medical care. Such problems can then lead to punishment for breaches of regulations (Medical Justice 2015).

Experiences of loss of agency and powerlessness are key to the consequent risk of further harm in detention, rather than the specific identity of the perpetrators (Cleveland et al. 2018). The extent of state responsibility for their experiences of serious harm may not be the determining factor in the impact of those experiences on their mental health, but for some victims of torture and ill-treatment the effect is to make it very difficult for them to trust state officials thereafter, even in a country separate from where the abuse occurred.

It can thus be difficult for HCPs in the detention setting, where they are likely to be viewed by the victims of torture as agents of the state, to engage in a trusting therapeutic relationship with the victims of torture, and the victims of torture may specifically avoid going to healthcare while in detention as they do not trust anyone in the detention setting to help them. Further, a key feature of PTSD is avoidance of the reminders of the trauma, which can contribute to avoidance of seeking healthcare and a general lack of help-seeking behaviour (Iverson et al. 2010).

This establishes barriers for those offering healthcare – the victims of torture may be in need of help but will not engage, resulting in exacerbation of the condition due to lack of treatment. Maintaining positive relationships is recognised as critical to helping children – for adults who are victims of torture, it is also key to achieving good healthcare outcomes (Dibben and Lean 2003). HCPs assessing a victim of torture whether as their treating clinician or as a forensic expert writing a medicolegal report have the same overarching duty of care to offer advice or take action if they discover unmet treatment needs.

Professional responsibility

HCPs working in detention settings have a number of duties which they must observe, laid out in Mandela Rules 24–35, (UNODC 2015) and shown in Table 11.2.

Outcomes

The Quality Standards for Healthcare Professionals Working with Victims of Torture in Detention (FFLM 2019) have the aims listed in Table 11.3.

A template based on that laid out in some quality standards in other settings issued by the National Institute of Health and Care Excellence (NICE) has been used for the 12 quality standards.

Table 11.2 Duties of HCPs working in detention settings (Mandela Rules 24–35, UNODC 2015).

- Identify, document, and report victims of torture and ill-treatment at the earliest opportunity
- Identify their healthcare needs
- Facilitate a trusting relationship
- Respect their patient's autonomy
- Respect the need for informed consent and confidentiality
- Determine how best they should be treated
- Report if a person is unfit for detention or for the processes required by the detaining authorities
- Fulfil their ethical obligations to detained patients, retaining their independence, and if there is a conflict, putting the needs of their patient above the requirements of a third party
- Clearly identify their role if carrying out assessment for processes required by the detaining authority or other third party

Source: Modified from UNODC (2015).

Table 11.3 Aims of Quality Standards for Healthcare Professionals Working with Victims of Torture in Detention (FFLM 2019).

- Identification, documentation, and reporting of victims of torture
- Improve treatment of health conditions for victims of torture and ill-treatment in detention
- Reduce the frequency of adverse outcomes such as self-harm and suicide attempts
- Improve the quality of life for victims of torture and ill-treatment in detention
- Empower healthcare professionals to maintain their ethical obligations to their patient if in conflict with the requirements of the detention authorities
- Reduce vicarious traumatisation of healthcare professionals
- Give patients a positive experience of care

Source: Modified from FFLM (2019).

Each standard has five sections:

- **Statement** – the purpose of the standard is defined
- **Rationale** – an explanation is given of the standard
- **Quality measures** – the key elements of the standard
- **Quality standards** – the ways in which the measures are assessed
- **Implications for the four main stakeholders** – commissioners, service providers, HCPs, and service users

Table 11.4 identifies the 12 Quality Standards for Healthcare Professionals Working with Victims of Torture in Detention.

Table 11.4 Twelve Quality Standards for Healthcare Professionals Working with Victims of Torture in Detention.

1. **Identification**

   Detained victims of torture are identified so that torture can be reported and their healthcare needs can be met
2. **Ethical obligations**

   Healthcare professionals working with detained victims of torture understand their ethical obligations
3. **Consent and confidentiality**

   The principles of medical information management are maintained by healthcare professionals working with detained victims of torture

(*continued*)

Table 11.4 (Continued)

4. **Communication**

   Healthcare professionals ensure that accurate communication is facilitated for the detained victims of torture in all clinical assessments for those not fluent in the primary language of the area in which they are detained or with other communication challenges

5. **Mental capacity**

   Detained victims of torture whose autonomy may be compromised receive appropriate assessment

6. **Access to healthcare**

   Pending release, detained victims of torture can access appropriate services or treatment equivalent to that available in the community

7. **Vicarious traumatisation**

   Healthcare professionals working with detained victims of torture receive support to prevent vicarious traumatisation and burnout and promote self-care

8. **Training**

   Healthcare professionals who work with detained victims of torture have the required training and competence

9. **Assessment required by detention processes**

   Victims of torture required to go through specific detention processes receive appropriate assessment of their vulnerability

10. **Children**

    Healthcare professionals understand their responsibility to safeguard the well-being in detention of children and young people who are victims of torture

11. **Mental health**

    Detained victims of torture receive appropriate assessment so that their mental healthcare needs can be met

12. **Sexual violence**

    Detained victims of torture who have past experiences of sexual violence receive appropriate assessment so that their healthcare needs can be met

In the first standard, identification, the rationale for this is set out with reference to the Mandela Rules on minimum standards for the treatment of prisoners (UNODC 2015) and the importance of facilitation of disclosure for the HCP to be able to identify, report, and meet the needs of the victims of torture. Identification of torture experiences and their impact on physical and mental health enables the obligation to report torture to be met and, for so long as the victim of torture remains in detention, the delivery of effective and responsive care, including secondary referral or support from allied professionals, if indicated. A care plan is advocated to summarise the mental, physical, and sexual healthcare needs; list investigations and referrals required; and outline the treatment plan as well as record the risk assessment, the outcome of considerations the detainee may be unfit for specific detention processes or unfit to fly, and to specify the frequency of review required to monitor for any deterioration of health.

The quality measures specified include evidence that there are processes to identify victims of torture and make a report, evidence of local measures to assess the impact on their health, and evidence of care plans being both made and completed.

Pathway following disclosure of torture:

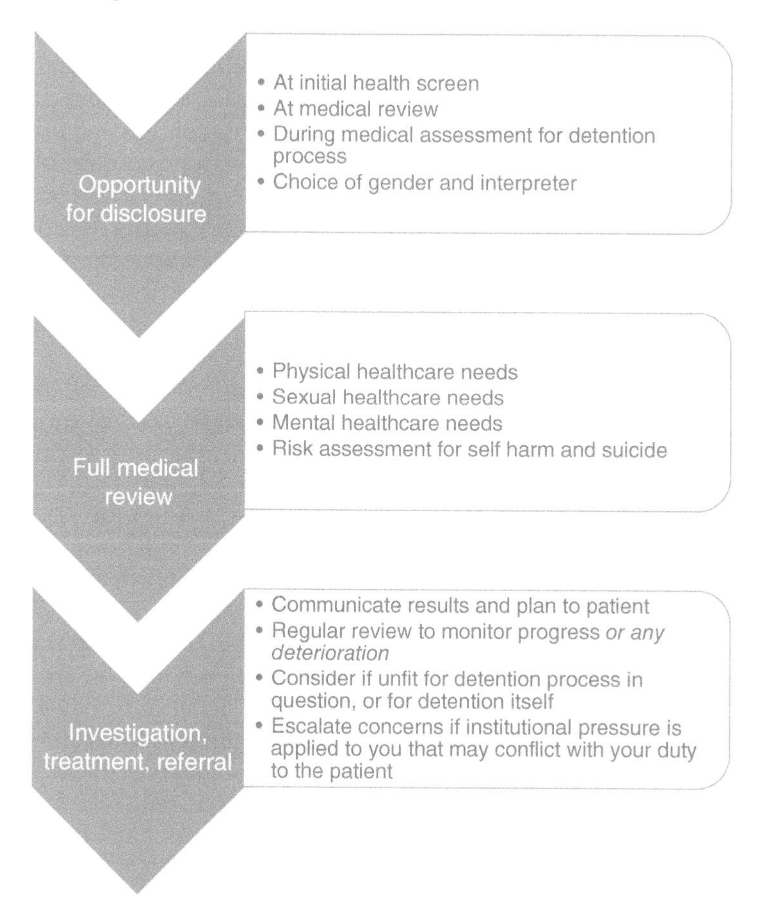

The second standard discusses the ethical obligations and reminds HCPs that detained patients are entitled to a standard of care equivalent to that in the community, that HCPs must take no part in torture or ill-treatment, and that adults in detention are vulnerable until proven otherwise. Most importantly, if the needs of their employer and patient conflict, the HCP must put the needs of their patient first. If things go wrong, or services are unethical, abusive, inadequate, or pose a potential threat to the detainee's health, they must speak up and support their colleagues to do the same. The associated quality standards show how quality measures of transparency, independence, equivalence of care, safeguarding, confidentiality, non-participation, and critical challenge can be met.

The third standard explains how for a victim of torture in detention, issues of consent and confidentiality require extra attention due to the patient's vulnerability and loss of autonomy. Information sharing for non-clinical purpose or examination for detention rather than a primary healthcare purpose may be requested and consent must be informed, and by a person with mental capacity for the specific issue in question.

Standard four is concerned with communication and the challenges for those without fluency in the primary language where they are detained or with a hearing difficulty or other communication challenge. Since a full assessment of healthcare needs cannot be made without clear communication, this standard is of primary importance. Patients need to be assured of confidentiality and offered a choice as to the gender of their interpreter.

The fifth standard considers mental capacity assessment and identification of those where their mental capacity for specific decisions and detention processes may be in question. Mental health disorder is relatively common in detention but does not of itself mean a person lacks mental capacity, although it should raise a concern that capacity may be affected. Relevant disorders in victims of torture include mood disorders such as depression and PTSD, psychotic symptoms, cognitive deficits arising post-head injuries, and intellectual disability. Deficits in decision-making capacity in victims of torture may not be recognised due an unrecognised language barrier, absence of an interpreter, fluctuations in decision-making capacity, behaviour being misinterpreted as attention seeking, and HCPs working outside their training and competence. A capacity assessment may be needed for a number of contexts, not only around medical assessment and treatment but also in cases of food and fluid refusal and for participation in legal processes. A screening tool is provided and standards explained for mental capacity assessment.

The sixth standard discusses access to healthcare both within the place of detention and for secondary care and the importance of maintaining continuity of care and good communication with the detainees and external HCPs involved in their care. Victims of torture have a right to rehabilitation, which is not achievable in detention, and detention itself may adversely affect their health, with the chance of this becoming greater the longer they are detained, so they must be monitored for signs of deterioration in detention. Evidence of their being unfit for detention or deteriorating in detention must be reported to the detention authorities.

Vicarious traumatisation is addressed in the seventh standard with the importance of recognition, training, and proactive management discussed. Hearing about the traumatic experiences of others, and witnessing their distress, can affect the healthcare professional's ability to empathise and to make objective decisions. It may bring up memories of one's own past trauma, leading to burnout, cynicism, and secondary trauma, and a consequent deterioration in professional performance.

Training needs are further elaborated in the eighth standard, reflecting the demanding nature of this type of clinical work and the GMC stipulation that doctors must work within their training and competence. The quality measures list the specific elements required, and standards show how they can be demonstrated.

In the ninth standard, assessment requests by the detention authorities of a person's fitness to undergo various detention processes are discussed. Ethical guidance directs a doctor to consider specifically if a person has any medical condition that makes them unfit for the process in question, since it is not ethical for a doctor to participate in or sanction any procedure that may cause a patient harm. For a victim of torture in detention, this may include continued detention in itself and also processes such as control and restraint and segregation. A past experience of torture may have contributed to the need for the process itself and increase the risk of harm resulting from it. The doctor needs to be able to make a full physical and mental health assessment, consider the patient's mental capacity and autonomy, and the impact of any institutional pressure from the detaining authorities on their assessment and clinical decision. A flowchart is provided demonstrating the pathway for such assessments and any necessary report or escalation of concern.

A 10th standard is on the specific responsibilities of HCPs in regard to children and young people who may be both victims of torture and in detention. It describes the universal vulnerability of all children, their right to the best possible healthcare, and the requirements for proper assessment of how their age affects their experience and understanding of their detention. The core significance of relationships and an understanding of child development as well as associated safeguarding responsibilities are specified. The risks of various forms of maltreatment include the significantly increased risk of exploitation with its attendant long-term risks to the child or young person.

Appropriate mental health assessment is the 11th standard. Important principles from the Royal College of Psychiatrists are set out, that '*people with a mental disorder should only be subjected to immigration detention in very exceptional circumstances*', and that 'detention itself is likely to precipitate a deterioration in mental health, that detention is not an alternative to in-patient care, and that all care should be equivalent to NHS standards of care'. Victims of torture are at higher risk of developing mental health disorders, and the impact of detention on them may be worse than on others, with the inevitable reminders of previous trauma when they were detained, powerless, and under the control of others. Risk assessments for self-harm and suicide must be made and regularly reviewed. The measures and standards demonstrate how HCPs can demonstrate appropriate assessment of mental healthcare needs.

The final and 12th standard looks at the needs of those who have suffered sexual violence, whether recent or historic. Difficulties in disclosure can particularly affect this group of patients, so they have not sought help or received treatment. Shame, stigma, and difficulty in trusting authorities may compound this problem. Gender choice of HCP and interpreter is highly important here, as is communication, so that each patient receives both verbal and written information in a form they can understand.

Throughout the document, the 12 quality standards are internally cross-referenced, so that the relevant links can be quickly made between them. The document benefitted greatly from both the multi-disciplinary nature of the working party and the external review by the survivors of torture, the Faculty of Law at Oxford University, Physicians for Human Rights, the International Rehabilitation Council for Torture Victims, Dignity Institute, the International Red Cross, the Secure Environments Group of the Royal College of General Practitioners, NHS England, and the Ethics Committee of the British Medical Association.

Conclusions

All HCPs working in places of detention should be aware of these standards and the principles underlying them. HCPs should have an understanding of the various methods of torture and ill-treatment and by application of the Quality Standards be able to identify, record, and initiate appropriate actions and interventions to minimise harm to vulnerable patients and ensure that such actions are not repeated.

References

Bosworth, M. (2016). *The Impact of Immigration Detention on Mental Health: A Literature Review. Review into the Welfare in Detention of Vulnerable Persons, Cm 9186*. London: HSMO Criminal Justice, Borders and Citizenship Research Paper No. 2732892.

Cleveland, J., Kronick, R., Gros, H., and Rousseau, C. (2018). Symbolic violence and disempowerment as factors in the adverse impact of immigration detention on adult asylum seekers' mental health. *Int J Public Health* 63: 1001–1008. https://link.springer.com/content/pdf/10.1007%2Fs00038-018-1121-7.pdf (accessed 25 January 2021).

Council of Europe (2010). The CPT standards 2010. https://www.refworld.org/pdfid/4d7882092.pdf (accessed 25 January 2021).

Dibben, M. and Lean, M. (2003). *Achieving compliance in chronic illness management: illustrations of trust relationships between physicians and nutrition clinic patients. Health Risk Soc* 5: 241–258. https://www.tandfonline.com/doi/full/10.1080/13698570310001606950?scroll=top&needAccess=true (accessed 25 January 2021).

Faculty of Forensic and Legal Medicine (2019). Quality Standards for healthcare professionals working with victims of torture in detention. https://fflm.ac.uk/resources/publications/summary-quality-standards-for-healthcare-professionals-working-with-victims-of-torture-in-detention/ (accessed 1 August 2021).

Higson-Smith C. (2016). Updating the estimate of refugees resettled in the United States who have suffered torture. https://www.cvt.org/sites/default/files/SurvivorNumberMetaAnalysis_Sept2015_0.pdf (accessed 25 January 2021).

Iverson, A., van Staden, L., Hughes, J.H. et al. (2010). Help-seeking and receipt of treatment among UK service personnel. *Br J Psychiatry* 197: 149–155. https://www.cambridge.org/core/journals/the-british-journal-of-psychiatry/article/helpseeking-and-receipt-of-treatment-among-uk-service-personnel/02D742DF4E06B2DFE5EE901D9607E2E1 (accessed 25 January 2021).

Kalt, A., Hossain, M., Kiss, L., and Zimmerman, C. (2013). *Asylum seekers, violence and health: a systematic review of research in high-income host countries. Am J Public Health* 103 (3): e30–e42. https://doi.org/10.2105/AJPH.2012.301136. https://www.ncbi.nlm.nih.gov/pmc/articles/PMC3673512/ (accessed 25 January 2021).

Medical Justice (2015). A secret punishment- the misuse of segregation in immigration detention. http://www.medicaljustice.org.uk/wp-content/uploads/2016/05/MJ_Segregation_report_WEB_version_04_11_15-1.pdf (accessed 25 January 2021).

Miller, N. and Najavits, L. (2012). *Creating trauma-informed correctional care: a balance of goals and environment. Eur J Psychotraumatol* 3: 1. https://doi.org/10.3402/ejpt.v3i0.17246 https://www.ncbi.nlm.nih.gov/pmc/articles/PMC3402099/ (accessed 25 January 2021).

OHCHR (2004). Istanbul Protocol, the UN Manual on the Effective Investigation and Documentation of Torture and other cruel inhuman or degrading treatment or punishment. UN OHCHR 2004. http://www.ohchr.org/Documents/Publications/training8Rev1en.pdf (accessed 25 January 2021).

OHCHR (2005). Basic Principles and Guidelines on the Right to a Remedy and Reparation for Victims of Gross Violations of International Human Rights Law and Serious Violations of International Humanitarian Law (Resolution adopted by the UN General Assembly on 16 December 2005). https://www.ohchr.org/en/professionalinterest/pages/remedyandreparation.aspx (accessed 25 January 2021).

Royal College of Psychiatrists (2016). Definition of torture in the context of immigration detention policy. PS07/16. https://www.rcpsych.ac.uk/pdf/PS07_2016.pdf (accessed 25 January 2021).

Royal College of Psychiatrists (2022). Position Statement on detention of people with mental disorders in Immigration Removal Centres. https://www.rcpsych.ac.uk/docs/defaultsource/improving-care/better-mh-policy/position-statements/position-statement-ps02-21---detention-of-people-with-mental-disorders-in-immigration-removal-centres---2021.pdf?sfvrsn=58f7a29e_6#:~:text=It%20is%20the%20view%20of,a%20return%20into%20immigration%20detention.

Shaw, S. (2016). Review into the welfare in detention of vulnerable persons. https://assets.publishing.service.gov.uk/government/uploads/system/uploads/attachment_data/file/490782/52532_Shaw_Review_Accessible.pdf (accessed 25 January 2021).

Sweeney, A., Clement, S., Filson, B., and Kennedy, A. (2016). *Trauma-informed mental healthcare in the UK: what is it and how can we further its development? Mental Health Rev J* 21 (3): 174–192.

UNCAT (2012). UN CAT general comment 3 https://www.refworld.org/docid/5437cc274.html (accessed 25 January 2021).

UNODC (2015). UN Standard Minimum Rules for the treatment of Prisoners (Nelson Mandela Rules). www.unodc.org/documents/justice-and-prison-reform/GA-RESOLUTION/E_ebook.pdf (accessed 25 January 2021).

Von Werthern, M., Robjant, K., Chui, Z. et al. (2018). *The impact of immigration detention on mental health: a systematic review. BMC Psychiatry* 18: 382. https://www.ncbi.nlm.nih.gov/pmc/articles/PMC6282296/ (accessed 25 January 2021).

WMA Tokyo Declaration of the World Medical Association (1975). https://www.wma.net/policies-post/wma-declaration-of-tokyo-guidelines-for-physicians-concerning-torture-and-other-cruel-inhuman-or-degrading-treatment-or-punishment-in-relation-to-detention-and-imprisonment/ (accessed 25 January 2021).

12 A forensic approach to intimate partner homicide

Jane Monckton Smith
University of Gloucestershire, Cheltenham, Gloucestershire

Introduction

Intimate partner abuse (IPA) is a pattern of violence and maltreatment that receives a lot of research attention, but despite this, it is still popularly considered a natural output of intimate relationships that are volatile or have deteriorated (Monckton Smith 2020). This suggests that the violence is largely situational and produced through the particular dynamics of those in the relationship. Similarly, intimate partner homicide (IPH) is often characterised as the result of arguing taken too far, where the fatal outcome was accidental, passionate, or merely reckless. There may be some value in both perceptions, depending on how IPA is defined and categorised. Johnson (2017) argues that it is important to recognise that domestic abuse is not a monolithic concept, and there are different types or categories of abuse that may be more or less dangerous or harmful, with differing motivations. The type of IPA considered to be the most dangerous in terms of links with IPH is related to power and control and may be referred to as coercive control (Stark 2009) or intimate terrorism (Johnson and Ferraro 2000). These typologies of IPA are significantly linked to male perpetrators and female victims, though it is also observed in same sex relationships (Frankland and Brown 2014) and some female to male IPA or IPH. It is also suggested that this form of IPA is most strongly associated with formal help seeking by victims (Stark 2009) and in preceding IPH (Monckton Smith 2020).

One of the stark consistencies in IPH is that men are overwhelmingly the perpetrators, and women the victims. The UN Global Homicide Report (UNODC 2019) shows that overall men commit 90% of *all* homicides and also form 80% of its victims. The victim numbers are reversed in the (IPH) category where women make up 82% of victims, but men still dominate as perpetrators (UNODC 2019). In same sex relationships, men are more likely to suffer homicide at the hands of a male partner (Ibrahim 2019), and victim precipitation (where the perpetrator of abuse initiated the violence against themselves through their abuse) is common. Women are underrepresented in this category as assailants, forming less than 10% of killers (ONS 2016). However, it has been found in research that in some female-perpetrated IPH, where they were not responding to violence against themselves, female killers can follow a very similar behavioural pattern to male killers (Monckton Smith 2020).

Current Practice in Forensic Medicine, Volume 3, First Edition. Edited by John A.M. Gall and J. Jason Payne-James.

Femicide (the killing of women because they are women) is estimated to have killed 87000 women in 2017 (UNODC 2019), and numbers are rising across the world, with over 50% of that number being domestic homicides. It is argued by the United Nations Development Programme that female inequality and patriarchal systems are a consequence and cause of male violence against women (UNDP 2018). Inequality is an important concept in exploring why IPH is dominated by female victims. Most cultures normalise and justify hierarchies within intimate relationships, with men commonly sitting above women and children. Strains of this have historically permeated public systems of law, and civil and criminal justice nearly everywhere, and in some places this is still the case. This notion of sexed hierarchies is important, especially as one of the key triggers for an IPH is attempts by female victims to challenge the control over them through these domestic chains of command and leave the relationship (Monckton Smith 2020; Dobash and Dobash 2015; Stark 2009). Research, however, has shown that personality disorder is observed in a number of these cases and may also provide a link to intense feelings of entitlement in both male and female perpetrators.

Feminism, which is the only historically consistent challenge to female inequality, is often represented as a purely political movement, positioning it as ideologically driven and thus diminishing some of its arguments. Even though women are gaining more representation in some places, that representation does not necessarily come with increased power. Although male privilege through women's inequality may arguably benefit all men, the specific interpretation of that privilege by individual men is not uniform. Some men, for many reasons, may internalise the privilege so that their sense of entitlement to it is intense. There are also cultures and societies that inscribe the inequality to such an extent that whole communities will support it, police it, and justify it. Johnson (2008) also argues that cultural, societal, medical, and legal messaging may also impact an individual's sense of identity and status which can be experienced through their intimate relationships.

Unlike many other homicide categories where the killers are considered worthy of forensic attention, for example serial killers or mass killers, intimate partner killers are often perceived as ordinary men who 'lost it' one day, rather than someone with a trackable history of behaviours and characteristics that might predict use of fatal violence. In media and legal proceedings for example, they are more routinely assessed as husbands or fathers, rather than as dangerous predators. Any threat they pose is more often considered temporary and confined to their current partner, even to the extent of discounting threat or risk to their children. But children often witness homicides, and are also targeted, or may become collateral damage. One study found that in this category, children witnessed 35% of actual homicides and 62% of the attempted homicides, and that children discovered the bodies of their mothers in 37% of IPH cases and 28% of the attempted homicide cases (Lewandowski et al. 2004). It was also found that in 572 cases of homicide, 44 children were killed (Fawcett 2010), and that the most likely scenario was that children would witness the killing of their mother (Chanmugam 2014:79). Katz (2014) found that the homicide will not be the first time the children have witnessed violence towards their mother. It is estimated that for every completed homicide, there will be three attempts (Lewandowski et al. 2004), and as these are not recorded formally, the scale of the issue is underestimated.

This 'crime of passion' narrative has historically dominated explanations for IPH across nearly all knowledge-producing institutions, including law, religion, medicine, psychiatry, and criminal justice. This has given the explanation a powerful footprint in all respected and authoritative environments. The 'crime of passion' explains IPH as spontaneous and unpredictable, with killers who are of little forensic interest presenting minor ongoing risk to themselves or others (Monckton Smith 2020). However, IPH has been found to be preceded by patterns of IPA and coercive control, with elements of planning in nearly all cases (Monckton Smith 2020; Juodis et al. 2014). The violence and intimidation, rather than being spur of the moment, is used instrumentally to maintain control, and those controlling individuals are the highest risk for offending and the most difficult to treat (Day and Bowen 2015). Patterns of coercive control also have strong links to other forms of homicide with predictors and histories for both being very similar (Iratzoqui and McCutcheon 2018; Felson and Lane 2010) and with IPA directly linked in 54% of mass homicides (Brandt and Rudden 2020). It is argued that domestic violence histories could be considered as a risk marker when forming policy for violence reduction more generally.

The powerful influence of the 'crime of passion' narrative and beliefs about IPA are deep in the cultural bone marrow (Websdale 1999), yet it has little to support it in the research literature. Its strong footprint in all the powerful knowledge-producing institutions could make it appear as if there is an evidence base that supports the position, and its use in forensic settings after IPH may be misleading. Media reports repeat what are often in fact defence narratives in IPH trials, and this has been found to have negative effect where they reinforce some of the myths. In November 2020, for example the BBC broadcast a documentary focused on the life of Oscar Pistorius, a champion sprinter who is also an amputee. Pistorius had been convicted of murdering Reeva Steenkamp, his girlfriend, in 2015. Despite this murder conviction, in a promotional piece for the documentary it was stated that Pistorius' story was *inspirational* and that he had '*found himself*' at the centre of a murder investigation (BBC 2020). There was a significant backlash from some, and the trailer was withdrawn by the BBC (BBC 2020). The concern centred on the lionising of Pistorius with a suggestion that he had stumbled into the circumstances of the killing of Reeva rather than being the architect of them. Such reporting and representation of IPH is a concern, and in the United Kingdom, there is now a code of conduct for journalists when reporting on domestic homicides supported by the UK Independent Press Standards Organisation (IPSO 2019) after campaigning by the charity Level_up. This code addresses not only the dignity and respect that should be afforded victims of homicide, but the accuracy and positioning of the headlines and stories. In IPH it is common to identify a proximate provocation to explain a spontaneous response (Monckton Smith, 2012). Lee and Wong (2020) report of their media study that 'domestic homicides were portrayed as isolated incidents and perpetuated the notions of victim blaming and offender excusing, rather than as connected to a pattern of domestic violence'.

The 'crime of passion' discourse

The crime of passion narrative is an historical artefact. It is a story of fatal violence from a husband to his wife. In this narrative, the husband is faced with an intolerable provocation and responds immediately and passionately. The acceptable provocations are culturally situated and constructed, and link to biological determinism, and heteronormative and patriarchal belief systems. The most common are revealed in defence narratives, and themed by Lees (1997) from her research observing homicide trials at the Old Bailey, as nagging, alleged infidelity, or challenge to the husband's seniority in the relationship. Other research has found that notions of passionate love are also used to defend killers, and where the killer could argue they loved the victim, they can receive reduced charges and lower sentences (Monckton Smith 2012). The killer is by this assessment a good man pushed to his limits by a provocateur; alternatively, he has trouble controlling his excessive, but fundamentally normal, urge to violence, dominance, and anger. The narrative suggests that there are inherent qualities in men that mean they will act impulsively and violently if provoked. This belief puts pressure on victims to make sure they do not provoke violence or abuse and can lead to what is called 'victim blaming'. Barthes (2001) makes an interesting case for the utility of this process of societal mythmaking, where acknowledging a small amount of badness can actually protect that myth from further scrutiny. He says that:

> 'One immunises the contents of the collective imagination by means of a small inoculation of acknowledged evil; one thus protects it against the risk of generalised subversion' (2001:123).

It has proved difficult for experts and researchers to challenge the idea that IPA and IPH are the patterns of dangerous individuals with histories of IPA and coercive control, rather than the patterns of ordinary men who are passionate husbands and lovers, when their clear criminal activities are justified through such powerful tropes. Even with the evidence base, and the persistent challenges put forward in critical and theoretical work, the myth remains. Courts are slow to accept expert evidence on the known patterns and motivations that are observable in IPA and IPH and the repeat and serial nature of the offending.

Coercive control discourse

The defining of coercive control has created what is described as a paradigm shift in conceptualisation of IPA and IPH (Hanna 2009). It reconstructs the motivations of the IPA perpetrators and the responses of the victims into a narrative that explains how and why the abuse is happening in a context of what Stark (2009) calls a liberty crime. A public consultation in the United Kingdom in 2012 found that coercive control was the best framework for understanding IPA, and the pattern was criminalised in England and Wales in the Serious Crimes Act (2015), and many other jurisdictions have followed. Coercive control positions the abuse as creating a 'hostage-like' situation removing choice and agency from victims, often in a culture of fear. This model captures the difficulties victims experience in attempting to separate from their abusers (Stark 2009). It shows how victim responses to the abuse are more rational than dysfunctional, and perpetrator patterns and behaviours are both functional and effective for them. This is in direct contrast to the crime of passion discourse.

Although mental illness may result from the trauma of domestic abuse, coercive control shows how victim responses to professionals, and their situation, are not the *result* of mental illness, they are explained as rational strategies to maintain a safe environment for themselves or their children (Stark 2009). Mental illness and trauma can result from the abuse, but it is not the sole cause of the victim response. Coercive control constructs IPA perpetrators as determined actors who repeat their patterns of behaviour with specific aims and predictable outcomes. Stark (2009) characterises coercive control as denying basic human rights to its victims and trapping them in relationships with their abusers. This model presents the common and repeated behaviours of the perpetrators as rooted in entitlement and privilege, encouraged, excused, and facilitated through structural inequalities. The broad range of controlling tactics, which are designed to subjugate victims and keep them compliant, are deployed with some purpose. They are not intermittent, rather the control is an unrelenting campaign that can be dynamic and vary in its intensity and tactics. The tactics can often include violence and sexual violence, which is an effective method of control, and also may include financial control, psychological manipulation and abuse, threats, and monitoring. The overt abuse is often mediated through periods of so-called love bombing (Monckton Smith 2020). Victims often respond with compliance, and this is underpinned through fear of the consequences of failing to comply with perpetrator demands and expectations. As Stark (2009) notes 'the perpetrator sets in the mind of the victim the price of their resistance'. For example, questions such as 'why doesn't she just leave?' are answered in this discursive approach by demonstrating the entrapment it creates. This becomes especially important in a legal setting. Victim behaviours can come under scrutiny by many, including defence counsel, children's services, and police, in suggesting they are partly culpable, or feigning fear, because they did not leave. There is little opportunity for the victim to defend themselves against such accusations, especially as expertise in IPA and IPH is not readily sought in a legal setting. It is often the case that a medical diagnosis is the only acceptable expert evidence to explain why victims behave as they do. This pathologising of the victim and their responses diminishes the more universal and normal human response to such abuse. The victim is 'othered' and this in turn makes the abuser and their abuse less visible as a pattern.

Medical narratives and discourse

In a forensic or legal environment, explanations for both victim and perpetrator behaviours are often rationalised using medical diagnoses, especially psychological or psychiatric conditions. Assessments of IPH killers in criminal trials that cannot be explained wholly through passionate responses are often situated in a medical narrative, diagnosing mental illness (especially temporary mental illness and depression) to reduce charges or culpability, and this is in part due to the requirements of the legal defences available. The legal system may require that female victims are diagnosed as suffering from conditions such as battered women's syndrome to explain their behaviours and responses to IPA, behaviours that may be more accurately explained as rational responses to chronic fear. It has been argued that 'battered women's syndrome' for example, is a dated concept, and there are more appropriate conceptualisations for victim responses to IPA (Ferraro 2004). These medical and mental health assessments have more status than expert testimony of IPA and IPH.

The medicalising of the perpetrator and victim behaviours suggests that those cases that reach court and/or a full trial are extraordinary rather than representative, and explaining the outcomes through medical diagnoses individualises responses and behaviours, rather than making links to known patterns of control. IPA is a pattern of behaviour not a mental illness. It is a pattern that can exist independent of mental illness or in parallel with it. The abuse may be exacerbated by mental illness, personality disorder, or substance misuse, for example, but it is widely agreed that these things are not causal. Chantler et al. (2019) found in their study of domestic homicide reviews that perpetrator mental health, largely characterised by anxiety and depressive disorders, formed most explanations for IPH. Medical assessments, especially after homicide, can focus on a wide range of mental illnesses ranging from mild depression to psychosis. Psychosis is a condition that is represented in IPH killers but has been found to be present only in a minority of cases (Chantler et al. 2019). This medical approach has been resisted in feminist scholarship, and the reasons for this are understandable. If mental illness or conditions are used to explain IPA, the structural inequalities that facilitate it are made invisible. Many researchers have found that structural inequalities and the expectations that some men have about hierarchies within relationships are key to what motivates IPH (Stark 2009). It is not argued that there is no place for psychological or psychiatric diagnoses of killers or victims, but situating mental illness as the context and cause of the abuse or homicide is not helping in preventing future homicide or helping with other preventative and safeguarding practices more generally. It has been argued, for example that antisocial personality disorder (ASPD) and psychopathy are more likely associated with IPA patterns and spousal violence that precede homicide. Juodis et al. (2014) state that 'psychopathy has been shown to be a strong predictor of persistent and severe violent recidivism against female intimate partners'. It is also argued that some controlling patterns evolve through dependence (Johnson 2008). Day and Bowen (2015) argue that it may be beneficial to suggest that specific consideration of the patterns and behaviours employed in coercive control could improve the outcomes of the current perpetrator behaviour change programmes.

IPH and IPA as expert knowledge

Better knowledge around coercive control in forensic and other professionals may be more effective in addressing perpetrator risk, threat, and desistance (Roberts 2007). Roberts (2007) argues that prediction of the duration and potential lethality of domestic abuse is one of the most critical issues in forensic mental health and social work, and that the civil and criminal justice systems rely on clinicians to advise them. However, there is often no requirement that expert witnesses have specific expertise in IPA and IPH and their patterns. Kennedy and Scriver (2016) note that despite high-profile calls for the inclusion of sensitive medical and forensic topics such as domestic and sexual abuse in the curricula of medical schools, they are frequently omitted, and this can result in a serious knowledge gap. Antovic and Stojanovic (2017) argue that the '*education and training of physicians of all specialties in recognizing the specific elements of DV abuse, as well as application of medical protocols to the treatment of DV victims, are necessary for a better understanding of the health hazards related to this field*' (p. 232). Arguing for better knowledge of domestic abuse, especially for forensic professionals, requires recognition of the status of IPA and IPH as specific areas of expertise, and that experts in related areas cannot rely on 'common sense' and extrapolate IPA or IPH knowledge from that. This argument permeates all levels of the criminal justice and legal systems, and repeated calls for specialist training of all manner of professionals dominate homicide reviews.

Forensic expertise is also crucial in identifying evidence of abuse. Reijinders et al. (2008) completed a comparative study investigating whether emergency room physicians and nurses were competent in describing, recognising, and determining abuse-related injuries. They found that it was forensic physicians, perhaps unsurprisingly, who scored significantly better than other emergency room staff. They concluded that training for all professionals involved in such assessments should be mandatory. However, criminal Justice and family court thinking around IPA and IPH expertise is disjointed. There is partial acceptance that the perpetrators are identifiable and that they have common histories and risk markers, and that specific knowledge and expertise can be crucial, but this is often those practices that come before a trial. In policing responses, for example risk identification checklists (RICs) are in common, if not blanket usage in some places (the United Kingdom and the United States especially). The use of psychological assessments of intimate or rejected stalkers is growing in a domestic context, while recognising that personality disorder rather than mental illness is more common in this category (Mullen et al. 2009). There are also tools built around temporal sequencing that track escalation and threat based on the notion that there is a common journey to IPH (Monckton Smith 2020). The use of these approaches indicates some acceptance that the patterns are consistent with identifiable motivations, and the outcomes of those patterns are predictable. Predicting escalation towards homicide cannot be used to prove homicide, but the disconnect between the more scientific approaches and the unscientific approach of the crime of passion after a homicide is contradictory.

Response practices

Specialist knowledge of domestic abuse and coercive control plays a crucial role in identifying threat and risk (Juodis et al. 2014). The known and common risk behaviours established in the literature include history of controlling patterns; history of violence and IPA; use of non-fatal strangulation; use of violence; sexual violence and aggression; threats to suicide; stalking; jealousy; possessiveness; fearful victims; threats to kill; and pregnancy (Adams 2007; Humphreys 2007). The use of these checklists is a semi-actuarial approach to assess the risk of homicide, the principle being that the more characteristics present, the more likelihood of homicide or serious harm. This approach, however, has been found to be ineffective where there is no corresponding knowledge of the patterns and motivations in IPA. It is argued that without good knowledge of IPA and IPH, the checklists give a probability of 'little better than chance' of identifying imminent risk of homicide (Turner et al. 2019). The suggestion is that it is adequate knowledge that makes the checklist more or less effective. The effectiveness of knowledge and expertise in IPA and IPH specifically is then widely argued to be crucial.

Stark and Hester (2019) state that RICs, rather than being used mainly for predicting harm, more often help 'ration' the allocation of scarce justice resources according to a woman's relative risk as determined at the scene (p. 84). This practice of determining evidence and risk based on the incident or the 'scene' inhibits recognition of the patterns that exist in this context and that are in fact written into the legal definition of coercive control and stalking offences in the United Kingdom. Given that IPA perpetrators can be broadly considered a 'type', notwithstanding the complexities in this statement, it is not surprising that they repeat their patterns over and again, they are both repeat and serial offenders. Scene or incident assessment, although dominant in police practices for all offending, is unhelpful in IPA and IPH. It is found that repeat domestic abusers are unlikely to be sanctioned by police in a system that focuses on incidents rather than patterns of offending. Hester (2006) found that criminal sanction was no more likely after the 50th offence, than the first, and Women's Aid (Women's Aid 2021) found that victims will suffer, on average, 35 assaults before reporting. Policing has historically focused on crisis management in this context and calming the 'incident' rather than approaching the situation as caused by a potentially dangerous and predatory individual. More attention is focused on the behaviour of victims who fail to exit dangerous relationships or support prosecutions. Responsibility for acting on the resultant information from a risk assessment is often largely left with the victim. The 'risk score' is not perceived as a societal threat and as such does not always engender wider concern. But those who go as far as to kill are known to sometimes kill more than once. The case of Theodore Johnson in the United Kingdom in 2018 is a case in point. Johnson killed three intimate partners between 1983 and 2016. He was convicted each time, but the danger he presented to women more generally, including any future partners, was minimised, and he was given very short sentences and minimal oversight in the community on release each time (BBC 2018). Johnson is now unlikely to be released having been convicted of murder rather than manslaughter with the mandatory life sentence for his third killing (BBC 2018). He is not the only example, and this case shows that IPH killers are perceived very differently to those who kill strangers.

However, post-relationship abuse and stalking (now recognised in the Domestic Abuse Bill (2021) in the United Kingdom) strongly suggests that leaving does not end the abuse, and research shows that separation does in fact escalate the potential risk of serious harm and homicide (Monckton Smith 2020). Stalking research has revealed that the most common form of stalking, which makes up over 50% of cases, occurs after a separation where there has been IPA (Mullen et al. 2009). Even in the face of this, a need for neutrality is a common argument put forward by police when attending IPA incidents, an approach that can impede an arrest, with that neutrality strengthening the abuser's hand and creating more barriers for victims in attempting to leave (Bancroft 2002). Removal of the perpetrator is argued to be the most effective protective measure (Brandt and Rudden 2020), and where there is effective criminal justice intervention, the outcomes are much better (Stark 2009). However, it has been found that outcomes in terms of recidivism can vary depending on the type of perpetrator (Johnson and Goodlin-Fahncke 2015). Hanna (2009) states that 'violence against women is not only a personal crisis for individual women but also a political crisis that the law has a deeper responsibility to remedy' (p. 1459). The relatively low status of domestic abuse and its victims is a part of most cultures, and the required social change will take effort and resources (Rodriguez et al. 2021). Rodriguez et al. (2021)

argue that '*trained staff can make a significant contribution and be the drive with their skills and knowledge of the significant shift in mindsets that are needed to achieve the ultimate goal of eradicating domestic violence in all its forms*' (p. 548).

There are aspects to the perpetrators that are unique in some respects to the context, for example the obsessive nature of stalking and coercive control; the access they have to the victims; and the power they have over the victims choices and liberty. Training in IPA and IPH although argued to be crucial in most domestic homicide reviews does not seem to have the status afforded by many other forms of professional training. However, despite the dominant beliefs that domestic abuse is produced through individual dynamics between specific couples, police responses are dominated by risk assessment processes that reinforce the importance of recognising the danger of the patterns and the serial nature of the offending. These risk processes, and the embracing of evidence-based tools, are completely divorced from the prosecution and family court processes, which often rely on outdated and 'common sense' notions of domestic abuse that can be detrimental to the victim (Monckton Smith et al.).

More recently, the method of 'temporal sequencing', which is an established method for tracking risk escalation in many categories of homicide, has been applied to IPH (Monckton Smith 2020). This method produced a timeline consisting of eight clear stages in a perpetrator's journey to homicide. The data was drawn from 400 case studies of IPH using multiple sources, the most important of which was published domestic homicide reviews (DHRs). These statutory reviews were introduced in the United Kingdom in 2011 as part of the Domestic Violence, Crime and Victims Act 2004. These reviews are quite unique in their method in that they are supposed to centre the victim's voice and present events through the victim's eyes (Home Office 2016). This gives information and detail that can be missing in other types of death review. The introduction of the reviews was an attempt to gather information to prevent future homicides.

The timelines produced through temporal sequencing illustrate predictability and culpability, rather than unpredictability and lack of culpability seen in the crime of passion model. Previous research has associated the most significant increase in risk of homicide with a triad cluster of characteristics that form a mini sequence: presence of coercive control–presence of violence– -and a separation. This cluster is reported to increase the risk of homicide by 900% (Stark 2009). The presence of (non-fatal) strangulation assaults increases risk by between six and seven times (Monahan et al. 2019) (see also Chapter 5), and the links to homicidal ideation are clear, given the context in which non-fatal strangulation is used as a method of creating threat, fear, and control. Feminist scholars argue that IPH happens when the protagonist 'changes the game from trapping a woman in a relationship, to punishing her for leaving it' (Dobash and Dobash 2015). The method of temporal sequencing establishes that in many homicide categories there is a psychological journey that progresses sequentially towards a fatal assault, especially after a separation. If there is an identifiable journey, then there are potential opportunities for intervening and stopping the homicide (Stanton 2016). Luckenbill (1977), for example, suggests that at any stage the protagonists in a male confrontational homicide have the choice to halt the progression. Similarly, Stanton shows that each stage in the progression towards genocide precedes the next stage, and each stage gives opportunities to halt the progress. Predicting risk is not a precise science (Shapiro and Noe 2015). All risk assessment methods and models will be broadly inductive in nature, giving only a statistical probability, or even possibility, of a homicide occurring. However, unlike some other homicide categories, the antecedent and high-risk behaviours of potential IPH killers are often made known to professionals beforehand and in a context where focused intervention could be effective. However, with risk processes being used as a form of resource management, their full potential is yet to be recognised. Most potential IPH perpetrators, whether or not they actually kill, are somewhere on the timeline at any point in time, even if they never reach stage eight, which is a homicide. In fact, most perpetrator's exit the sequence around stage five, but their patterns of abuse can last well after the end of the relationship. The following is a summary of the eight-stage progression documented in the research:

Stage one: history

IPA perpetrators are repeat and serial offenders, and this is a key risk marker noted on RICs. The importance of history cannot be overstated. The domestic violence disclosure scheme (DVDS) allows police in the United Kingdom to proactively inform new partners of a perpetrator's history. The fact that the DVDS exists is an acknowledgement that the history of IPA is a high-risk marker, and that offenders are likely to repeat past patterns. The research also tells us that repeat perpetrators also often share personality and behavioural characteristics in many cases: possessiveness, jealousy, controlling patterns, thin skinned, violent, or narcissistic.

Stage two: early relationship

This stage is when a controlling person meets a potential partner. This stage is often characterised by the speed at which the relationship develops and the search for a 'commitment' from the victim. There may be early declarations of love, early cohabitation, and pregnancy, for example. The search for a commitment is linked to possessiveness and jealousy. IPH killers are more likely to see relationships as practices in possession, demanding abject loyalty and a lot of control.

Stage three: relationship

This stage is where a relationship is formed (at least from the perpetrator's perspective), and it is dominated by controlling patterns. It may, or may not, include obvious violence. Perpetrator's may carry on multiple relationships at the same time but will display jealousy, possessiveness, and control in all of them.

Stage four: trigger

This is a stage when there is some challenge to the control the perpetrators have over their relationship or life. The single most significant trigger found in research is separation or its threat. Life changes such as redundancy, retirement, bankruptcy, and illness are also associated with potential separation and loss of control of the victim (and children).

Stage five: escalation

Controlling patterns escalate in an attempt to regain the perceived or real loss of control over the victim, the relationship, or their life. Stalking patterns may begin here; however, stalking, monitoring, and tracking are common within controlling relationships.

Stage six: homicidal ideation

The controlling persons begin to decide how they want to resolve the situation, especially where control appears irretrievable. For some, homicidal ideation begins here, though most will exit at stage five.

Stage seven: planning

The homicide is planned. Opportunities to carry it out may be sought. For example, some will plan very quickly and may purchase a weapon and immediately use it, and some will plan over a much longer period.

Stage eight: homicide

Could be homicide of primary victim; could be homicide/suicide; could involve killing of children or others; could be a missing person; and could be a staged suicide or staged accident.

The sequence is being used in many contexts, not simply in assessing risk, but in assessing homicides, requesting protective orders, and designing safeguarding interventions for example. The development of theoretical and practical approaches brings with it a requirement for specialism or expertise in their use, or they have little value. The development of a forensic and scientific approach to IPA and IPH should perhaps apply to all legal contexts, and knowledge of that literature should be part of the expertise of any forensic practitioner when considering IPA and IPH.

Conclusions

There is a growing acceptance that IPH is worthy of focused forensic attention, and the use of evidence-based tools and forensic expertise is a testament to that. However, there is also ongoing resistance to any challenge to the crime of passion discourse at every level of criminal, civil, and social justice. On one hand, there is acceptance of predictability and aberrant motivation, serial and repeat offending, and recognisable dangerous patterns, and on the other hand, there is a belief that IPH is about the dynamics between two people that is unpredictable and merely dysfunctional with links to mental illness. Reconciling these divergent discourses is going to take recognition that domestic abuse is worthy of focused forensic attention that is based on the domestic abuse and homicide literature, and that knowledge cannot be assumed because domestic abuse is a 'human relationship' problem.

References

Adams, D. (2007). *Why Do They Kill? Men Who Murder Their Intimate Partners*. Nashville: Vanderbilt University Press.

Antovic, R.A. and Stojanovic, J. (2017). Mediocolegal characteristics of domestic violence. *Srp Arh Celok Lek* 145 (5–6): 229–233.

Bancroft, L. (2002). *Why Does He Do That? Inside The Minds of Angry and Controlling Men*. New York, NY: The Penguin Group.

Barthes, R. (2001). (i) Operation Margerine; (ii) Myth Today. In: *Media and Cultural Studies Key Works*. Chapter 6 (ed. M.G. Durham and D.M. Kellner), 121–128. Oxford: Blackwell Publishing.

BBC (2020). Oscar Pistorius: BBC removes documentary trailer after backlash. https://www.bbc.co.uk/news/entertainment-arts-54706858 (accessed July 10th 2021).

BBC (2018). Triple killer Theodore Johnson jailed for 26 Years. https://www.bbc.co.uk/news/uk-42583114 (accessed July 10th 2021).

Brandt, S. and Rudden, M. (2020). A psychoanalytic perspective on victims of domestic violence and coercive control. *Int J Appl Psychoanal Stud* 17: 215–231.

Chanmugam, A. (2014). Social work expertise and domestic violence fatality review teams. *Soc Work* 59 (1): 73–79.

Chantler, K., Robbins, R., Baker, V., and Stanley, N. (2019). Learning from domestic homicide reviews in England and Wales. *Health Soc Care Community* 28 (2): 485–493. https://doi.org/10.1111/hsc.12881.

Day, A. and Bowen, E. (2015). Offending competency and coercive control in intimate partner violence. *Aggress Violent Behav* 20: 62–71.

Dobash, R.E. and Dobash, R.P. (2015). *When Men Murder Women (interpersonal violence)*. Oxford: Oxford University Press.

Fawcett, J. (2010). *Up to Us: Lessons Learned and Goals for Change After Thirteen Years of the Washington State Domestic Violence Fatality Review*. Seattle: Washington State Coalition Against Domestic Violence.

Felson, R.B. and Lane, K.J. (2010). Does violence involving women and intimate partners have a special etiology? *Criminology* 48: 321–338. https://doi.org/10.1111/j.1745-9125.2010.00186.x.

Ferraro, K.J. (2004). The words change, but the melody lingers: the persistence of the battered woman syndrome in criminal cases involving battered women. *Violence Against Women* 9: 110–129.

Frankland, A. and Brown, J. (2014). Coercive control in same sex intimate partner violence. *Multicultural Research on Intimate Partner Violence* 29: 15–22.

Hanna, C. (2009). The paradox of progress: translating evan stark's coercive control into legal doctrine for abused women. *Violence Against Women* 15 (12): 1458–1476.

Hester, M. (2006). Making it through the criminal justice system: attrition and domestic violence. *Soc Policy Soc* 5 (1): 1–12.

Home Office (2016). Guidance for the conduct of Domestic Homicide Reviews. https://www.gov.uk/government/publications/revised-statutory-guidance-for-the-conduct-of-domestic-homicide-reviews (accessed 14 August 2019).

Humphreys, C. (2007). Domestic violence and child protection: exploring the role of perpetrator risk assessments. *Child Fam Soc Work* 12: 360–369.

Ibrahim, D. (2019). Police reported violence among same-sex intimate partners in Canada 2009–2017 statistics Canada. https://www150.statcan.gc.ca/n1/pub/85-002-x/2019001/article/00005-eng.htm (accessed July 10th 2021).

IPSO (2019). IPSO blog: reporting on domestic violence. https://www.ipso.co.uk/news-press-releases/blog/ipso-blog-reporting-on-domestic-violence/.

Iratzoqui, A. and McCutcheon, J. (2018). The influence of domestic violence in homicide cases. *Homicide Stud* 22 (2): 145–160.

Johnson, M. (2017). A personal social history of intimate partner violence. *J Fam Theory Rev* 9: 150–164.

Johnson, M. (2008). *A Typology of Domestic Violence: Intimate Terrorism, Violent Resistance, and Situational Couple Violence.* Boston: Northeastern University Press.

Johnson, R. and Goodlin-Fahncke, W. (2015). Exploring the effect of arrest across a domestic batterer typology. *Juv Fam Court J* 6 (1): 15–30.

Johnson, M. and Ferraro, K. (2000). Research on domestic violence in the 1990s: making distinctions. *J Marriage Fam* 62: 948–963.

Juodis, M., Starzomski, A., Porter, S., and Woodworth, M. (2014). A Comparison of domestic and non-domestic homicides: further evidence for distinct dynamics and heterogeneity of domestic homicide perpetrators. *J Fam Violence* 29: 299–313.

Katz, C. (2014). The dead end of domestic violence: spotlight on children's narratives during forensic investigations following domestic homicide. *Child Abuse Negl* 38 (12): 1976–1984.

Kennedy, K.M. and Scriver, S. (2016). Recommendations for teaching upon sensitive topics in forensic and legal medicine in the context of medical education pedagogy. *J Forensic Leg Med* 44: 192–195.

Lee, C. and Wong, J.S. (2020). 99 reasons and he ain't one: a content analysis of domestic homicide news coverage. *Violence Against Women* 26 (2): 213–232.

Lees, S. (1997). Ruling Passions: Sexual violence, reputation and the law. Buckingham: OpenUniversity Press.

Lewandowski, L.A., McFarlane, J., Campbell, J.C. et al. (2004). 'He killed my mommy!' Murder or attempted murder of a child's mother. *J Fam Violence* 19: 211–220.

Luckenbill, D.F. (1977). Criminal homicide as a situational transaction. *Soc Probl 25* (2): 176–186.

Monahan, K., Purushotham, A., and Biegon, A. (2019). Neurological implications of nonfatal strangulation and intimate partner violence. *Future Neurol* 14: 3.

Monckton Smith, J. (2020). The homicide timeline: using foucauldian analysis to track an eight stage relationship progression to homicide. *Violence Against Women* 26 (11): 1267–1285.

Monckton Smith, J., Williams, A., and Mullane, F. (2014). *Domestic Abuse, Homicide and Gender: Strategies for Policy and Practice.* Hampshire: Palgrave Macmillan.

Monckton Smith, J. (2012). *Murder, Gender and the Media: Narratives of Dangerous Love.* Hampshire: Palgrave Macmillan.

Mullen, P.E., Pathe, M., and Purcell, R. (2009). *Stalkers and their Victims.* Cambridge: Cambridge University Press.

ONS (2016). Office for national statistics. https://www.ons.gov.uk/peoplepopulationandcommunity/crimeandjustice/bulletins/domesticabseinengandandwales/yearendingmarch2017.

Reijinders, U.J.L., Giannakopoulos, G.F., and de Bruin, K.H. (2008). Assessment of abuse-related injuries: a comparative study of forensic physicians, emergency room physicians, emergency room nurses and medical students. *J Forensic Leg Med* 15: 15–19.

Roberts, A.R. (2007). Domestic violence continuum, forensic assessment and crisis intervention. *Fam Soc* 88 (1): 42–54.

Rodriguez, L., Power, E., and Glynn, E. (2021). Introduction to domestic violence, abuse, and coercive control for counselors: an evaluation of the impact of training. *Gend Work Organ* 28 (2): 547–557.

Shapiro, D.L. and Noe, A.M. (2015). *Risk Assessment: Origins, Evolution and Implications for Practice. SpringerBriefs in Behavioural Criminology*. London: Springer.

Stanton, G. (2016). 10 stages of genocide. http://genocidewatch.net/genocide-2/8-stages-of-genocide/ (accessed 2 May 2019).

Stark, E. (2009). *Coercive Control: How Men Entrap Women in Personal Life*. Oxford: Oxford University Press.

Stark, E. and Hester, M. (2019). Coercive control: update and review. *Violence Against Women* 25 (1): 81–104.

Turner, E., Medina, J., and Brown, G. (2019). Dashing hopes? The predictive accuracy of domestic abuse risk assessment by police. *Br J Criminol* 59 (5): 1013–1034.

UNDP (2018). Violence against women a cause and consequence of inequality. https://www.undp.org/blogs/violence-against-women-cause-and-consequence-inequality (accessed July 10th 2021).

UNODC (2019). Global study on homicide. United nations office on drugs and crime. https://www.unodc.org/unodc/en/data-and-analysis/global-study-on-homicide.html (accessed 1 June 2021).

Websdale, N. (1999). *Understanding Domestic Homicide*. Boston: Northeastern University Press.

Women's Aid (2021). Facts and figures. https://womens-aid.org.uk/domestic-abuse/facts/#:~:text=Women%20are%20on%20average%20assaulted,adjacent%20room%20when%20violence%20happens (July 10th 2021).

13 Non-lethal physical abuse in the elderly

John A.M. Gall[1,2,3]
[1] Era Health, Melbourne, Australia
[2] Department of Paediatrics, The University of Melbourne, Melbourne, Australia
[3] Victorian Forensic Paediatric Medical Service, The Royal Children's Hospital and Monash Medial Centre, Melbourne, Australia

Elder abuse is not something new but has been with us for centuries. As life expectancy increases, the potential for elder abuse also increases, and according to the World Health Organisation (WHO 2020), about one in six older people experience some form of abuse. The actual rates of physical abuse are difficult to determine due to the very limited published data. A meta-analysis of elder abuse in the community indicates a rate of 2.6% (Yon et al. 2017). There is no information as to the extent of abuse within aged care facilities and hospitals. Unfortunately, the rates of abuse may be higher for older people living in institutions than in the community. It is not infrequent that press reports appear outlining some form of abuse within nursing homes and aged care facilities. The WHO predicts an increase in elder abuse as many countries experience rapidly growing and ageing populations.

What then is non-lethal physical abuse? There is no universal definition, but one can be formulated utilising a modification of the WHO definition of elder abuse. For the purposes of this chapter, non-lethal elder physical abuse is a single or repeated act, occurring within any relationship where there is an expectation of trust, resulting in non-lethal physical injury or injuries. These injuries may be intentional or unintentional and include bruises, abrasions, lacerations, stabs, incisions, fractures, burns (thermal, chemical, and radiation), asphyxiation, and internal injuries. Decubiti have not been included as these are either nursing care complications or neglect-associated conditions.

Elder abuse is not dissimilar to child abuse. Despite its existence for centuries, it was only about 50 years ago that child abuse became topical and a subject for research. Developmentally, the concept of elder abuse is at least 50 years behind child abuse and, not surprisingly, has been poorly investigated. Similar to child abuse, there are difficulties with the diagnosis complicated by concept, communication, the ageing process, and mimickers of injury.

This chapter provides an overview of non-fatal/non-lethal physical abuse in the elderly with consideration being given to the ageing process and the associated medical conditions, the mimickers of abuse, and how these affect the assessment of injury

Current Practice in Forensic Medicine, Volume 3, First Edition. Edited by John A.M. Gall and J. Jason Payne-James.

including at the time of autopsy. The issues of psychological, financial, and toxicological abuse, neglect, and sexual assault are not discussed.

Failure to diagnose

There are similarities between child abuse and elder abuse in relation to failure to diagnose. Both children and the aged may be unable to articulate that abuse has occurred. If they are able to report abuse, they may be too afraid of the consequences to report it. Those consequences may be punishment of some form including ongoing and perhaps even more severe abuse. They may not be believed perhaps due to concurrent mental health issues and/or senility (real or perceived). It may be because it is too inconvenient for the receiver of the report to action the report particularly where mandatory reporting (with associated penalties for not reporting) is not in existence or applicable. Also, the aged person may be so isolated that the only person that they may report to is, in fact, the perpetrator of the abuse.

For health professionals, there may be additional barriers for failing to report and acting upon perceived elder abuse. The first of these is not considering abuse as part of the differential diagnosis. In other words, being unsuspecting of abuse. If it is not a consideration when making an assessment of injuries, it will never form a diagnosis. Denial is another barrier where there is knowledge of the family, and a belief that the healthcare professional, facility, employee, family, or a family member could not possibly engage in abuse. Even if abuse is considered, there may be a reluctance by the healthcare professional to become involved for a variety of reasons. Finally, there may be a failure to act due to the perceived futility of reporting based on past experiences. The end result of this is that the aged person remains at the mercy of the perpetrator of abuse.

There is one factor that usually does not feature in child abuse but may in elder abuse: that reporting abuse does not matter. In this instance, a diagnosis is considered immaterial. People become old, and their contribution to society is deemed to be less with increasing age to a point where they become totally dependent on family, relatives, friends, and the society. Their death becomes beneficial, a convenience, and financial gain, both for the estate and the state.

The ageing process

The types of injuries identified in elder abuse are not necessarily different to those seen in other age groups. The extent and ease with which some of these injuries may occur, however, may be altered due to the ageing process. The features of ageing are of multi-system deterioration that lead to a gradual debility, and this process may be affected in individuals by superimposed degenerative diseases (Harrison and Fleming 2008). Ageing is the end result of both the genetic and environmental factors, and the overall process will vary from person to person.

From a forensic perspective, the most common injuries seen in live patients are skin based and include bruises, abrasions, lacerations, incisions, bites, and burns. Fractures may also occur and become more prevalent with age due to loss of bone strength. As skin and bone injuries are the more common findings in live patients, it is important to briefly consider the structure of these two tissues, the healing process, and the effect that age may have upon them.

Skin

The skin is the largest organ in the human body and plays a protective role. Beyond about 70 years of age, there are changes in its strength, thickness, and elasticity that subsequently result in a reduced functional capacity (for review, see Blume-Peytavi et al. 2016). Anatomically, the skin consists of an outer layer, the epidermis, which is covered by a layer of keratin, an underlying layer of avascular keratinocytes, with a variable number of melanocytes, situated on a basement membrane. Beneath this is the dermis that contains an extracellular matrix, supporting blood vessels, lymphatics, and nerves.

Injury to the skin results in an inflammatory response, the extent of which will depend on the extent of injury sustained. Whether the surface of the skin is breached or not there will be an acute inflammatory reaction of variable extent that may exhibit the classical signs of redness, heat, swelling, pain, and, perhaps, loss of function. Where there is a breach of the surface of the skin, healing occurs either by primary or secondary intention union. From a forensic perspective, the dating of open skin injuries may be very important in making an assessment about a case. In many circumstances, a knowledge of the healing process and of the factors that may affect that process may enable comment to be made as to the age of a particular injury. In brief, primary intention occurs in clean, incised wounds that have good apposition of the edges such as may occur post-surgery. The process includes (Burt and Fleming 2008; Govan et al. 1991):

- *An immediate phase:* the site of injury cleft is filled by blood clot and fibrin;
- *A few hours post-injury:* early inflammatory changes to close the edges of the wound with initial movement (regrowth) of a basal layer of epithelial cells across the cleft of the wound and formation of mild hyperaemia at the edges;
- *2–3 days*: ongoing movement of growing epithelium across the wound cleft with macrocyte activity to remove the blood clot and any other debris and fibroblastic activity within the underlying connective tissue/subcutaneous tissue;
- *10–14 days*: the epithelial layer now covers the complete wound; and
- *weeks*: the presence of hyperaemic scar tissue which eventually undergoes devascularisation and remodelling of the underlying collagen – after many months, the scar is minimalised and may merge with the surrounding tissues.

Healing by secondary intention occurs in open wounds where there has been loss of tissue, tissue damage, necrosis, and/or infection. The sequence of healing is not dissimilar to that of primary intention healing, but the process takes longer and includes (Burt and Fleming 2008; Govan et al. 1991):

- *An immediate phase:* where the injury cavity fills with blood clot and fibrin, and an acute inflammatory response commences at the junction between the area of injury and the living tissue;

- *A few days post-injury:* the scab dries out, and a basal layer of epithelial cells begin to grow across the injury site beneath the surface debris. At the base of the injury, new capillaries are being formed which facilitate access to the area by macrophages, polymorphs, and fibroblasts;

- *About one week post-injury:* the epithelium continues to grow across the area of injury. Granulation tissue (capillary loops and fibroblasts) forms in the base of the wound with the laying down of collagen; and

- *2 weeks onwards:* depending on the width of the injury, the proliferating epithelial cells may now cover the previous area of injury. The capillaries at the base of the wound become less prominent, and there is a reduction in inflammatory cells. With time, there is resolution of underlying tissue with formation of a full thickness epithelium and the development of a collagenous scar which, with time, becomes less vascular.

With ageing, the process of primary and secondary intention wound healing does not differ from that seen within younger healthy people. The healing process itself, however, may be affected (see the following discussion) due to ageing changes and concomitant medical and nutritional factors. This may complicate any effort to age a particular injury.

Both the epidermis and dermis are affected by age. The epidermis becomes atrophic with a concomitant reduction in its function as an environmental barrier. Its ability to heal following injury also decreases (Lavker et al. 1987; Choi et al. 2007). Changes also occur within the dermis with atrophy of the extracellular matrix and a reduction in the deposition of collagen types I and III (Lovell et al. 1987; Pearce and Grimmer 1972). Elastic fibres, which confer compliance and recoil to the skin, similarly undergo disintegration (Braverman and Fonferko 1982). The end result is an atrophic, less-elastic, and more friable protective barrier that is more susceptible to injury and that heals at a potentially slower pace. Falling oestrogen levels have been shown to be the principle mediator of some of these ageing effect (Bolognia et al. 1989; Emmerson and Hardman 2012).

There are a number of environmental factors that contribute to the development of an 'aged' skin and that may also contribute to healing occurring at a potentially slower pace. Principal among these is chronic sun exposure (ultraviolet light) that causes solar elastosis, which is the deposition of abnormal elastin within the dermis (Dawber and Shuster 1971). Smoking similarly causes elastosis (Martires et al. 2009; Kadunce et al. 1991). Medical conditions and medications may also contribute to the friability of the skin and/or the healing process. These are discussed below.

Bone

Bone consists of a collagenous matrix (type 1 collagen within a mucopolysaccharide ground substance including some small amounts of non-collagenous protein mainly in the form of proteoglycans and bone-specific proteins) with minerals (mostly calcium and phosphate) and cells (mostly osteoblasts, osteocytes, and osteoclasts). With respect to the cells within the bone, the osteoblasts (originating from mesenchymal stem cells) are involved with bone formation and appear as rows of small cuboidal cells on the surfaces of trabeculae and the Haversian system where new bone is formed. Osteoblasts produce type I collagen and non-collagenous bone proteins and are involved in the mineralisation of the bone matrix. The osteocytes are essentially mature osteoblasts that are found encapsulated within the mineralised bone matrix. Osteoclasts are multi-nucleated cells involved in bone resorption.

Bone is categorised into two different forms, woven and lamellar. Woven bone is immature bone and in adults is mainly found in reactive and neoplastic conditions including in the early stages of fracture healing. It consists of collagen fibres that are arranged haphazardly without any specific cellular orientation. Lamellar bone is mature bone in which the collagen fibres are arranged in parallel layers enclosing osteocytes. There are two different structural forms of lamellar bone: cortical bone, which is a dense compact bone made up of compact haversian units, and cancellous bone, which has a honeycomb appearance. The former is found lining the outer walls of most bones and in particular the shafts of the long bones. The latter is found within the interior of all bones.

Bone is not a static structure but a dynamic organ that undergoes remodelling during adult life. This is undertaken in an orderly and specific sequence whereby osteoclasts excavate an area (cavity) of bone. They are subsequently replaced by osteoblasts that fill the cavity with new bone. This is a slow process and takes some months to occur (Blom et al. 2018).

As with skin lesions, the ageing of a healing fracture may be of forensic significance. Most fractures are accompanied by some degree of tissue damage and haemorrhage. There are three phases involved in the healing process. These are an inflammatory phase, repair, and bone remodelling, and the process includes (Burt and Fleming 2008):

- *Inflammatory reaction (lasting up to one week):* early organisation with the development of capillaries and an inflow of fibroblasts associated with phagocytosis of debris and necrotic tissues;

- *A repair phase:* during which there is early bone regeneration followed by the development of callus. After about one week following the fracture, osteoblastic activity is present beneath the disrupted periosteum and at the end of the bone fracture with development of a provisional callus bridging the gap between the two bone fragments. Initially, osteoid tissue is laid down and then woven bone. From about three weeks onwards, a well-formed callus is present joining the fractured ends. Osteoblastic and -clastic activity continue to occur, which leads to the third phase of the repair process;

- *Remodelling*: remodelling of callus takes weeks to months and involves ongoing osteoblastic and -clastic activity, resulting in the eventual formation of dense lamellar bone and eventual reduction in osteoblastic and -clastic activity.

Changes in bone associated with ageing are well known. A reduction in sex steroids, particularly in post-menopausal women, and also in males, results in a loss of bone mass. This loss of bone mass commences from about the third decade of life, and as age increases, the balance between the formation of new bone and the removal of old bone favours a loss of bone mass, resulting in osteoporosis (Almeida et al. 2007; Manolagas and Parfitt 2010; Gibon et al. 2017). This has been shown to be associated with a reduced proliferation, differentiation, and osteogenic capability of the osteoblast precursor, the mesenchymal stem cells (Sethe et al. 2006). From the onset of menopause and for about the next 10 years, the rate of bone loss in women is about 3% per year. This rate of bone loss gradually slows for those aged 65–70 years and becomes about 0.5% per year by the age of 75 years. Men experience a similar phase of rapid bone loss, but this occurs some 15–20 years later than for women. In both cases, osteoporosis develops (Blom et al. 2018).

Inflammation, although an essential part of bone repair and healing, may lead to a delay in fracture healing and the development of complications such as non-union (Mountziaris and Mikos 2008; Marsell and Einhorn 2010; Claes et al. 2012). Unlike the inflammation associated with systemic inflammatory conditions, the inflammation associated with fracture healing is limited and regulated (Mountziaris and Mikos 2008). It stimulates angiogenesis and promotes the formation of osteoblasts from the stem cells. Inflammation is also associated with the release of cytokines essential for the healing process (Thomas and Pueo 2011; Claes et al. 2012). Inhibition of these inflammatory pathways adversely affects bone formation. Cyclooxygenase-2 and prostaglandin E2, for example play an important role in healing, and their inhibition by nonsteroidal anti-inflammatory drugs and selective COX-2 inhibitors adversely affects healing (Gerstenfeld et al. 2003; Pountos et al. 2012; Lu et al. 2017). Chronic inflammation is similarly detrimental to bone healing as evidenced by a variety of diseases (e.g. chronic obstructive pulmonary disease, diabetes mellitus, rheumatoid arthritis, and systemic lupus erythematosus) characterised by osteoporosis (Claes et al. 2012).

A number of other factors in relation to the regulation of bone turnover deserve consideration particularly in the ageing patient (for an overview see Blom et al. 2018). Bone consists of a number of minerals including calcium, phosphorus, and magnesium. The more important of these is calcium. The majority of the body's calcium (about 99%) and phosphorus are stored within the bone. Calcium has an important role in cell function including nerve conduction and muscle contraction. Dietary calcium absorption occurs within the small intestine and is mediated by active vitamin D metabolites (especially calcitriol). In bone, the active vitamin D metabolites in conjunction with parathyroid hormone promote osteoclastic bone resorption. In cases, therefore, where there is a fall in serum calcium levels below normal, there is a concomitant increase in the secretion of parathyroid hormone. In the immediate phase, this increased secretion causes increased renal tubular reabsorption of calcium followed by an increased intestinal absorption of calcium. With prolonged hypocalcaemia, increased osteoclastic activity results in bone resorption and subsequent release of calcium from the bone. In situations where there is a prolonged hypocalcaemia, the potential to develop osteomalacia exists. A number of conditions may lead to hypocalcaemia, and these include excessive intake of phosphates, hyperparathyroidism, vitamin D deficiency, malabsorption disorders of the bowel, renal failure, pancreatitis, calcium channel blocker overdose, rhabdomyolysis, tumour lysis syndrome, and the use of corticosteroids and biphosphonates – all conditions that are seen in the elderly.

There are a number of other factors that influence bone metabolism and these include:

- *Parathyroid hormone:* this has a significant effect on increasing bone turnover and osteoclastic bone resorption;
- *Oestrogen:* with age, a reduction in oestrogen results in a reduced bone density;
- *Glucocorticoids:* in excess, glucocorticoids cause suppression of bone formation;
- *Thyroxine:* in excess, thyroxine increases bone turnover;
- *Mechanical stress:* mechanical loading of bone may act as an osteogenic stimulus; and
- Diet and especially calcium intake.

Injury types

One of the challenges in forensic medicine is ascertaining whether an injury is accidental or inflicted, and with the latter, whether it has been deliberately inflicted by another person or the result of self-harm. Witnessed acts of abuse generally do not present to the forensic physician and not always to the forensic pathologist. Abusive acts may be dealt with directly through the legal system. Sometimes, the patient is not able to provide a history as to what has occurred either because they are prevented from doing so by third persons or because of mental health or neurological issues. The challenge, therefore, is to ascertain how an injury may have occurred, when it may have occurred, and by whom.

Elderly patients may present with signs of abuse including:

- bruises, abrasions, lacerations, bite marks, incisions, stab wounds, burns, and gunshot wounds;
- pressure sores;
- missing or fractured teeth;

- eye injuries;
- hair loss (traumatic alopecia);
- limited or loss of function of a limb, suggesting either an underlying soft tissue injury (including ligamentous and muscular injury; sprains and strains) or recent healing or old fracture, subluxation, or dislocation;
- areas of tenderness or pain due to underlying internal organ damage;
- bleeding from an orifice, haematuria, otorrhoea, haematemesis, and haematochezia;
- oedema/swelling;
- general disability; and
- loss of consciousness.

The mechanisms whereby these may be caused include:

- pushing, shoving, and rough handling;
- kicking, hitting, punching, and slapping;
- biting;
- burning;
- cutting or stabbing;
- shooting;
- physically restraining; and
- hair pulling.

There are a number of external injuries that may be identified as potential markers of abuse. These include bruises, abrasions, lacerations, bite marks, loss of teeth, hair loss, incisions, stab wounds, bites, burns, and marks from restraint. Gunshot wounds do not necessarily reach the attention of forensic physicians and tend to be addressed by the legal system with the assistance of the treating physicians and surgeons in cases where the patient is alive. Where death has occurred, the matter is addressed by the forensic pathologist and legal system. Gunshot injuries, therefore, will not be considered further in this chapter. Other injuries that may also be markers of abuse but may not always be immediately evident include musculoskeletal injuries and internal organ damage. Each injury type will be briefly discussed below with consideration being given to the influence of age-related changes and of medical conditions that may affect their presentation. More detailed information regarding these injuries may be found in the various textbooks of clinical forensic medicine (Gall et al. 2003; Madea 2014; Byard and Payne-James 2016; Payne-James and Jones 2019) and forensic pathology (Saukko and Knight 2016).

Oedema, erythema, and tenderness

Oedema, erythema, and pain or tenderness in a particular area may be associated with trauma. There are, however, many other causes for these signs, particularly in the elderly, and their presence needs to be considered with caution to distinguish injury from medical condition.

Bruises

A bruise (contusion) is a focal area of discolouration due to leakage of blood from ruptured veins and small vessels in the tissue. Bruises are usually, but not always, the result of blunt trauma. The colour of bruises varies from red to purple to grey to yellow and may show a variety of colours within one bruise. Bruises occurring at the same time on the body may appear with different colouring. Although it would seem logical that the colour of bruises would represent the associated biochemical breakdown of haemoglobin, this does not occur in reality. The ageing of bruises is limited, and the only factor currently understood with respect to the visual ageing of a bruise is that a bruise with a yellow discolouration is 18–24 hours old or older (Langlois and Gresham 1991), but that may also be challenged as there is variability in the perception threshold for yellow in the general population and that a subject's ability to perceive yellow in a bruise declines with age (Hughes et al. 2004).

There are a number of factors that may affect the appearance of a bruise:

- *Depth of injury:* an injury may occur in a deeper anatomical plane, which results in a delayed presentation of the bruise due to the time required for migration of the haemorrhaged cells towards the skin surface. For example, a significant injury to the thigh may not result in visible bruising for several days post-injury.

- *Laxity of tissue:* as the extent of bruising is dependent on the available space into which blood from the damaged vessels is able to escape into the surrounding connective tissue, lax tissues such as the periorbital area will appear to bruise more easily than those where there is dense fibrous tissue such as the palm of the hand and sole of the foot.

- *Site of bruising:* the extent of the underlying vasculature varies in density in differing parts of the body which in turn affects the extent to which haemorrhaging and thus a bruise may occur.

- *Underlying tissues:* some parts of the body are more prone to bruising than others due to the nature of the underlying tissue. For example, skin overlying bone tends to bruise more easily than the more resilient areas such as the abdominal wall.
- *Force of impact:* the force of impact and implement used will influence the extent of bruising. That being said, it is not possible to determine purely on the size and extent of a bruise how much force was used to cause the bruise.
- *Migration:* a bruise, or more accurately the blood elements within the bruise, may migrate from the site of impact to another area due to gravity (Figure 13.1). For example, an injury to the forehead may result in periorbital bruising and later appear in remote sites such as the cheeks and neck.
- *Skin pigmentation:* the visibility of a bruise is dependent on the individual's skin pigmentation.
- *Treatment:* the appearance and length of time a bruise is visible may be influenced by treatment particularly with the use of ice in the early phase of injury.
- *Age, diseases, and medications:* these are discussed below.

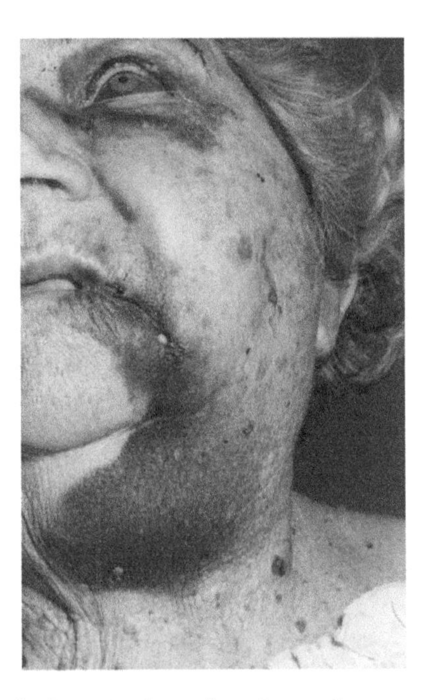

Figure 13.1 Migration of a bruise onto the neck post-assault.

Some bruises have characteristic features that may assist in determining the cause:

- *Petechial bruising:* Petechial bruises are pinpoint (up to approximately 2 mm in diameter) areas of bruising resulting from vascular damage resulting from a raised intravascular pressure (Figure 13.2). Petechiae are characteristic findings in cases where there has been some form of compression of the upper part of the body (neck or chest). They tend to be evident on the upper part of the body and particularly on the face and conjunctiva. Persons grabbed about the neck, where there is neck compression, may develop petechiae affecting areas such as the face and conjunctiva. Where there has been significant chest compression (e.g. a vehicle running over a person's chest) may also result in similar petechiae appearing above the line of compression. A special form of petechial bruising is the suction mark left following a 'love bite' or 'hicky' (Figure 13.3).

- *Imprints:* When hit with a particular object that has a specific pattern (e.g. a belt buckle and soul of shoe), the subsequent bruise that forms may be a mirror image, or imprint, of the object (Figure 13.4). Similarly, when blunt force is applied over an object located between the applying force and the skin, an imprint bruise may also develop (e.g. a fabric). Another form of imprint is where a person is hit with an open hand, and the outline of fingers and palm may be evident within the bruising pattern.

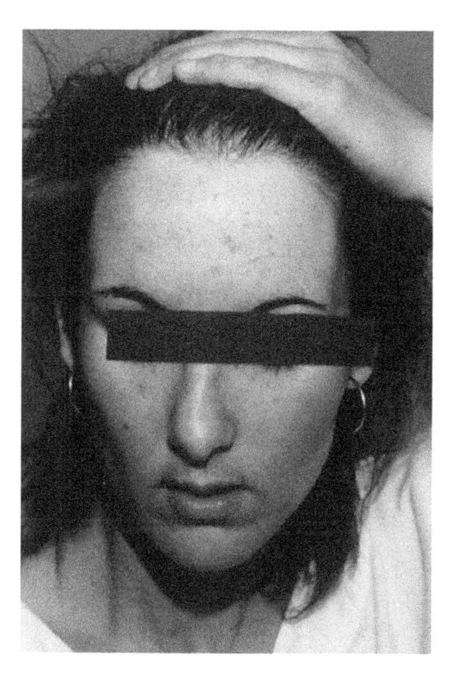

Figure 13.2 Petechial haemorrhages/bruises on the forehead post neck compression. *Source:* Gall et al. (2003), Figure 37/With permission of Elsevier.

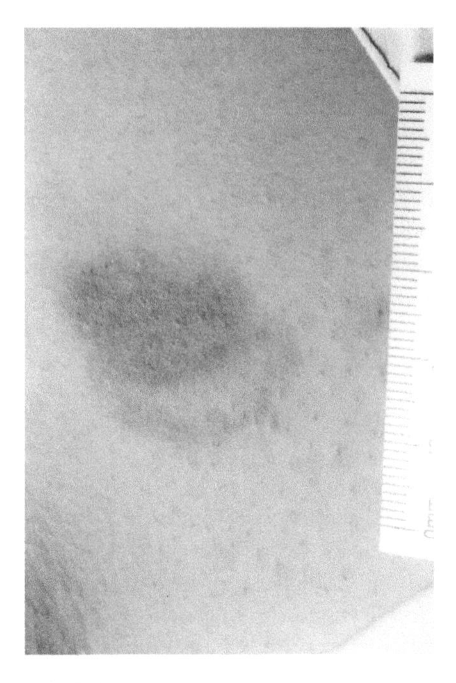

Figure 13.3 A 'love' bite or 'hicky'.

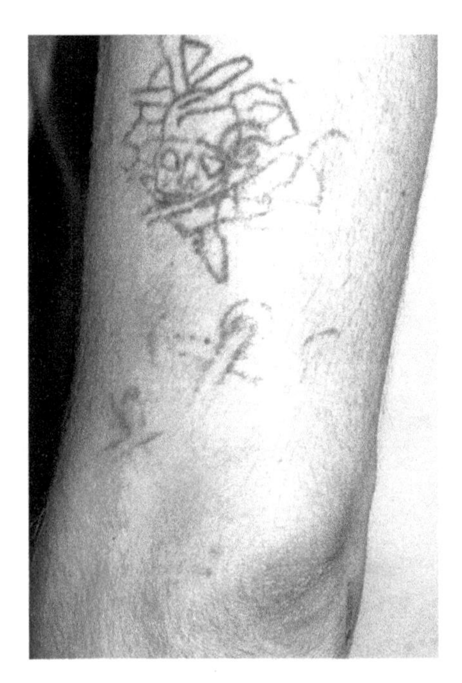

Figure 13.4 Patterned bruise on the forearm following being hit with a belt buckle. *Source:* Gall et al. (2003), Figure 38/With permission of Elsevier.

- *Fingertip and grip marks:* These injuries are more frequently identified in child abuse cases but may also be seen in adult assault cases. Grip marks may be present in the elderly, and the cause may be either accidental (friable skin) or deliberate (excessive grip force). The principal site for fingerprint bruising tends to be about the arms.

- *Tramline bruising:* Following forceful contact with an object that has essentially parallel edges such as a rod and cane, 'tramline' bruising may result (Figure 13.5). The name arises due to the resultant essentially parallel linear bruising with sparing of the intervening skin. The mechanism for the development of this appearance is that with the contact of the skin by the implement there is a stretching of the edges (corresponding to the edges of the implement), resulting in damage to the adjacent underlying dermal blood vessels beneath the longitudinal edges or sides of the implement. With release of the implement force, blood flows from those damaged vessels causing the characteristic tramline appearance, while the vessels directly beneath the compressing implement remain intact.

- *Bite mark:* Bite marks may result in damage to the skin beyond bruising. In circumstances where bruising occurs, the bruising tends to be located in an oval or curved arrangement with some central sparing (Figure 13.6).

Medications, medical conditions, and skin changes associated with ageing may result in the development of bruises as may some toxins that may or may not be associated with trauma (Neutze and Roque 2016). Tables 13.1 and 13.2 provide a

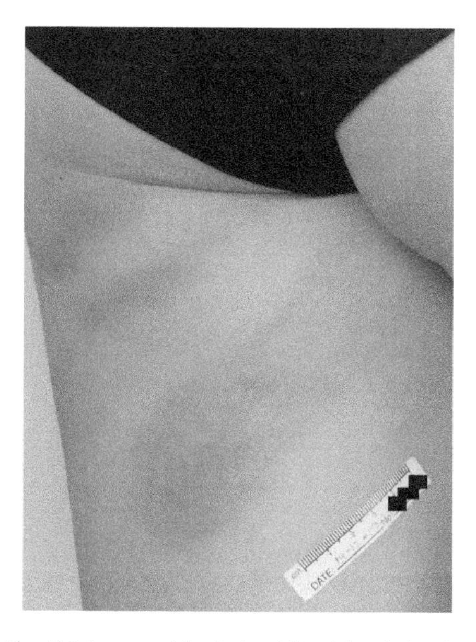

Figure 13.5 Bruising to the thigh caused by being hit with a belt. At one end, tramline bruising is shown.

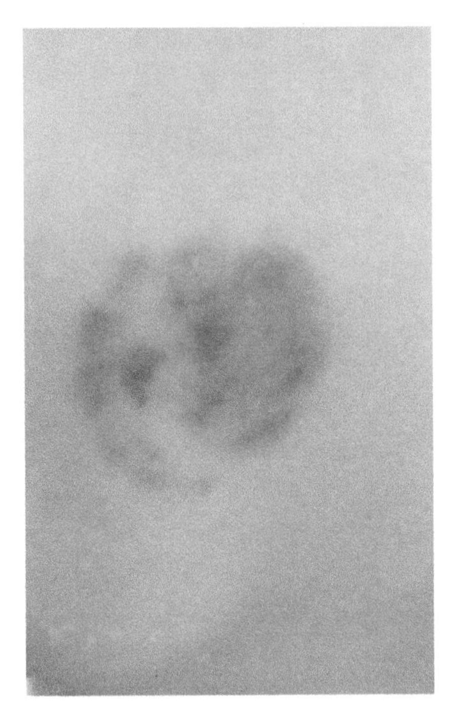

Figure 13.6 Bite mark on the arm.

Table 13.1 Medical conditions associated with easy bruising (this list is not exhaustive).

Bleeding disorders	Other conditions
Disseminated intravascular coagulation	Ageing skin
Factor VIII deficiency	Alcoholism
Factor IX deficiency	Aplastic anaemia
Haemophilia	Cancer including metastatic carcinoma
Von Willebrand disease	Ehlers–Danlos syndrome
Vitamin K deficiency	Leukaemia
	Liver disease
	Myelofibrosis
	Renal failure
	Scurvy
	Solar skin damage

brief list of medical conditions and medications that may be associated with easy bruising. Neither list is exhaustive. With any consideration of the potential non-accidental cause of bruising, a careful and thorough history is necessary with consideration of all medical conditions and medications being taken. Poisons also need to be considered.

Table 13.2 Some medications that may cause easy bruising (this list is not exhaustive).

Medication	Mechanism of action
Acenocoumarol; apixaban; betrixaban; dabigatran; edoxaban; enoxaparin; heparin; phenindione; phenprocoumon; rivaroxaban; warfarin	Inhibition of coagulation/ vitamin K antagonists
Corticosteroids	Collagen degradation
Aspirin; cangrelor; cilostazol; clopidogrel; dipyridamole; non-steroidal anti-inflammatory drugs; selective serotonin reuptake inhibitors; ticlopidine	Platelet aggregation inhibition
Alcohol; antibiotics (many); carbamazepine; some diuretics; H_2 blockers; interferon; methotrexate; quinine; thiazide diuretics; sodium valproate	Thrombocytopenia

Given the ageing changes in the skin, whether or not there are associated medical conditions and/or medications that may contribute to the friability of the skin, the presence of bruises in the elderly patient may be expected particularly due to their propensity for falls and knocks against firm surfaces. Bruises commonly occurs following a fall, and the most common cause of injury in the elderly are falls (Burns et al. 2016). There are, however, certain sites where bruising may not normally be expected. These include the face and neck, chest wall, abdomen, and buttocks (Figure 13.7) (Crane 2000; Saukko and Knight 2016). Excessive and extensive

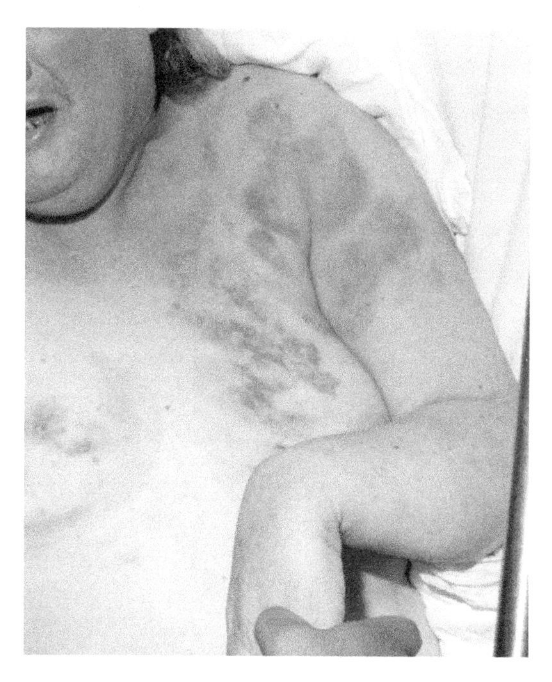

Figure 13.7 Excessive bruising and bruising on the chest and breasts should raise suspicions of possible abuse.

bruising, and bruising in multiple sites, is reason for suspicion of non-accidental injury.

Abrasions

An abrasion (graze or scratch) is the result of blunt trauma to a particular part of the body with an associated frictional element directed across the surface of the skin causing disruption to the surface layers of the skin. Technically, it is disruption to the epidermis only, but there may be bleeding associated with this due to disruption of the vascular element contained within the papillary dermis which lies immediately beneath the epidermis in a convoluted manner. Abrasions are important injuries as it is possible to determine the direction of force due to the presence of skin tags. Bruising may be associated with an abrasion.

Abrasions are commonly associated with falls and may be seen in association with bites and with being pulled along the surface causing frictional or carpet burns (Figure 13.8). There is one site where abrasions may be of particular importance and that is in attempted efforts to strangle a person. Not only does the possibility of petechial haemorrhages on the face and conjunctiva occur, but due to the pressure about the neck, the victim may sustain linear abrasions to their neck resulting from injury from their own fingernails whilst trying to remove the compressive item about their neck (Figure 13.9).

As with bruising, because of the fragility of ageing skin and the propensity to bump against firm surfaces, abrasions are not uncommon in the elderly, particularly on the anterior leg and knee and the forearms.

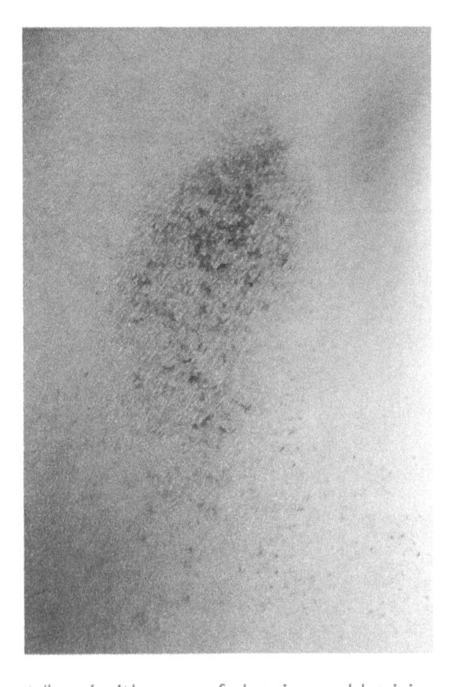

Figure 13.8 Healing carpet 'burn' with areas of abrasion and bruising.

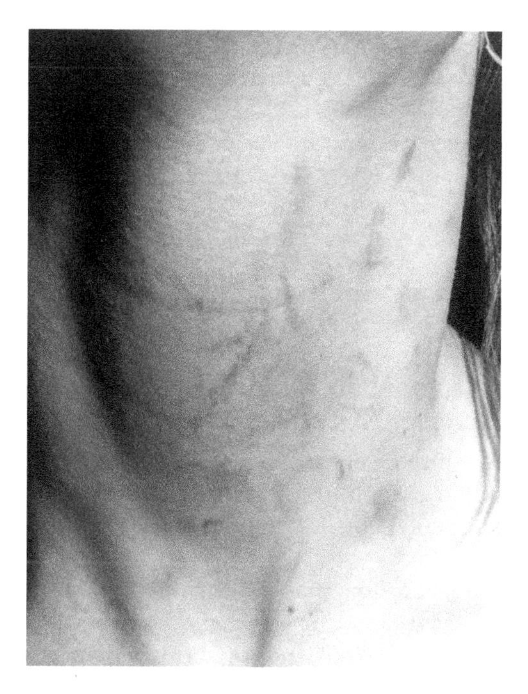

Figure 13.9 Abrasions caused by the subject's fingernails in an attempt to release pressure from around her neck during an assault.

Lacerations

Lacerations are the result of blunt trauma causing a tear or split in the skin of variable depth. This tear extends beyond the epidermis (or overlying capsule in internal organs) and into the underlying tissues. Lacerations may also occur as a result of blunt trauma affecting the internal organs. Lacerations are a different entity to incisions. They have specific features regarding their appearance that assist in their diagnosis (Table 13.3). Because they are the result of blunt trauma and may involve crushing and or tearing forces, the edges tend to be ragged, crushed, abraded,

Table 13.3 Comparison of features of lacerations and incisions.

Feature	Lacerations	Incisions
Margins	• Ragged • Crushed • Abraided • Inverted • Bruised	• Regular • Not crushed • Usually not abraided • Everted • Usually not bruised
Foreign material	Often	Not usually
Tissue bridging strands	Yes	No

inverted, and bruised. There may be foreign material within the wound together with bridging strands. These bridging strands extend between the edges of the wound and may consist of fibrous tissue, blood vessels, and nerves.

There are many causes for lacerations including falls, being hit with a firm implement, and coming into contact with a firm surface such as in motor vehicle accidents. Lacerations are not an uncommon finding in the elderly, but they usually occur on the forearms and legs as a result of falls or knocking against a firm surface.

Incised wounds

By comparison with lacerations, incised wounds (cuts and slashes) are the result of sharp injury caused by implement such as scalpels, knives, and similarly sharp implements. These injuries extend beyond the surface layer of the skin (or capsule of an organ) into the subcutaneous tissue and deeper structures of skin or parenchyma of an organ. By contrast with lacerations, the margins of these wounds tend to be regular, they are not crushed, they are usually not abraded, they are inverted, and usually there is no associated bruising (Table 13.3). Foreign material is not normally found within these wounds, and the bridging strands, a feature of lacerations, are not present.

Incised wounds may be the result of an accident, an assault, or be self-inflicted. Ageing does not necessarily increase the risk of sustaining an incision unless the person has motor skill difficulties or there are problems with their vision. In cases of accidental injury, the more common injuries are identified to the hands and fingers sustained whilst using a knife or similar implement in the kitchen or workplace. Incisions on other parts of the body may arrive accidentally as a result of dropping a sharp object onto a part of the body. In cases of assault, the location of incised wounds will depend on the events in the assault. Incisions resulting from an assault may also be sustained on the ulnar aspect of the forearm and palmer surface of the hands and fingers. These locations are indicative of defensive injuries.

Stab wounds

A stab wound is a penetrating injury where the depth is usually greater than its width (Figure 13.10). These injuries may result from sharp (e.g. a knife) or blunt (e.g. a screwdriver) implements. The appearance of the wounds will vary depending on the nature of implement used and any movement that may occur of the victim's body or of the weapon whilst the weapon is being inserted or withdrawn. The stab wound edges may have associated irregular margins, and the margins may be bruised and abraded depending on whether the hilt of the blade or handle hits the skin. If a knife is used, the appearance of the stab wound is often elliptical or slit-like. By examining the wound width, it may be possible to estimate whether the knife used was double or single edged/bladed. If the knife has a single cutting edge with a blunt back edge, the wound may have an acutely angled corner that corresponds to the cutting edge

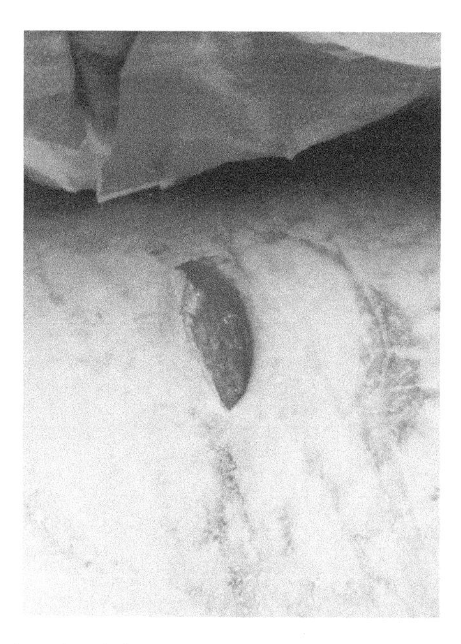

Figure 13.10 Stab wound to the abdomen.

but a split opposite corner corresponding to trauma created by the blunt back edge. A double cutting-edge knife may have two acutely angled corners without any evidence of splitting.

Stab wounds may be accidental but are usually associated either with an assault or as a result of deliberate self-harm. In cases of assault, defensive injuries may be present. These are discussed further below. In cases of non-fatal self-harm, the injuries tend to be on a part of the body easily accessible to the subject. In cases of suicide, the injuries again are within the reach of the individual and often involve the anterior chest wall and abdomen. Accompanying these injuries may be more superficial injuries that are referred to as tentative or hesitation injuries.

Bites

Bites may be inflicted by either humans or animals. Human teeth are usually arranged in a 'U' shape, whereas animals have a more 'V'-shaped arrangement (Clement 2011). Bites may be associated with bruising, abrasions, lacerations, and, less commonly, amputations. Penetrating wounds are more associated with animal bites. Human bites appear as an irregular, crescentic area of injury (bruising, abrasion, and/or laceration) usually with an opposing similar image and central sparing (Figure 13.6). One area of crescentic injury may be more prominent than the other. Bite marks are not a normal finding in any person and indicates intentional injury perpetrated either by another person or, less frequently, as a result of self-harm. In cases where either definite bite marks or potential bite marks are identified, the engagement of a forensic dentist to assist in the analysis of the injury is recommended.

Burns

Burns result in tissue damage and may result from thermal, chemical, or radiation sources. Thermal burns tend to be the most common type encountered in forensic medicine. The depth, extent, and configuration of a burn will depend on the agent causing the burn, its temperature (if relevant), and the time of exposure. Variables associated with this will depend on the nature of any protective clothing that may be worn and of the site on the body where the burn occurs. Burns associated with elder abuse tend to occur on the palms of the hands, soles of the feet, and hips/buttocks (Quinn and Tomika 1997).

Fractures

Fractures are the result of blunt trauma. The force may be directed either directly to the site of fracture or indirectly at a point distant to the fracture site. Associated with the fracture will be a degree of soft tissue injury. Direct force generally results in a transverse or comminuted fracture. An indirect force may cause an oblique or spiral fracture but may also damage the nearby joints including the formation of an haemarthrosis. For example, the twisting of a limb may result in a spiral fracture. A combination of twisting, bending, and compression may cause an oblique or spiral fracture.

A brief overview of the changes in bone associated with ageing, disease, and medications has been given above. Within the elderly, principally osteoporosis and also osteomalacia need to be considered, and the medical conditions, medications, and other conditions that contribute to osteoporosis are listed in Table 13.4. Elderly patients have bones that are more fragile and fracture prone than do younger patients. Although younger patients also may sustain pathological fractures particularly as a result of cancer, the incidence of cancer-related pathological fractures increases with age. Fractures as a result of medical-related conditions may occur at any site in the body although tend to occur mostly in the long bones. Fractures resulting from falls tend to occur in the hips, wrists, and forearms. Although fractures at these sites may be in keeping with the fall, their presence does not indicate that the fall was accidental rather than the result of a deliberate push or trip. Suspicious fractures include:

- mid and lower facial fractures, fractures of the mandible and zygoma (Fenton et al. 2000); and
- spinal and truncal fractures (Fanslow et al. 1998).

Spiral fractures are also considered to be highly suspicious of nonaccidental injury (Dyer et al. 2003).

Self-harm

Self-inflicted or self-harm injuries occur within all age groups (Figure 13.11) (Gall et al. 2003, 2011). The difficulty with these injuries is ascertaining whether

Table 13.4 Conditions associated with or that lead to osteoporosis (this list is not exhaustive).

Nutritional causes	Medical and other conditions	Neoplastic conditions	Medications/ toxins
Eating disorders	Ageing	Carcinomatosis	Alcohol
Malnutrition (esp. calcium)	Ankylosing spondylitis	Leukaemia	Corticosteroids
Malabsorption syndromes	Chronic renal disease (renal osteodystrophy)	Multiple myeloma	Heparin
Scurvy	Chronic inflammatory conditions		
	Chronic obstructive pulmonary disease Cushing syndrome		
	Diabetes mellitus Gonadal insufficiency		
	Hyperparathyroidism		
	Immobilisation		
	Osteogenesis imperfecta (some subclinical forms)		
	Systemic lupus erythematosus thyrotoxicosis (hyperthyroidism)		
	Tuberculosis		
	Rheumatoid arthritis		

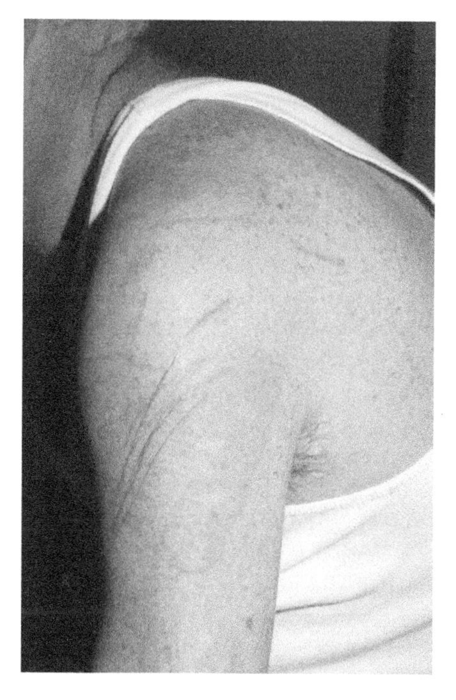

Figure 13.11 Self-harm by a right-handed person causing abrasions and lacerations to his left shoulder and arm.

they are the result of an accident, deliberate harm by another person, or the result of self-infliction. Self-harm injuries include bruises, abrasions, lacerations, incisions, stab wounds, gunshot wounds, bites, and burns. Some of these injuries result in death, and some were intended to cause death, but the attempt was unsuccessful. Others may be cries for help or an indicator of frustration. There are specific features regarding these injuries that assist in identifying their potential of having been self-inflicted in the living (Gall et al. 2003, 2011). These include:

- all injuries tend to be of a similar injury type (e.g. all bruises and all abrasions) although they generally consist of abrasions or incisions rather than bruises or lacerations;
- all injuries tend to be of a similar severity (e.g. all abrasions are of a similar severity) and in most cases tend to be superficial;
- the injuries are in accessible sites for the individual;
- the injuries tend to occur on the non-dominant side of the body (i.e. a right-handed individual will create injuries affecting the left-hand side of the body);
- the injuries are often multiple and of similar orientation (e.g. parallel);
- the injuries generally do not involve vital structures such as the eyes but tend to involve non-vital structures such as the face, chest, abdomen, arms, and legs;
- are beneath the overlying clothing which is usually spared; and
- there is an absence of defensive injuries.

Defensive injuries

When making an assessment of anybody who had sustained any injury, the presence of defensive injuries should be considered. Defensive injuries are injuries that result from defending one's self. They occur when the victim attempts to ward off a weapon or assailant. Defensive injuries consist of one or more of bruises, abrasions, lacerations, incisions, and stabs. In the elderly, given the potential for a fall during the assault, fractures may also occur. Defensive injuries generally involve the upper limbs, thighs, and neck. When being attacked, it is natural to use the forearms and hands to protect oneself. If being kicked or hit, the victims may position their body so that more posterior part of the body is used to protect the anterior part of the body. Defensive injuries tend to occur on the extensor and ulnar aspects of the forearms and hands. In cases where an assault is made with a knife, sometimes attempts are made to grab the weapon, resulting in incisions or other injuries to the fingers and hands.

Other injuries

There are a multitude of other injuries that may occur. Blunt force trauma to the mouth may result in loss of teeth, tooth fracture, and associated injuries to the gums, mouth mucosa, and lips. Eye injuries may occur from both blunt and sharp trauma. Traumatic alopecia may occur in addition to alopecia due to ageing. Ligature marks (bruises and/or abrasions) about the ankles and/or wrists may indicate restraint. Loss or limited function of a limb may occur due to the underlying joint, and soft tissue injuries (i.e. ligamentous and muscular injury, sprains, strains, subluxation, and dislocation) may be associated with medical, accidental, and non-accidental causes. Internal injuries may also occur as a result of both blunt and sharp trauma.

Mimics of elder abuse

There are many mimics of elder abuse including chronic diseases, the use of various medications, poisons, and injuries sustained due to causes other than non-accidental injury. Chronic diseases are difficult to differentiate from abuse (Hunsaker 2014). When assessing injuries in elderly patients, a thorough assessment including a full medical history is necessary to determine whether the injuries identified are the result of accidental, non-accidental, or self-harm origin.

How to assess and avoid errors in the interpretation of injuries

There are many reasons as to why a diagnosis of elder abuse may be missed (Swagerty et al. 1999). When assessing injuries in elderly patients, just as in children, consideration of an injury being the result of a non-accidental cause needs always to be considered. In elderly patients, the coexistence of chronic diseases and the concomitant use of medications complicates the assessment. For every patient who presents with physical injuries, it is essential that the health practitioner consider the type of injury, how it may have been caused, and the potential for a non-accidental cause. The approach to the patient may be difficult due to cognitive impairment and/or reluctance to provide information for fear of retaliation by the abuser. Ideally, the patient should be interviewed and examined in the absence of any carer. As with any medical assessment, a full medical history needs to be taken including information as to how any injury identified may have occurred. If the patient is unable to provide these details, this information should be sourced from independent and non-carer sources in cases where injuries and suspicions of physical abuse exist. In addition to a thorough medical history and medical examination, care needs to be taken to ensure that errors in the interpretation of injuries are minimised (Gall and Payne-James 2011).

Acknowledgement

Figures 13.2 and 13.4 are reproduced by kind permission from Gall, J.A.M., Boos, S.C., Jayne-James, J.J., Culliford, E.J. (Eds). (2003). *Forensic Medicine Colour Guide*,

Edinburgh, Churchill Livingstone, Figures 37 and 38, Copyright Elsevier Science Ltd (2003).

References

Almeida, M., Han, L., Martin-Millan, M. et al. (2007). Skeletal involution by age-associated oxidative stress and its acceleration by loss of sex steroids. *J Biol Chem* **282**: 27285–27297.

Blom, A., Warwick, D., and Whitehouse, M.R. (ed.) (2018). *Apley & Solomon's System of Orthopaedics and Trauma*, 10e. Boca Raton, FL: CRC Press. Kindle Edition.

Blume-Peytavi, U., Kottner, J., Sterry, W. et al. (2016). Age-associated skin conditions and diseases: current perspectives and future options. *Gerontologist* **56**: S230–S242.

Bolognia, J.L., Braverman, I.M., Rousseau, M.E., and Sarrel, P.M. (1989). Skin changes in menopause. *Maturitas* **11**: 295–304.

Braverman, I.M. and Fonferko, E. (1982). Studies in cutaneous aging. The elastic fiber network. *J Invest Derm* **78**: 434–443.

Burns, E.R., Stevens, J.A., and Lee, R. (2016). The direct costs of fatal and non-fatal falls among older adults - United States. *J Saf Res* **58**: 99–103.

Burt, A.D. and Fleming, S. (2008). Cell injury, inflammation and repair. In: *Muir's Textbook of Pathology*, 14e (ed. D.A. Levinson, R. Reid, A.D. Burt, et al.), 47–76. London: Hodder Arnold.

Byard, R. and Payne-James, J. (ed.) (2016). *Encyclopedia of Forensic and Legal Medicine*, 2e. Amsterdam: Elsevier.

Choi, E.H., Man, M.Q., and Xu, P. (2007). Stratum corneum acidification is impaired in moderately aged human and murine skin. *J Invest Derm* **127**: 2847–2856.

Claes, L., Recknagel, S., and Ignatius, A. (2012). Fracture healing under healthy and inflammatory conditions. *Nat Rev Rheumatol* **8**: 133–143.

Clement, J.G. (2011). Bite marks. In: *Current Practice in Forensic Medicine* (ed. J. Gall and J. Payne-James), 291–308. Wiley Blackwell: Chichester.

Crane, J. (2000). Injury interpretation. In: *A Physician's Guide to Clinical Forensic Medicine* (ed. M.M. Stark), 99–116. Totowa, NJ: Humana Press.

Dawber, R.H. and Shuster, S. (1971). Scanning electron microscopy of dermal fibrous tissue networks in normal skin, solar elastosis and pseudo-xanthoma elasticum. *Br J Derm* **84**: 130–134.

Dyer, C.B., Connolly, M.T., and McFeeley, P. (2003). The clinical and medical forensics of elder abuse and neglect. In: *Elder Mistreatment: Abuse, Neglect, and Exploitation in an Aging America* (ed. T.J. Bonnie and R.B. Wallace), 339–381. Washington, DC: National Academies Press.

Emmerson, E. and Hardman, M.J. (2012). The role of estrogen deficiency in skin ageing and wound healing. *Biogerontology* **13**: 3–20.

Fanslow, J.L., Norton, R.N., and Spinola, C.G. (1998). Indicators of assault-related injuries among women presenting to the emergency department. *Ann Emerg Med* **32**: 341–348.

Fenton, S.J., Bouquot, J.E., and Unkel, J.H. (2000). Orofacial considerations for pediatric, adult, and elderly victims of abuse. *Emerg Med Clin North Am* **18**: 601–617.

Gall, J.A.M., Boos, S.C., Payne-James, J.J., and Culliford, E.J. (2003). *Forensic Medicine. Colour Guide*. Edinburgh: Churchill Livingstone.

Gall, J., Goldney, R., and Payne-James, J.J. (2011). Self-inflicted injuries and associated psychological profiles. In: *Current Practice in Forensic Medicine* (ed. J. Gall and J. Payne-James), 273–290. Wiley Blackwell: Chichester.

Gall, J. and Payne-James, J.J. (2011). Injury interpretation: possible errors and fallacies. In: *Current Practice in Forensic Medicine* (ed. J. Gall and J. Payne-James), 239–271. Wiley Blackwell: Chichester.

Gerstenfeld, L.C., Thiede, M., Seibert, K. et al. (2003). Differential inhibition of fracture healing by non-selective and cyclooxygenase-2 selective non-steroidal anti-inflammatory drugs. *J Orthop Res* **21**: 670–675.

Gibon, E., Lu, L.Y., Nathan, K., and Goodman, S.B. (2017). Inflammation, ageing, and bone regeneration. *J Orthop Translat* **10**: 28–35.

Govan, A.D., Macfarlane, P.S., and Callander, R. (1991). *Pathology Illustrated*, 3e. Edinburgh: Churchill Livingstone.

Harrison, D.J. and Fleming, S. (2008). Normal cellular functions, disease and immunology. In: *Muir's Textbook of Pathology*, 14e (ed. D.A. Levinson, R. Reid, A.D. Burt, et al.), 11–29. London: Hodder Arnold.

Hughes, V.K., Ellis, P.S., and Langlois, N.E.I. (2004). The perception of yellow in bruises. *J Clin Forensic Med* **11**: 257–259.

Hunsaker, J.C. (2014). Violence against the elderly. In: *Handbook of Forensic Medicine* (ed. B. Madea), 761–775. Wiley Blackwell: Chichester.

Kadunce, D.P., Burr, R., Gress, R. et al. (1991). Cigarette smoking: risk factor for premature facial wrinkling. *Ann Intern Med* **114**: 840–844.

Langlois, N.E. and Gresham, G.A. (1991). The ageing of bruises: a review and study of the colour changes with time. *Forensic Sci Int* **50**: 227–238.

Lavker, R.M., Zheng, P.S., and Dong, G. (1987). Aged skin: a study by light, transmission electron, and scanning electron microscopy. *J Invest Derm* 88 (3 Suppl): 44s–51s.

Lovell, C.R., Smolenski, K.A., Duance, V.C. et al. (1987). Type I and III collagen content and fibre distribution in normal human skin during ageing. *Br J Derm* **117**: 419–428.

Lu, L.Y., Loi, F., Nathan, K. et al. (2017). Pro-inflammatory M1 macrophages promote osteogenesis by mesenchyme stem cells via the COX-2-prostaglandin E 2 pathway. *J Orthop Res* **10**: 28–35.

Madea, B. (ed.) (2014). *Handbook of Forensic Medicine*. Wiley Blackwell: Chichester.

Manolagas, S.C. and Parfitt, A.M. (2010). What old means to bone. *Trends Endocrinol Metab* **21**: 369–374.

Marsell, R. and Einhorn, T.A. (2010). The biology of fracture healing. *Injury* **42**: 551–555.

Martires, K.J., Fu, P., Polster, A.M. et al. (2009). Factors that affect skin aging: a cohort-based survey on twins. *Arch Derm* **145**: 1375–1379.

Mountziaris, P.M. and Mikos, A.G. (2008). Modulation of the inflammatory response for enhanced bone tissue regeneration. *Tissue Eng B Rev* **14**: 179–186.

Neutze, D. and Roque, J. (2016). Easy bruising and bleeding. *Am Fam Physician* **93**: 279–286.

Pearce, R.H. and Grimmer, B.J. (1972). Age and the chemical constitution of normal human dermis. *J Invest Derm* **58**: 347–361.

Payne-James, J.J. and Jones, R.M. (2019). *Simpson's Forensic Medicine*, 14e. CRC Press.

Pountos, I., Georgouli, T., Calori, G.M., and Giannoudis, P.V. (2012). Do nonsteroidal anti-inflammatory drugs affect bone healing? A critical analysis. *Sci World J* **2012**: 606404.

Quinn, M.J. and Tomika, S.K. (1997). *Elder Abuse and Neglect*, 2e, 49–179. New York: Springer.

Saukko, P. and Knight, B. (2016). *Knight's Forensic Pathology*, 4e. Boca Raton, FL: CRC Press.

Sethe, S., Scutt, A., and Stolzing, A. (2006). Aging of mesenchymal stem cells. *Aging Res Rev* **5**: 91–116.

Swagerty, D.L., Takahashi, P.Y., and Evans, J.M. (1999). Elder mistreatment. *Am Fam Physician* **59**: 2804–2808.

Thomas, M.V. and Pueo, D.A. (2011). Infection, inflammation, and bone regeneration: a paradoxical relationship. *J Dent Res* **90**: 1052–1061.

World Health Organisation (2020). Elder abuse (fact sheet). http://www.who.int/mediacentre/factsheets/fs357/en/ (accessed 12 January 2021).

Yon, Y., Mikton, C.R., Gassoumis, Z.D., and Wibe, K.H. (2017). Elder abuse prevalence in community settings: a systematic review and meta-analysis. *Lancet Glob Health* **5**: e147–e156.

14 Physical intervention and restraint

Anthony Bleetman and Stanislav Lifshitz*
**University of Warwick Medical School, Coventry, UK*

Introduction

Physical intervention and restraint are integral parts of behaviour and conflict management in agencies who may be called upon to manage non-compliant, challenging, threatening, or violent behaviour. The need to give staff training in physical intervention and restraint extends to many agencies including the police, prisons, the acute and mental health sectors, government agencies, night clubs, airlines, public transport, sports venues, children's homes, and elderly care facilities, among others. The use of behavioural interventions and de-escalation techniques is paramount to all programmes aimed at reducing aggression and injury. Trained physical skills need to be selected carefully in order to assure the safety of all parties involved when previous measures fail. Certain physical interventions incur additional risk of injury or death, including by positional asphyxiation, and should be avoided if at all possible. The use of behavioural and physical interventions should be captured in a reporting system in order to improve the effectiveness of training and application.

The organisational approach to managing challenging behaviour, aggression, and violence

In the United Kingdom, to be lawful, any physical intervention must be reasonable, proportionate, and necessary and as such should be integrated and incorporated into a comprehensive set of communication, behaviour management, and non-contact skills. Each organisation should have a policy dealing with behaviour management for all subjects in their care, and this should consider the specific needs and vulnerabilities of the population they need to manage. A robust use-of-force reporting system serves to identify the specific threats faced by staff and their subjects and allows the organisation to select those skills that staff are likely to need. The reporting system will also monitor the efficacy and safety of interventions, allowing the organisation and its trainers to ensure that the training syllabus is the best possible

Current Practice in Forensic Medicine, Volume 3, First Edition. Edited by John A.M. Gall and J. Jason Payne-James.

for the operational or clinical environment in which their staff operate. Skills that are not used often, fail to work reliably, or that result in excess injury are thus identified and replaced with better skills. The physical fitness and health requirements of staff expected to deliver physical intervention need to be determined and monitored by the organisation and managed through occupational health arrangements.

To minimise the use of force and reduce the risk of injury to all parties, it is helpful to adopt a strategic 'top-down' approach to the prevention and management of violence in any organisation or environment in which staff are expected to operate. This should start with an organisational policy that considers the role, purpose, and function and the subject group that staff will be expected to manage. The organisation's reporting process should identify the nature of the challenging behaviour that staff face and monitor the efficacy of behavioural and physical interventions. Identifying the type of threat through the reporting system will allow trainers to identify the skills that staff need to acquire to manage the specific threats within their operational or clinical environment. For example, the skills that a police officer requires for working in an inner city environment are likely to be far different from the skills required by staff working in a secure home for adolescent autistic patients or those working in a secure psychiatric intensive care unit.

The skills that staff require will include communication skills and de-escalation strategies that are the most appropriate for their working environment. For those working in a closed environment, the organisation should consider optimising the environment to promote cooperative behaviour and should also identify triggers and antecedents of challenging behaviour in each individual service user so that non-physical behavioural management strategies can be tailored for each and obviate the need for physical intervention.

Beyond communication, optimising the environment, and identifying behavioural and physical interventions, the organisation should identify if there is a need for restraint, seclusion, restraint devices, lesslethal weapons, or access to rapid tranquillisation. Staff need to deploy skills when necessary to maintain safety to all parties and never as a punishment or in an attempt to 'educate' service users. Doing so breaches the legal requirements of reasonableness, proportionality, and necessity that are required at law. Furthermore, inappropriate use of force may serve to psychologically damage subjects in care and irreversibly damage the therapeutic for mental health patients or those in care and is ethically unacceptable.

Having identified the intervention skill set required, the organisation should determine the minimum fitness levels for staff destined to undergo training to acquire those skills and also to be aware of the educational constraints of the staff group. In general terms, an adult learner will struggle to acquire more than six new physical skills. It is imperative therefore, that the organisation chooses only those skills that the staff member is likely to need and deploy, often at times of great stress with little warning.

Having determined the skill set required for its staff, the organisation should write a skills manual and supporting material to underpin the policy and training material. The organisation's policy and training material should be offered for legal, tactical, and medical reviews on a regular basis, supported by data from the organisation's reporting process. This level of review will serve to enhance the organisation's resilience in the event of adverse outcome, complaint, litigation, injury, or death.

Minimising the risk of injury and death

Any and all physical interventions may result in injury. Injury to the staff member, service user, or third party can occur from: slips; trips and falls; failure of the skill; escalation of violence; obstacles and hazards within the operational environment; the nature of the skill; and any specific vulnerabilities or conditions of both the staff and the service user.

Injury will inevitably occur in some operational situations where there is a need to intervene to prevent imminent violence or to terminate a violent episode. The skills selected for these situations should have the best possible safety profile in comparison to any other skills that might be executed in the same situation for the same purpose.

A number of medical conditions may precipitate unintentional aggression and violence. These include but are not limited to: epilepsy; diabetes; drug effects; head injury; sepsis; cancer; alcohol; thyroid disease; dehydration and other metabolic disorders; and a number of psychiatric and behavioural conditions. Staff require training in this area.

Staff may be vulnerable to injury in both the training and operational environments due to: individual constitution; fitness; musculoskeletal disorders; obesity; cardiovascular and neurological disorders; stature; gender; psychological vulnerability; individual personal history; physical hazards in the operational environment; pregnancy; and recent injury. They may also be injured if the operational situation escalates or if the skills fail.

Subjects may be vulnerable to injury due to: stature; gender; physical and mental constitution; age; development (physical and psychological); mental illness; special needs; recent injury; extremes of BMI; extremes of age; musculoskeletal, cardiovascular, and neurological disorders; individual previous history; physical disability; exhaustion; effects of medications; connective tissue disorders; recent injury; previous orthopaedic surgery; eating disorders; chronic lung disease; asthma; long-term use of steroids, anti-coagulants, and beta blockers; heart disease; diabetes; and epilepsy. There may also be psychological or behavioural vulnerabilities or a history of previous exposure to abuse or violence.

Restraint skills may be relatively safe at the moment of deployment, but if the subject is not adequately monitored, there is a danger of restraint-related injury or death. Service users may require restraint following excited delirium, malignant neuroleptic syndrome, serotonin syndrome, or any other form of acute behavioural disturbance. Staff need very specific training in this respect.

Some physical interventions will result in unintentional discomfort to the subject. Some organisations teach staff to deliver the so-called paincompliance skills. These are skills designed to terminate a subject's behaviour through distraction caused by the deliberate application of pain. This is more common in the policing and prison environments but is frowned upon in any therapeutic environment. Police officers are taught several pressure point skills and joint stretching techniques that will cause pain. These skills are useful in rescue from a stranglehold and other similar high-risk and dangerous scenarios. Pain compliance skills are not routinely taught to staff working in therapeutic environments.

Use of force in therapeutic environments

The use of force in any clinical setting is problematic from both the ethical and practical perspectives. Staff need to be trained to perform manoeuvres that cause the least harm to the subject while remaining effective and preventing harm to themselves, the patient, and the third parties. This training should be delivered on the background of de-escalation skills and with the understanding that use of force should be limited to these occasions where it is the only option available. Training staff to the right level and with the most appropriate skills remains a challenge (Cowman 2017).

Most literature regarding the management of violence in the clinical setting is non-empirical and does not clarify how to translate training into practice (Price and Baker 2012). The common conception is that steps that lead to violence and the de-escalation techniques are similar and divide the situations into roughly two stages: pre-agitation where verbal de-escalation, non-restrictive seclusion, and relaxation techniques are the primary interventions and agitation where help should be required, other patients should be relocated, and force may be used (Rubio-Valera 2016). There is no clear evidence that training in de-escalation techniques reliably has the desired effect, and no strong conclusions can be drawn about the reduction of assaults or injuries from these studies (Price 2015), and a Cochrane database search has found no good evidence of the efficacy of non-pharmacological interventions to contain disturbed or violent behaviour. Current practice is thus based on evidence not derived from any quality trials or studies (Muralidharan 2006) but rather from experiential learning.

When identifying specific risks to vulnerable groups the literature is also scarce. The population most often covered is elderly people who often suffer from osteoporosis, easy bruising, difficulty breathing, and reduced muscle mass (Gastmans 2006). Other populations often suffer some or all of these risk factors, but without proper evidence extrapolation is problematic.

The most severe risk is death, which is often due to asphyxiation (Berzlanovich 2012; Rubin 1993), and this risk is higher in the elderly (Gastmans 2006), small children (Rubin 1993), the obese, and those with limited pulmonary reserve and low muscle mass. These deaths often happen while the service user is mechanically restrained (Gastmans 2006; Berzlanovich 2012; Rubin 1993), but physical restraint that applies pressure to the thorax could conceivably produce the same effect. An understanding of positional asphyxia is essential.

The use-of-force hierarchy

In policing, It is useful to consider physical interventions as an incremental hierarchy starting at the lowest level with communication skills and escalating through low-level physical holding and guiding; strikes, punches, and kicks; batons; irritant sprays; handcuffs; restraint devices; conducted energy devices (e.g. Taser™); other less-lethal weapons, dogs; vehicles; and firearms (Table 14.1). As we ascend this hierarchy the risk of injury to all parties will increase. The challenge of an organisation or agency which needs to provide physical interventions training to its staff is to select those few skills that are likely to be needed, likely to succeed in their purpose, lawful, and that carry the lowest possible risk of injury to all parties.

Table 14.1 Use-of-force hierarchy for UK policing.

1. De-escalation/tactical communication from close quarters
2. Primary control skills (e.g. escorting)
3. Handcuffs
4. Physical restraint, strikes, kicks, ground restraints, etc.
5. Irritant spray
6. CED (Taser)
7. Dogs
8. Baton
9. Police vehicles as 'weapons'
10. Firearms

Organisational approaches to managing challenging behaviour and violence

The range of options currently available to police

Police officers receive training in managing violence and aggression including the use of force. The Police Personal Safety Manual (UK) is the high-level manual of guidance and exists for the guidance of chief officers in carrying out their duty to provide appropriate training and policies and for police officers and police staff who may be required to deal with conflict as part of their role.

The Police Personal Safety Manual provides general guidance on the use-of-force issues and includes a directory of techniques. Individual forces will be able to select techniques from the manual that may be required for specific policing roles.

The content of personal safety training typically includes legal issues; communication; conflict management; de-escalation; medical considerations (positional asphyxia, acute behavioural disturbance, and excited delirium); breakaway and restraint techniques, including pressure points, take-down and floor-restraint skills; strikes; edged-weapon defensive skills; team skills, vehicle skills, and SPEAR (spontaneous protection enabling accelerated response); baton use; handcuffing; mechanical restraints; and incapacitant spray.

In addition, specialist officers may be trained in the use of CEDs (Taser), dogs, cars as a weapon, and firearms. New less-lethal use-of-force modalities are evaluated by the Scientific Advisory Committee on the Medical Implications of Less Lethal Weapons (SACMILL), a group managed by the Surgeon General at the Ministry of Defence. New technologies to manage and terminate conflict are on the horizon. A recent example of this is the Bola Wrap; a Kevlar cord is fired from a hand-held device and wraps the subject from a distance. Other potentially promising technologies are under development. The driver for all of these is to immobilise a non-compliant (or violent) subject and terminate the episode with the lowest possible risk to all parties. Some of these in time will likely enter the use of force hierarchy.

In many situations it will be safer to control a person while maintaining a distance. Police officers already have options that enable them to control a person while maintaining a degree of distance, including batons, irritant spray, CEDs, dogs, and firearms.

Physical interventions in other (non-policing) environments

Prisons

The hierarchy of use of force in prisons and its underlying principles share a similarity with those of the police. Prisoners are more likely to be unarmed but may use improvised weapons and are also more likely to be violent than individuals encountered in routine policing. Security systems in prisons help separate prisoners, thus enabling the prison officers to target the specific prisoner or prisoners for physical intervention. Officers can use personal protective equipment and shields to mitigate injury in cases of riots and against prisoners armed with improvised weapons. The display of riot gear may be enough to de-escalate volatile situations.

Care homes

The population of care homes is varied in its mental and physical challenges, with patients having a wide range of medical and psychiatric conditions that can create potentially challenging situations for themselves and the staff. Patients can have severe dementia and be unable to understand or communicate with other people while retaining a significant amount of physical strength, while others can be fully aware but unable to perform the smallest tasks to aid in their own care.

Workers in care homes are not security or police personnel and are significantly more likely to be female than any other care demographic, enhancing the need for use-of-force techniques that are tailored to their needs and the needs of their patients. Care homes are unlikely to have armed or significantly dangerous residents and are more likely to have people requiring support, direction, and position adjustment. The carers are unarmed, with security staff being often far away or completely absent from the facility, and therefore, staff need to be taught appropriate techniques to achieve their goals and fulfil the needs of their patients. Focus is put on support, direction, soft restriction of movement, and skills that assist medical interventions such as placing IV lines or nasogastric (NG) tubes.

Some use of force of a higher degree may be required to restrain movement in people with severe dementia who retain physical strength or partial mobility and may injure themselves and care staff by getting out of secure beds and falling or (often unintentionally) assaulting care staff. Patients may experience delirium and require physical or chemical restraints until medical intervention can be implemented. The focus on preventing trips and falls necessitates techniques for safe body weight support and redirection, and the large amount of medical conditions necessitates even more attention to safe and pain-free joint manipulation, positional asphyxia risks, and soft restraints when necessary.

Mental health

'Psychiatric patients' is an umbrella term that encompasses a wide range of patients with mental health disorders. Despite public misconception, most of the patients pose little threat to staff and each other, and only a minority are explosive, dangerous, or aggressive.

Patients in psychiatric settings can have widely differing levels of cognition and understanding, are often under effect of psychiatric medication that may alter perception or consciousness, and are prone to trips, falls, and intentional, or accidental self-injury. Psychiatric patients have an increased risk of acute behavioural disturbance (aka 'excited delirium'), a condition that can put caretakers at risk and may be lethal to the patient unless quickly managed in an acute medical setting.

Medical interventions are much more likely to be available and utilised in a psychiatric setting owing to the presence of medical doctors and specialised pharmaceuticals. The use of medication to potentially reduce injury rate from use of force needs to be balanced against the increased risk of apnoea and aspiration due to deep sedation, while the use of physical restraint devices should be limited to short durations and acute need only.

Restraint devices used on patients for prolonged periods of time have been shown to negatively affect the parameters of well-being and the perception of the caretakers and facility. With growing scrutiny over apparent overuse of restraint devices and seclusion, it is imperative to create bespoke interventions that are tailored to specific patients and focus on cooperation, de-escalation, and physical touch that is not based on paincompliance. Individualised care plans are strongly recommended for every patient so that staff understand the triggers, antecedents, and optimal behavioural interventions for each and every person in their care. Doing so is likely to reduce the need for seclusion, physical intervention, and restraint.

In settings that house highly dangerous psychiatric patients, the use-of-force gradient can extend to the use of physical force, multi-person restraints, disarming techniques, and use of riot gear. These settings are specialised and often utilise specially trained security staff and lock-down systems not unlike those in high-security prisons.

Some psychiatric patients may be prone to rare but high-risk violent behaviours that are sudden and potentially unprompted. Contrary to public perception, these are a relatively rare occasion in regular psychiatric settings but must be considered when training the staff at such facilities. Techniques to deal with this specific threat involve breakaway and escape techniques, disengagement, and high-level physical restraint involving several members of staff. Focus must be put into quick creation and coordination of such staff teams with care to limit restraint times, use of unnecessary force, and ameliorate the risks of positional asphyxia.

Acute health

People mostly arrive to the acute health setting of their own volition and in need of care and help. They are often scared and alone or surrounded by worried loved ones. People rarely arrive to the acute care setting seeking conflict or violence. Despite this, conflict often arises due to tensions inherent to the perceived or actual acuity or severity of their presenting illness or injury, and the limited emotional and physical resources of the staff working in these settings. Poor communication skills from staff may trigger angry outbursts and occasionally violence in patients who are already having a 'bad day'.

There is evidence that a significant portion of conflicts and subsequent use of force can be prevented with improved communication, de-escalation, and rapport building with a small minority of unavoidable conflict stemming from criminal activity, altered mental states due to acute conditions of drugs, or mental disability. Hospital designers have attempted to secure the perimeter, controlling access into sensitive clinical areas, and there have been efforts to ameliorate stress and anxiety through the provision of information to patients and their visitors.

The acute care setting is the preferred location for the treatment of acute behavioural disturbance (sometimes used interchangeably with the term 'excited delirium' and use of which remains controversial), with specialised tranquillisation medication and equipment to help maintain the basic life functions of a deeply sedated patient. Treatment of such patients requires physicians to be knowledgeable in this uncommon but life-threatening condition and its potentially extreme acuity and severity.

The acute care setting often has security personnel on site for the protection of medical staff and employs panic buttons to quickly summon additional team members and security.

The gradient of use of force is different in the caretaker and security populations, with the former focusing on de-escalation and communication techniques, alleviation of fears, and rare use of low-grade restraint techniques with additional breakaway training. The latter also focuses on communication but extends to restraint, transfer, and higher grade physical techniques to protect themselves and the staff. Occasionally, victims and perpetrators of criminal altercations can arrive in the acute health setting, presenting a unique challenge to the acute health staff. The possible risk of such altercations necessitates a broadening of the range of tools that acute health security personnel employ to include non-lethal weapons and firearms, the latter being reserved for extreme circumstances and environments.

The range and risks of physical interventions

Physical interventions are used across a very broad range of settings, from care homes to maximum security prisons, and must be tailored to the needs of each organisation or setting and, where possible, to the needs of specific individual within an organisation. Physical interventions are often grouped together but can be as mild and as safe as assisting an elderly person to get out of bed or as dangerous as subduing an armed assailant and pinning him to the ground.

To understand the full range of what physical interventions are used for, we must create a gradient of resistance and response and build a model that enables escalation of physical interventions when resistance increases and de-escalation when it decreases, while enabling the use of higher degree use-of-force techniques when necessary.

Alongside communication skills, lowest on this gradient are support techniques, which focus on assisting those who are unstable due to physical illness, age, or intoxication or confused due to mental illness, dementia, or mental disability by creating a

Figure 14.1 A support technique. *Source*: Dr Tony Bleetman thegsa.co.uk.

comforting, stable, and secure support. These techniques focus on preventing the target from falling, self-injury, and wandering off, and are used on people who are largely compliant. An example would be a hand gently supporting the lower back with the second hand holding the elbow of the target individual (Figure 14.1). This enables the person performing the technique to carry part of the subject's weight in a way that does not break balance or puts strain on the back. It also enables the user to gently turn the subject to the desired location and lead them there. If the subject is intoxicated, delirious, or resisting for any other reason in a passive or non-threatening way, the user can increase the force used slightly, firmly leading by applying pressure on the lower back and holding the forearm in hold that is both secure and painless and limiting the movement of the forearm in a way that can harm the user if the situation escalates. When the subject is actively resisting and poses a risk to staff or caretakers, it is important to call for help and apply two-person techniques.

Single-person techniques are limited in scope and effectiveness against a subject who is as strong as the user or stronger and is resisting and should not be used in a situation that poses any risk to the staff member. They are occasionally used when taking care of children as the strength disparity is greater and the potential risk is lower, or as a last resort for lone operators and to contain a situation while help is on the way.

To demonstrate this, we might consider a scenario in which a person in a restaurant is belligerent towards staff and begins damaging property without responding to verbal de-escalation attempts (which should almost always be attempted before any force is used). The police are called, but trained staff can help contain the situation pending their arrival.

A single person attempting to stop the person from damaging the property will either risk being injured or use significant force, injuring the person. Bothof these outcomes are unacceptable in this situation.

Alternatively, two people approaching from the front can each restrain an arm on either side, without resorting to paincompliance, and with little risk to harm to either side. With restraint in place, de-escalation can be attempted, and removal techniques can be used to await the arrival of police (Figure 14.2).

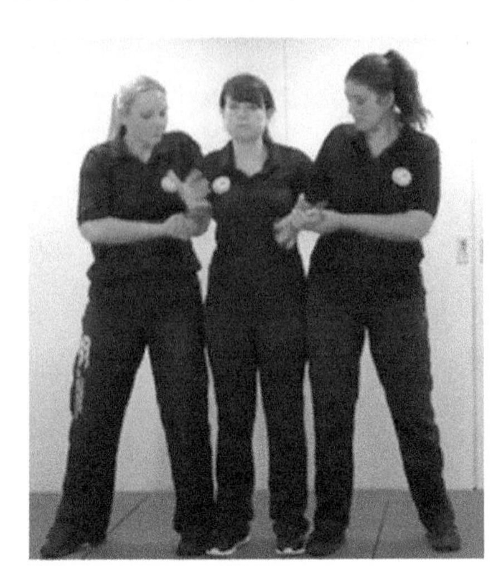

Figure 14.2 A two-person standing restraint without pain compliance. *Source*: Dr Tony Bleetman thegsa.co.uk

Should a person continue to struggle and be unresponsive to communication after being restrained, additional members of staff can assist in three- or even four-person restraints. It may be necessary to descend to the floor in a controlled fashion and hold the subject there until help arrives, with a person securing each arm, a person keeping the head safe and a person securing the legs to prevent kicking (Figure 14.3). These techniques must be taught and practiced and will not achieve their goals without appropriate training and regular refresher courses.

The correct implementation of these interventions, learning to work together as an adhoc team, and avoiding the pitfalls that can endanger all parties must be taught as part of a curriculum and practiced on a regular basis in refresher courses. Simulation training is useful.

One cannot assume that a care worker who went through a week-long course five years ago would be able to perform these interventions in a safe and effective manner. The techniques taught should be simple and easy to remember and perform if we are to expect people to perform them without regular practice. Restraint techniques

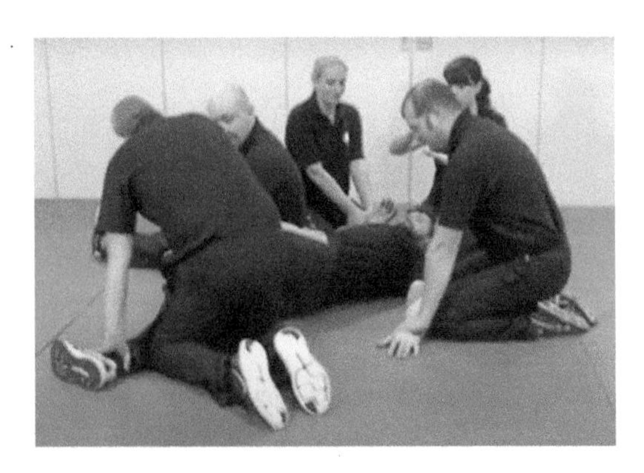

Figure 14.3 Securing a service user on the ground with two additional staff members. *Source*: Dr Tony Bleetman thegsa.co.uk

should target large joints, must utilise simple and intuitive movement patterns, and, most importantly, the fewest necessary techniques should be taught.

The human body has several physiological and anatomical limitations that we must consider when teaching or performing restraint techniques, with special attention being given to positional asphyxia, safe range of joint movement, and preventing head and face injury.

Positional asphyxia results when the ability of the chest and abdomen to inflate to allow air into the lungs is limited because of the position of the body or the outside pressure on it. Positional asphyxia risk is increased in sitting or prone bodies and further increased when bent forwards sitting or when pressure is applied to the chest or abdomen, restricting the physiological movement of the chest wall and the diaphragm. The most common cause of death due to physical restraint is asphyxiation of a person in the prone position that had some element of a restrainer's bodyweight pressing on the back.

The prone position additionally makes it harder to monitor the well-being of an individual being restrained and thus should be only used sparingly and for as short a time as possible or avoided completely. There should never be any pressure on an individuals' neck, chest, back, or abdomen, and the pressure from the sides should be minimal as well. A person restrained in the prone position should be monitored constantly for signs of asphyxia, and the position should be switched to supine or de-escalated out of as soon as feasible by transfer into a side or supine position if at all possible (Figures 14.4 and 14.5).

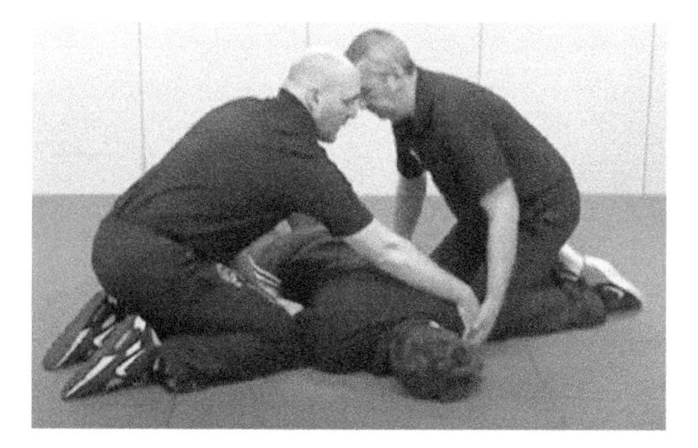

Figure 14.4 By avoiding pressure on the torso, the risk of positional asphyxia in the prone position is decreased. *Source*: Dr Tony Bleetman thegsa.co.uk

Figure 14.5 A case of prone restraint that allegedly lead to the death of the person restrained. *Source*: Dr Tony Bleetman thegsa.co.uk.

Death during restraint is not uncommon in an individual in a state of acute behavioural disturbance. Autopsy in many of these cases often fails to identify a cause of death. It is possible that these individuals succumb to progressive hypoxia and acidosis; others have postulated that these individuals develop a stress-induced cardiomyopathy triggering a fatal arrhythmia. For obvious reasons, this is an area which is difficult to research. Most agree, however, that rapid termination of violence displayed by an individual in an acute behavioural disturbance prior to rapid delivery of medical care is paramount and probably saves lives (Otahbachi et al., 2010; Savaser et al., 2013).

Joints in the human body need to sacrifice their stability to increase their range of motion, while their stability increases with size. Some joints, such as the elbow and finger joints, can only move on one axis and should never be moved beyond their range of motion or in any other axis. Joints with greater degrees of freedom, such as the wrist and shoulders, can be moved in all axes but not beyond their limits, which vary somewhat between individuals. The elbows, knees, and finger joints should never be hyperextended as this is painful and can cause lasting injury. Techniques that rely on holding fingers should be avoided due to complexity of execution in the heat of the moment and the risk of injury.

The wrist should not be bent more than 90° in any direction. The shoulders' range of motion enables a wide range of movement in all directions, although crossing the arm across the chest should be avoided to reduce asphyxiation risk. The shoulders' range of movement is also the most varied across individuals, and care must be taken when resistance is encountered. Internal rotation and backward extension of the arm increase the risk of shoulder dislocation and must be done with care.

Injury to the head is likely in descent to floor and should be controlled if descent is attempted. When transferring a restrained person through doorways, corridors, or into cars, a person should be controlling the head at all times. Such techniques should be done with no fewer than three restraining individuals.

Injury to the face can result in loss of eyesight and teeth and damage to physical appearance. Techniques should avoid the face and especially sensory areas, holding the head by its sides, back, or chin. Whenever the chin is held, breathing should be monitored due to increased risk of positional asphyxia. The teeth can be injured when a person attempts to bite another.

If a bite has connected, pulling away should be avoided to the risk of injury to the teeth of the biter and soft tissues of the bitten. The skin can break whenever any force is applied, and while the severity of injury is usually low, the minimal needed amount of force should be used to avoid unnecessary pain.

Prolonged pressure on an area can cause permanent nerve damage, so physical and mechanical restraints should always be applied for a limited time, and their necessity should be re-evaluated often (Figure 14.6).

There is an almost infinite variation of physical skills that can be used to restrain or defend against an attack, but they are mostly variations on several general moves that utilise important biomechanical principles. Those variations may not be interchangeable, as minor movements and slight changes in angle can create significant stress on joints, pain in tender spots, or impede breathing efforts.

Certain special populations might be more vulnerable to specific techniques, and their use should be avoided; for example, flexing the body forward slightly in a sitting

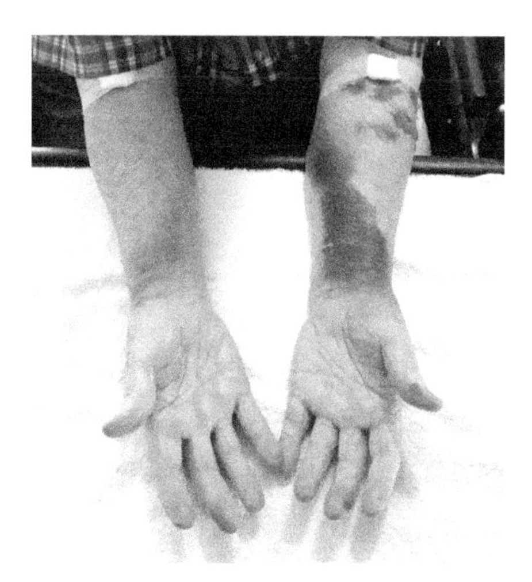

Figure 14.6 Forearm hematoma caused by mechanical restraint. *Source*: Dr Tony Bleetman thegsa.co.uk.

restraint may not limit the breathing of a healthy adult but may affect a person with severe obesity, scoliosis, or restrictive lung disease; extending the head backwards can injure a person with Down's syndrome or those with rheumatoid arthritis.

It is paramount to have the techniques reviewed and tailored to the population under care, and have a reporting system in place that allows adapting the techniques continuously, removing those that fail, are not used, or have shown to be dangerous and adding those that can deal with newly encountered situations, populations, and other needs.

Mechanical restraints

Long gone are the dark days of straightjackets and chaining patients to beds. In some situations, a violent subject may need to be restrained for prolonged periods to prevent injury. Manual (hands-on) restraint by staff members is unpleasant for all parties, demeaning and potentially dangerous to the restrained subject. In recent times, there has been a resurgence of mechanical restraints which serve to safely restrain an individual for long periods without the trauma and danger of prolonged hands-on floor groundpins. Some organisations use soft hand cuffs, soft body belts, and even modified bean bags. These are showing signs of reducing injury and obviating the need for prolonged traumatic restraints. Each organisation where restraint may be required should consider the careful introduction of mechanical restraints in situations where prolonged physical restraint may be required.

Conclusions

There are a multitude of methods of physical intervention and restraint that can be applied safely in a wide range of organisational settings. All those personnel using them must be trained to apply them, understand their risks, and use them in a planned and structured way appropriate to the relevant individual and the setting they are being applied. Lack of training and standardised approaches risks adverse effects and harms and at the extreme, but far from unknown, fatality.

Acknowledgement

The images reproduced in this chapter appear with the kind permission of the General Services Association.

References

Berzlanovich, A. (2012). Deaths due to physical restraint. *Dtsch Arztebl Int.* 109: 27–32.

Cowman, S. (2017). A descriptive study of violence management and priorities among psychiatric staff in mental health services, across seventeen European countries. *BMC Health Serv Res.*.

Gastmans, C. (2006). Use of physical restraint in nursing homes: clinical-ethical considerations. *J Med Ethics* 32: 148–152.

Muralidharan, S. (2006). Containment strategies for people with serious mental illness. *Cochrane Database Syst Rev* 19: CD002084.

Otahbachi, M., Cevik, C., Bagdure, S., and Nugent, K. (2010). Excited delirium, restraints and unexpected death: a review of pathogenesis. *Am J Forensic Med Pathol* 31: 107–112.

Price, O. (2015). Learning and performance outcomes of mental health staff training in de-escalation techniques for the management of violence and aggression. *Br J Psychiatry* 206: 447–455.

Price, O. and Baker, J. (2012). Key components of de-escalation techniques: a thematic synthesis. *Int J Ment Health Nurs* 21: 309–310.

Rubin, B.S. (1993). Asphyxia deaths due to physical restraint. A case series. *Arch Fam Med.* 2: 405–408.

Rubio-Valera, M. (2016). Qualitative study of the agitation states and their characterization, and the interventions used to attend them. *Actas Esp Psiquiatr.* 44: 166–177.

Savaser, D.J., Campbell, C., Castillo, E.M. et al. (2013). The effect of the prone maximal restraint position with and without eight force on cardiac output and other haemodynamic measures. *JFLM* 20: 991–995.

15 Medical and toxicological aspects of chemical warfare: the nature, classification, and management of chemical agents used in warfare

Stevan R. Emmett[1,2,3], *Mark Byers*[4], *and Peter G. Blain*[5]
[1] *CBR Division, Dstl Porton Down, Salisbury Wiltshire, UK*
[2] *Royal United Hospitals Bath NHS Foundation Trust, Bath, UK*
[3] *Medical School, University of Bristol, Bristol, UK*
[4] *Institute of Pre-Hospital Care, St Mary University London Great North Air Ambulance Service, London, UK*
[5] *Faculty of Medical Sciences, Translational, and Clinical Research Institute, Newcastle University, Newcastle upon Tyne, UK*

Introduction

What are chemical weapons?

One of the most robust definitions of a chemical warfare (CW) agent is that outlined by the Organisation for the Prohibition of Chemical Weapons (OPCW) – the implementing body of the Convention on the Prohibition of the Development, Production, Stockpiling, and Use of Chemical Weapons and on their Destruction – otherwise known as the Chemical Weapons Convention (CWC) (OPCW 2021). This three-part definition includes (i) the toxic effects of the chemical, or chemical precursor itself, and their action to incapacitate, cause permanent harm or death to animals or humans, (ii) the incorporation of toxic chemicals into munitions or devices, and (iii) the use of equipment specifically in connection with munitions of devices. Currently, the use of chemical incapacitants, such as CS spray (CS), by law enforcement agencies in a civilian context is permitted by the CWC.

Any toxic chemical or precursor can be classed as a chemical weapon, should there be intent for its use to incapacitate, harm or cause death. There have been numerous historical instances where individuals, groups, and non-state and state actors have deployed such approaches to achieve their aims.

The physicochemical properties of chemicals, precursors, or toxins combined with the toxicity profile and a delivery system can dictate how such agents might be deployed, whether as a chemically laced object for assassination or a large release of toxic gas for large population effects.

Current Practice in Forensic Medicine, Volume 3, First Edition. Edited by John A.M. Gall and J. Jason Payne-James.

OPCW and control and schedules

The OPCW has three schedule lists of toxic chemicals and their precursors to remove the use and encourage non-proliferation of CW agents. These schedules aid the implementation of the Convention signed by 193 Member States covering some 98% of the global population. Each schedule has specific definitions, so the chemicals or precursors can be catalogued into the appropriate section and listed alongside the CAS registration number. This chapter explores some of the more prevalent CW agents and considers the relevant medical and pathological aspects, should casualties be encountered.

Hazard/threat assessment

Recognition of chemical weapon use

The population-based *risk* (probability of harm) from CW agents is very low but is very diverse and ranges from targeting of individuals (e.g. Kim Jong-nam alleged assassination with VX in 2017) (Campbell 2017), the collateral casualties from accidental spread, accidental release (e.g. chlorine), through to terrorist mass casualty events (e.g. Aum Shinrikyo, Tokyo, 1995) (BBC News 2011) and state terrorism/warfighting (e.g. use of nerve agents (NAs) in Halabja, 1988) (US Department of State 2003).

The *threat* (capability of harm) from CW agents is different to risk and reflects the capability of a perpetrator to undertake a human/environmental (e.g. denial of ground) exposure and depends on numerous factors. These factors include the availability of precursor chemicals, the chemical agents themselves, the relative ease of synthesis, a manufacturing/preparation facility, a satisfactory delivery system, plans to deploy them, and a motivation to do so; just some examples that can establish threat.

Environmental indicators and detection overview

Should a CW agent be released into the environment, typically a non-specific acrid odour would be noticed, but for some agents there are distinct, recognisable odours that may indicate use (e.g. hydrogen cyanide: bitter almonds, phosgene: freshly mown hay, and soman: vapour rub/camphor). There may also be environmental clues to use with evidence of dead small animals or the presence of clues such as clouds of gas and presence of sticky liquids.

Some basic detection equipment can aid in environmental identification of chemical agents. There are chemical detector papers that can demonstrate the presence of NA (e.g. G and V series) or blister agent compounds (e.g. mustards, H, HD, and HT) from liquid/droplet contamination. In the forward location, where a chemical agent release may have occurred, other environmental detection is also possible with person or vehicle/vessel-mounted equipment that may utilise colorimetric assays, flame spectrophotometers, mass spectrometry, through to Raman portable spectrometers. Commercial equipment provided to government agencies and national infrastructure organisations can readily detect NAs, blister agents, blood agents, toxic industrial chemicals, and other pharmaceutical-based agents such as the potent opioids.

These detector systems based in the forward environment at the point of agent release or where casualties are witnessed are highly effective in understanding which class of chemicals may have been used and therefore bring significant benefit in establishing incident management approaches, decontamination processes, and help direct which clinical interventions may be required. The disadvantage of such forward-based equipment lies in low sensitivity and false positive alarms for other environmental chemicals.

Bioanalytical detection overview

The ability to detect and identify *biological markers* from those that have been exposed to toxic compounds serves not only to aid in the immediate and appropriate medical management of casualties but also to undertake forensic attribution for the alleged use. Bioanalytical assays may also be employed for occupational health monitoring in laboratory/industrial workers (e.g. arsenic), the military, and those entering regions of unknown risk such as United Nation/OPCW weapons inspectors (Savelieva 2021).

In the absence of clear clinical symptoms, signs, or the triggering of environmental detection, early clinically diagnostic tools may have significant advantage, so medical countermeasure therapies can be targeted, optimised, or adjusted promptly. One example is following organophosphate exposures where re-inhibition of acetylcholinesterase (AChE) has occurred in patients despite initial optimum therapy; biological monitoring may re-initiate therapies or influence dosing decisions and therefore reduce the need for escalation of clinical care. Also, where a casualty's signs and symptoms do not fit a recognised toxidrome, or the clinician may have limited experience for recognition. Similarly, where the threat picture does not raise an index of CW use, detection and diagnostic testing may be the *key* trigger for therapeutic intervention.

The physical size of the bioanalytical/diagnostic equipment, its requirements (e.g. power and reagents), ruggedness, and run-time to a result all dictate its practicality and usefulness in different situations. In the pre-hospital environment, battery-powered lightweight devices with simple instructions and result interpretation are desirable to support early clinical decision-making (e.g. Securetec). While a highly sensitive, accurate, robust, and reproducible, often large, equipment array is desirable for toxicokinetic/forensic purposes, such equipment will take much longer to generate a result, requires calibration, the use of reference standards, and maintenance to recognised international standards (e.g. ISO/IEC 17025) as the outputs may support legal cases.

The identification of toxic chemicals, precursors, metabolites, or toxin adducts broadly fall into three categories: direct compound, precursor, or metabolite detection; enzymes as biomarkers; and surrogate markers of toxic use.

Direct compound, precursor, or metabolite detection

In toxicology, the most direct route to biological analytical measurement of a chemical agent is through direct identification of the offending parent molecule. Typically, gas chromatography–mass spectrometry (GC–MS) or high-performance liquid chromatography–mass spectrometry (HPLC–MS) techniques are effective in the identification of relatively small molecules and are the gold standard since well-defined libraries and databases exist. A number of biological matrices (e.g. blood, plasma, urine, and hair) may be used when previously validated against a standard, but urine is particularly helpful since excreted toxin and metabolites do not require lengthy sample processing. Tandem mass spectrometry is highly selective and lends itself well to detection of chemical toxins and metabolites. Some toxins and metabolites, particularly those that are highly polar (ionic), are present in biological matrices for a short period of time and may require other techniques of identification (see adducts).

Enzymes as biomarkers

Blood or tissue can be used as a reservoir to identify that endogenous enzymes have been modified by the presence of a toxic agent. With respect to CW, inhibition of blood cholinesterase, particularly AChE on red blood cells, using the colorimetric method developed by Ellman (1961) is the primary example and can be undertaken in a variety of environments. These assays are diagnostically sensitive and show specificity to the class of agent (i.e. organophosphorus (OP) compound) but not the chemical compound *per se* (e.g. VX). Non-enzymatic methods are required to identify the compound. Another example where enzyme activity can identify toxic chemical use is that of cyanide inhibition of cytochrome C oxidase (Ikegaya et al. 2001), but it is of limited utility in acute exposure; metabolic acidosis (lactataemia) with anion gap is more useful.

In the case of death and autopsy following chemical toxin exposure, enzyme histochemical staining can also prove useful. Should irreversible enzyme inhibition occur in tissues, following death or biopsy, as is the case with NAs and AChE, histochemical staining for the enzyme may show unexpected low activity in certain tissues. AChE histochemical staining of rectal mucosa has been a historical means of excluding Hirschsprung's disease (increased AChE expression is seen in the extrinsic nerve fibres in the aganglionic rectum found in Hirschprung's disease), and the same principle applies. Furthermore, should fresh tissue be available, quantitative histochemical enzyme staining can be correlated to biochemical enzyme activity to support evidence of toxic exposure (Jiri Bajgar et al. 2010; Patrick et al. 1980).

Surrogate markers of toxic use

When direct identification of a chemical toxin parent molecule or metabolites is inconclusive, it can be useful to look for protein adducts of the chemical as these may be present in biological samples for a longer period of time. Using techniques such as high-resolution tandem mass spectrometry, human serum albumin adduct biomarkers can be identified to prove the use of CW vesicants such as sulphur mustard (Steinritz et al. 2016). Since protein adducts can survive for much longer as bioanalytical samples, such protein targets with a covalently bound parent toxin or metabolite can be a valuable identification route. The OP NAs (e.g. sarin, soman, and VX) also bind to albumin and have been used for adduct identification and also bind irreversibly to cholinesterase, making this protein an option (Bao et al. 2012). It has also been reported in the open literature that Novichok NAs can be identified using a nonapeptide technique and high-resolution LC–MS and applied for verification purposes by the OPCW (Noort et al. 2021). With the nonapeptide technique, butyrylcholinesterase (BuChE – also known as pseudocholinesterase) is isolated from human plasma, then enzymatically digested with pepsin, which results in a nonapeptide fragment of active site NA moiety. Presence of such a moiety indicates the clear use of a NA in a casualty's plasma (Brimijoin et al. 2016).

With an increasing role for proteomics in both research and clinical diagnostics, spheres mapping of the proteome following a toxic exposure could create a hallmark footprint with which to correlate defined CW agents. Furthermore, host transcriptomic analysis for early changes seen in gene expression could not only provide insights into the mechanisms following exposure but also identify defined gene expression patterns for specific toxic insults. Future work in this area will provide alternative routes to the identification of known and unknown toxic insults, aiding forensic attribution and support counter proliferation efforts of the international community.

Classes of chemical weapons and casualty management

Organophosphorus nerve agents

Classification of organophosphorus nerve agents

The synthesis of OP NAs dates back to 1820 with the identification of compounds such as triethylphosphate. Similar chemicals, such as those reported by Willy Lange in 1932, were noted to be "strongly aromatic in odour and cause marked pressure in the larynx and breathlessness with disturbances of consciousness and painful hypersensitivity in eyes." Prior to this, in the 1920s, during prohibition, Jamaican Ginger (Ginger Jake) had been sold as a medicine, but it contained tricresyl phosphate, an organophosphate, which led to muscle pain, weakness, and neuropathy, and affected some 20 000 people.

The OP NAs themselves were first synthesised by German chemists in the 1930s and were identified as highly toxic during early work to develop novel insecticides. Gerhard Schrader in 1934 synthesised Tabun, which, as an organophosphonate, was an extremely potent and irreversible inhibitor of AChE and was recognised to have uses outside its intended development.

The synthesis of Tabun, during the unstable lead up to the World War II, meant there was military interest in its use as a weapon. Further research, with this in mind, led to other compounds, which would latterly become defined as the G-series (German) agents, and included sarin named after *S*chräder, *A*mbros, *R*itter, and van der L*in*de (GB) and soman (or GD, EA 1210, Zoman, PFMP, systematic name: O-pinacolyl methylphosphonofluoridate). Despite the availability of these NAs during WWII, it is believed that Adolf Hitler, who had been seriously affected by mustard gas during the first war, never sanctioned their use, fearful of the consequences of allied retaliation.

During the 1950s, further pesticide development was underway in the chemicals industry, which led to the synthesis of Amiton in 1952, which was even more potent but had different biophysical properties. Details of this compound, termed VG, and an analogue VX, were explored by the UK War Office and weaponised by allied nations. Stocks were subsequently destroyed under the terms of the CWC.

In 1992, a book entitled State Secrets authored by Vil Mirzayanov reported that the Soviet state had developed "newcomer" or "Novichok" NAs. Soman (GD) is an oxime-therapy-resistant NA, and the Novichok NAs are reportedly similar but significantly more toxic than the G- and V-series agents (Nepovimova and Kuca 2018). Novichok, alongside the other known NAs, are now included in the schedule of banned chemicals covered by the OPCW CWC (Mirzayanov 1992).

Pharmacology and toxicology

Organophosphates, and the NAs in particular, are some of the most toxic chemicals known. The main toxicological effects are driven through irreversible inhibition of AChE (see Emmett and Blain 2020), but they also bind to other serine hydrolases and enzymes expressed throughout the body, although the effects at these sites are poorly understood. The B-esterase enzymes include AChE, BuChE, carboxylesterase, chymotrypsin, and trypsin (Colović et al. 2013). BChE metabolises exogenous bioactive esters in the diet, the hormone ghrelin, and various medicines. It is important in the detoxification/metabolism of NAs, making this enzyme also an effective bioscavenger removing toxins before they reach the target tissues. Other OP detoxifying enzymes (e.g. A esterases, phosphotriesterases, paraoxonases, PON1, and DFP esterases) have also been identified, while other enzymes, such as cytochrome P450, can contribute to bioactivation of some pesticides (Costa 2006).

In general terms, OPs, NAs, and their metabolites are highly lipid solubles and so are readily distributed throughout the systemic circulation and accumulate in fatty tissue such as the central nervous system (CNS). Such ease of absorption and distribution, combined with the irreversible inhibition of AChE and relatively slow enzyme synthesis, contributes to the sustained symptomatology and slow recovery observed in OP poisoning (Hulse et al. 2019; Jokanović 2009; Eddleston et al. 2008; Singh et al. 2020).

The strong binding of NAs to AChE prevents enzyme hydrolysis of the neurotransmitter acetylcholinesterase (ACh), resulting in accumulation of ACh at synapses. Rates of enzyme inhibition vary depending on the OP, with the NAs showing the greatest affinity and hence toxicity (Worek and Thiermann 2013). Once the NAs bind to AChE, they phosphorylate the serine hydroxyl group at the enzymes' active site. Following this initial phosphorylation, the enzyme remains relatively unreactive to spontaneous hydrolysis and results in persistent inhibition that may only be reversed by nucleophilic attack from therapies such as the oximes. For some NAs, following

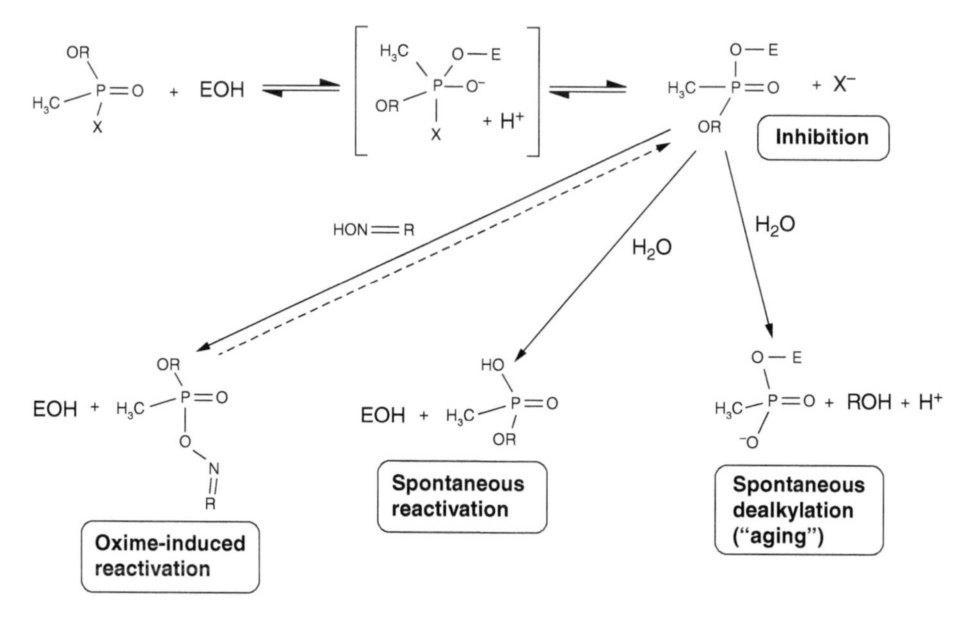

Figure 15.1 Scheme of the inhibition of AChE (EOH) by an organophosphonate and subsequent ageing steps. HON, R denotes a nucleophilic oxime. *Source*: Taken from Worek and Thiermann (2013).

phosphorylation, the enzyme may become dealkylated, resulting in the formation of an enzyme–ligand complex, termed an "aged" enzyme, which is truly irreversible. This is clinically significant as NAs, such as soman, rapidly result in a dealkylated enzyme–inhibitor complex that is not amenable to standard oxime therapies. Dealkylation enzyme ageing occurs at different rates (1–2 min through to hours), depending on the OP, NA, or the metabolite studied (see Figure 15.1) (Worek et al. 2016).

The distribution of AChE throughout the body is diverse and varies in molecular form, which can also have an impact on the OP affinity for the enzyme (Ogane et al. 1992; Taylor et al. 1994). Most AChE is expressed as a membrane-bound form, via a long hydrophilic tail, while other shorter forms may be extracellular. A readily quantifiable source of AChE is that expressed by red blood cells. However, other tissue sources, such as that of the neuromuscular junction, peripheral nervous system ganglia, and CNS cholinergic neurons, may be more representative of the physiological and functional effects of OP poisoning. The accumulating ACh in the synapse results in overstimulation of post-synaptic effector sites (nerves, muscles, or glands) and is discussed in more detail below.

Pharmacological impact of AChE inhibition

Given the systemic distribution of OPs and subsequent widespread inhibition of AChE that results in ACh excess, many cholinergic, and non-cholinergic, synapses will be affected. At the neuromuscular junction, direct motor endplate function will be altered through ACh overstimulation, resulting in prolonged depolarisation of post-synaptic nicotinic receptors (nAChR) located on striatal tissues, a phenomenon many anaesthetists will recognise as a depolarising neuromuscular blockade such as that seen with suxamethonium. Persistent stimulation of the nAChR at neuromuscular junction results in fasciculations and then paralysis as the inhibited AChE is unable to metabolise the ACh.

Upstream of the neuromuscular junction and somatic system, peripheral nervous system effects of cholinergic excess are widespread in the autonomic nervous system (ANS). Within the ANS, the two major components, parasympathetic and sympathetic, have opposing activities on the end organs. The parasympathetic system has a craniosacral output from the spinal cord with nicotinic acetylcholine receptors located at the ganglia, distally, and stimulation of these pathways results in ACh release at the effector organs, such as via the vagal nerve to the heart, bronchi, stomach, and intestines, that express post-synaptic muscarinic receptors (mAChRs). The cholinergic hyperstimulation from AChE inhibition tends to result in predominately parasympathetic effects where NA exposure results in hypersecretion (e.g. bronchorrhoea, lacrimation, vomiting, and diarrhoea), smooth muscle contraction (e.g. wheeze, abdominal pain, and cramps), bradycardia, and reduced blood pressure through increased release of nitric oxide. The parasympathetic system is crucial in the regulation of glycogen and other metabolic and endocrine pathways. Acute or chronic exposure to OPs or NAs can induce metabolic disruption (Li et al. 2019) and may result in type-2 diabetes mellitus and weight dysregulation (Lakshmi et al. 2020).

Within the sympathetic nervous system, nicotinic acetylcholine receptors (nAChRs) are found at the ganglia and in the adrenal medulla. Stimulation of these nAChR-based pathways results in the release of noradrenaline at the effector organ or adrenaline (predominantly) from the chromaffin cells of the adrenal medulla. Subsequent binding of the noradrenaline, or adrenaline, to adrenoreceptors then causes typical parasympathetic effects such as tachycardia and hypertension. The sympathetic control of many sweat glands is mediated by muscarinic cholinergic transmission, and NA poisoning results in profuse sweating.

Further upstream of the peripheral nervous system OP and NA-induced overstimulation of the muscarinic (mAChR) and nicotinic (nAChR) acetylcholine receptors in the brain can result in confusion, agitation, seizure, coma, and respiratory failure.

Historical Scenarios

Since the identification of NAs, a number of events have highlighted the toxic potential to individuals, small groups of people, and their deployment in wider state warfare.

In 1969, a US storage facility based in Okinawa, Japan, sustained an accidental release of VX, which led to the hospitalisation of 24 American soldiers (Asia Pacific Journal 2019). Other than this accidental event, it was not until 1984 that NAs were used on the battlefield during the Iran–Iraq War in which thousands of Kurds in northern Iraq died, the most notorious event being in Halabja.

In the 1990s, a terrorist group in Japan released sarin in Matsumoto (1994) and on Tokyo subway lines (1995), causing extensive civilian injuries and deaths. In addition to the Matsumoto Sarin Incident and the Tokyo Subway Sarin Attack (Okumura et al. 2005), the Aum Shinrikyo cult continued to synthesise sarin and VX from 1993 to 1995 and attempted to attack Daisaku Ikeda, the leader of a Buddhist organisation, using sarin in 1993. In 1994, Aum used VX in attempted attacks against the anti-Aum lawyer, Taro Takimoto, as well as a car park owner, and they killed an office worker in Osaka by administering VX intramuscularly. In 1995, Aum conducted a further attack by spraying and injecting VX on the chairman of the Aum Shinrikyo Cult Victims Association, Hiroyuki Nagaoka, which led to hospitalisation for several weeks.

It is, therefore, clear that since 1994 chemical weapons were no longer restricted to military applications on battlefields; rather, chemical agents can now be deployed in urban settings against normal civilians. One such example was the use of sarin in an attack on Damascus, Syria, in 2013, which killed more than 1400 people including 426 children. Sarin was also used in the 2017 Kahn Sheikhorn attack in Syria (Bellingcat 2017). VX was used to assassinate Kim Jong Nam at Kuala Lumpur International Airport in Malaysia in 2017, and several more recent assassination attempts involving Novichok, the newer class (fourth generation) of OP NAs, have also been perpetrated (Haslam et al. 2022). NAs are thus a clear and present danger to modern society, and the medical community should be mindful to consider their use in a toxicological differential diagnosis.

When the threat of NA use is high, a number of states around the world can consider instructing those at risk to take a pre-treatment, should they be faced with a release. Such preventative action, in the form of the carbamate compound pyridostigmine bromide, was deployed and used during the Gulf War of 1999. Pyridostigmine is a reversible cholinesterase inhibitor and licenced for medical use in the autoimmune disease myasthenia gravis and also as a pre-treatment for NA exposure. Pyridostigmine alone will not protect against a soman exposure but will enhance the effectiveness of post-exposure therapies such as atropine, oximes, and benzodiazepines (see below) (Marrs 2004; King and Aaron 2015; Hulse et al. 2019).

Symptoms and signs

As outlined, the classical reported toxidrome of OP inhibitors of cholinesterase reflects the increased cholinergic activity, a pro-cholinergic syndrome. The most predominant symptoms and signs are mediated through overstimulation of muscarinic receptors and reveal themselves clinically as miosis, excess secretions, increased gut motility (e.g. diarrhoea) bradycardia, and seizures. Nicotinic-mediated signs and symptoms can also be present but are less apparent (e.g. fasciculation, muscular weakness, and hypertension) and often require thorough clinical examination to be identified.

The biophysical properties of the OP or NA dictate not only the affinity for the cholinesterase enzymes, and whether chemical reversal is possible (see below), but also how they behave in the environment. The relationship between physical properties and environment (e.g. temperature) is important as this affects the predominant route of human exposure. At 25 °C, sarin is a volatile NA (22 000 mg/m^3), while VX is a liquid with the consistency of honey (10.5 mg/m^3), so their employed use will result in different exposure routes. It is noteworthy that exposure dose is a key factor affecting survivability, incapacitation, and the ability to seek early medical care, while exposure route remains a secondary consideration.

If the exposure dose is low, and casualties are ambulatory, they may report pro-cholinergic early symptoms in the form of vivid dreams, hallucinations, blurred vision, nausea, and marked changes in attention. Acute changes in multiple patients (i.e. >2) from the same geographical location should raise the index of suspicion of NA use, especially if there is a perceived threat.

The primary route of exposure is also important as it will affect the pattern of signs and symptoms early in clinical presentation until the toxin becomes ubiquitous and a global systemic toxidrome picture is seen. An inhalation exposure from a highly volatile NA is likely to initially result in rapid-onset respiratory effects including salivation, rhinorrhoea, and bronchorrhea/bronchospasm, while ingestion exposures have respiratory components with the addition of abdominal pain, cramps, and vomiting that merge into a systemic picture. Dermal exposures can be more insidious in onset with delayed but then rapid onset of systemic signs. The sensitivity of the short ciliary nerve and pupillary constrictor muscle, combined with the ease that organophosphates cross the blood–brain barrier (BBB), means that miosis is a sensitive, reliable, and prominent sign, should an exposure be suspected.

In the case of organophosphate exposures, an intermediate syndrome has been described, which usually presents 24–96 hours after poisoning but has *almost never* been reported with NAs. Clinical features include proximal muscle and neck flexor weakness which progresses to involve respiratory muscles. Figure 15.2 and the acronym consciousness, respiration, eyes, secretions, skin (CRESS) can be of help in determining whether a NA is responsible for the signs and symptoms.

A further type of rare OP and even rarer NA-induced paralysis is organophosphate-induced delayed neuropathy (OPIDN), which is predominantly a motor neuropathy characterised by limb flaccid paralysis with sensory sparing that occurs 2–3 weeks following exposure (Lotti and Moretto 2005). OPIDN is thought to occur via a proteolytic mechanism involving neurofilament cytoskeletal proteins including the phosphorylation and ageing of >70% of neuropathy target esterase (NTE) (Jokanović et al. 2011). The ageing of NTE is thought to be essential in the development of OPIDN and make NTE a useful biomarker, should clinical neuropathy be suspected.

Quick look (CRESS)					
Conscious	**Respiration**	**Eyes**	**Secretions**	**Skin**	**Other**
Unconscious seizures	Increased or decreased	Pinpoint	Increased	Sweating cyanosis	Bradycardia

Figure 15.2 Acute signs and symptoms of nerve agent exposure. *Source*: Based on Nato ATO Standard 2018.

Diagnostic aids

From the casualty perspective, cholinesterase activity measurements are by far the simplest way to aid diagnosis of OP NA exposure, and such tests are available in the pre-hospital or laboratory environment. It is not completely reliable as a number of other compounds, such as phosphine, fasciculin (a snake toxin (Green Mamba)), rivastigmine/galanthamine (drugs used in dementia), flavonoids from *Syzygium samarangense* (rose apple), and some coumarins, also inhibit cholinesterase activities. NAs have a stronger affinity for AChE, and erythrocyte expression of this enzyme is more stable than that of the circulating BuChE. BuChE has a turnover of approximately 50 days, while that of AChE is 120 days, and the former is affected by concomitant medicines use, acute infections, chronic disease, and liver and renal diseases. It should be noted that blood ChE is only a surrogate marker of exposure; generally, organophosphates preferentially inhibit plasma enzyme, while NAs inhibit erythrocyte- or membrane-bound AChE, and most casualties remain only mildly symptomatic (e.g. nausea) even at inhibitions <75–80%. However, during the Tokyo Sarin gas attack of 1994 (Nohara and Segawa 1996), there were symptomatic cases in the absence of any blood cholinesterase inhibition (Nohara and Segawa 1996).

Pathological hallmarks

Pathological clues to the use of NAs may be present in a variety of organ systems. The enhanced parasympathetic activity of cholinesterase inhibitors in the gut means that abdominal pain and diarrhoea are the commonly reported side effects of reversible inhibitors such as rivastigmine and galanthamine; this is also the situation for low-dose organophosphate or NA exposure. Enhanced smooth muscle tone tends to be more prominent in the ileum than jejunum (Mutafova-Yambolieva et al. 1993), an observation consistent that a gradient of cholinesterase activity exists along the GI tract (Nowak and Harrington 1985). Bowel intussusception has been reported following cholinesterase inhibition in animals and humans, but it is very rare.

The seizure activity reported following NA exposure is best understood following soman exposure and is characteristically clonus, which progresses to tonic–clonic activity and, as such, results in significant incapacitation. Left untreated, or with slow access to therapy, neuropathological damage occurs through glutamatergic excitotoxic mechanisms, most evident in the hippocampal pyramidal neurons.

Sarin and soman (de Araujo Furtado et al. 2012) can cause breakdown of the BBB in sensitive areas such as the thalamus, and this disruption likely plays an important role in the NA-induced cell death in sensitive brain regions. However, other authors (Grange-Messent et al. 1999) have indicated that soman-induced seizure increases the number of endothelial vesicles in the absence of structural changes in the endothelial tight junctions, suggesting that it is not an essential mechanism for BBB dysfunction, but vascular leakage could occur through a limited number of damaged tight junctions.

Management

Early decontamination is key following NA exposures to reduce the toxic load and remove any potential sources that may be providing continued off gassing (e.g. sarin-contaminated clothing) or ongoing absorption through skin (e.g. VX-contaminated skin). Removal of gross contamination or liquids may be achieved through dry decontamination process with substances such as Fuller's earth or active wet products such as reactive skin decontamination liquid (RSDL).

Basic and advanced life support measures, in personal protective equipment, should be the mainstay of casualty management following suspected NA exposure. In those with limited training this may be placing the casualty in the recovery position and administering an autoinjector countermeasure (see the following discussion). For those clinically experienced, clearing and maintenance of an airway, with the use of adjuncts (e.g. Guedel airway), should be prioritised alongside administration of immediate pharmaceutical interventions (e.g. intramuscular (IM) atropine, ± an oxime, and ± benzodiazepine).

Within the UK military, a forwardly deployable triple therapy autoinjector is available for self- or buddy–buddy administration. This contains 5 mg of atropine for its anticholinergic effects, pralidoxime for its AChE enzyme reactivation properties (see Figure 15.3), and avizafone, a prodrug of diazepam for its "neuroprotective" anticonvulsive effects. Commercial autoinjectors for NA therapy also exist in a variety of drug combinations and forms from single chamber atropine autoinjectors (Atropen) to dual therapy autoinjectors (e.g. DuoDote and Emergard) with adjuvant benzodiazepine devices (e.g. Seizalam and CANA). The aim of the combination therapy is to administer an antimuscarinic, an oxime, and an anticonvulsant.

Following immediate therapy with autoinjectors, higher echelon care can be more finessed through intravenous (IV)/intraoral (IO) administration of further NA therapies and supportive measures (e.g. fluids, inotropes, anaesthetics, measures to aid biological clearance, and neuroprotective drugs).

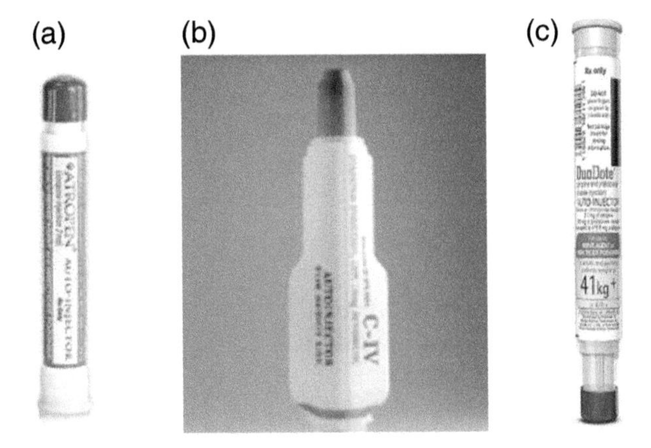

Figure 15.3 Example of nerve agent autoinjectors. (a) Atropen™, (b) CANA™, ™diazepam and (c) Duodote™, ™atropine and 2-PAM.

Atropine

Atropine is a nonselective competitive antimuscarinic compound that exerts its majority of benefit peripherally via muscarinic antagonism of NA-induced rise in ACh. As a competitive antagonist, dose escalation may be required to overcome muscarinic toxidromic signs such as bronchorrhea and bronchoconstriction and bradycardia. Dosing level, interval, and route can be adjusted to attain stable clinical parameters (e.g. a heart rate of >40 or reduction in secretions). Atropine also has early benefits in reducing NA-induced seizure.

In addition to atropine titration, s.c. hyoscine hydrobromide (scopolamine) can also be administered and brings the advantage of improved transport across the BBB. In NA intoxication, centrally acting antimuscarinics have been shown to lower seizure thresholds, cease seizures (McDonough et al. 2021), and may also support long-term neuroprotection. Alongside atropine, hyoscine should be used with caution, particularly in protracted use (days) and in the awake casualty, as it may increase agitation, anxiety, and hallucination.

Benzodiazepines and other neuroprotection

Acute NA exposures may result in focal or generalised seizures, and after a prolonged period or high exposure dose status epilepticus may occur, and in the absence of electroencephalogram (EEG) these may become sustained, resulting in neuronal damage or neuropathies. The use of benzodiazepines at the point of injury and known NA exposure may prevent long-term consequences. Both diazepam and midazolam have been shown to provide improved efficacy in acute animal models. Clinically, acute doses can be weaned over a number of days to maintain a degree of neuroprotection. Other pharmacological classes can also provide beneficial neuroprotection effects including the *N*-methyl-D-aspartate receptor (NMDA) antagonists and non-NMDA glutamate receptor antagonists, especially levetiracetam with its combined glutamatergic and GABAergic (gamma-aminobutyric acid) effects.

Other nerve agent therapies

Other therapies include bioscavengers, which bind to NAs in the systemic circulation, much like an antitoxin, reducing and removing the toxic load reaching AChE tissue effector sites. These compounds fall broadly into three categories: stoichiometric, pseudo-catalytic, and catalytic (Lenz et al. 2007). Butrylcholinesterase, purified or expressed, is stoichiometric and has been used in limited form as a prophylactic, or post-exposure therapy, to very good effect across a broad range of NA challenges.

Pulmonary agents: chlorine and phosgene

Classification

The pulmonary agents, otherwise known as lung-damaging agents, or more traditionally, "choking agents," are a group of CW agents which include phosgene (CG), diphosgene (DP), chlorine (CL), and chloropicrin (PS). Phosgene accounted for 80%

of all chemical fatalities in World War I, although chlorine was the first to be used (on 22 April 1915 at Ypres).

Pharmacology and toxicology

Pulmonary agents are gases that affect the respiratory tract. They are soluble in aqueous solution with their solubility determining which part of the respiratory tract they tend to injure; the less soluble affects the more distal airways.

The highly soluble agents, such as ammonia and formaldehyde, tend to affect the proximal respiratory tract, with the intermediately soluble agents such as chlorine mainly affecting the bronchi and bronchioles. The relatively insoluble agents such as phosgene, oxides of nitrogen, and perfluoroisobutene (PFIB) (see the following section) tend to penetrate deep into the peripheral lung tissue and are deposited within the alveoli.

On deposition, the agents dissolve in the tissue water; chlorine undergoes acylation and hydrolysis, leading to the formation of hypochlorous and hydrochloric acids as well as oxygen-free radicals. This causes cell damage to the alveolar-capillary membrane, interstitial fluid leakage, and the development of pulmonary oedema.

Phosgene (COCL) is relatively insoluble in water and aqueous solutions, which accounts for the delay between exposure and symptoms of around 24 hours, often preceded by a period of euphoria. However, once dissolved, it also undergoes acylation and hydrolysis, leading to the formation of carbon dioxide and hydrochloric acid.

The LCt_{50} of phosgene is believed to be around 3200 mg min/m^3, compared to the LCt_{50} of chlorine of 6000 mg min/m^3.

Historical scenarios

Chlorine is a ubiquitous chemical widely available due to its use in sanitation, disinfection, and antisepsis. There are over 15 000 commercially traded chlorine compounds, and chlorine is readily obtainable from industrial sectors. Although one of the first CW agents used in the modern era, it continues to be a weapon of choice, given its ease of production, distribution, and challenges in attribution. In the Iraq conflict between 2003 and 2011, chlorine-improvised explosives devices were regularly used by insurgents, and during the more recent Syrian War, all sides were accused of using chlorine as a weapon, in direct contravention of the CWC 1992. Following investigations by the OPCW, the Syrian Air force were accused of dropping chlorine barrels from helicopters, which on bursting spread chlorine gas across a wide area, causing a large number of casualties and deaths.

Environmental detection

Chlorine can be detected by its familiar odour, as in bleach, whereas phosgene has an odour of freshly mown hay. Chlorine and phosgene detectors are readily available, but whilst these may be present in industrial settings, they are unlikely to be readily available in the event of a terrorist release.

Table 15.1 Symptoms and signs of chlorine exposure

Symptoms and signs
Blurred vision
Burning pain, redness, and blisters on the skin if exposed to gas. Skin injuries similar to frostbite can occur if it is exposed to liquid chlorine
Burning sensation in the nose, throat, and eyes
Coughing
Chest tightness
Difficulty in breathing or shortness of breath. These may appear immediately if high concentrations of chlorine gas are inhaled, or they may be delayed if low concentrations of chlorine gas are inhaled
Fluid in the lungs (pulmonary oedema) that may be delayed for a few hours
Nausea and vomiting
Watering eyes
Wheeze

Symptoms and signs

The immediate signs and symptoms of dangerous concentrations of chlorine exposure may include a transient sensation of burning in the eyes, with lachrymation, a feeling of chest tightness, shortness of breath, and a cough secondary to interstitial fluid build-up (Table 15.1). The main clinical feature of most exposures is that of pulmonary oedema, with symptoms occurring after a variable latent period of between 20 minutes and 24 hours; this is dependent on the dose, the solubility of the gas, and also the physical activity of the casualty. Areas of patchy emphysema, atelectasis, and oedema of perivascular connective tissue precede the development of profound pulmonary oedema with death typically occurring between 24 and 48 hours after exposure, if the casualty does not recover.

High-dose exposures, which would be likely in a deliberate release, may result in the immediate onset of laryngospasm, reflex bronchorrhea and cough, and hypoxia progressing to respiratory arrest. It is also possible to see chemical burns, should there be aqueous solutions in the environment; such burns contribute to morbidity and mortality through interstitial fluid shifts.

There are no reliable biomarkers, and the measurement of free chloride ions is not reliable. Chest X-rays may show signs of chemical pneumonitis (Figure 15.4) that develop over time, but in the absence of medical intelligence of an exposure, this is a non-specific finding (Zellner and Eyer 2020).

Pathological hallmarks: At autopsy, cardiomegaly has been noted on a number of occasions alongside expected pulmonary oedema and vascular congestion.

Management

Treatment of chlorine exposure is generally supportive. Humidified oxygen should be used to treat all casualties initially, and β-adrenergic agents can be used to benefit patients with chest tightness and wheeze. Glucocorticoids show some benefit, although there

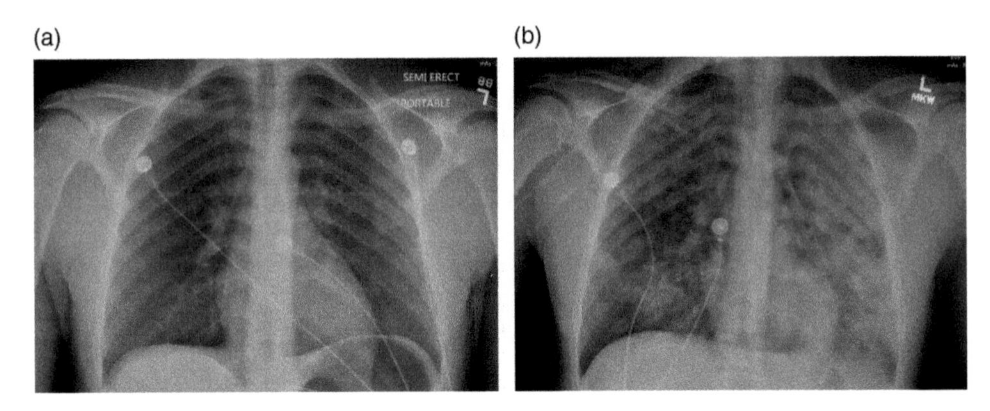

Figure 15.4 X-ray changes of chemical pneumonitis from inhalational chlorine exposure. (a) Initial point of injury, (b) 24 hours post-exposure. *Source*: From: https://learnem.org/inhalation-injury/ Accessed 21 December 2021.

are limited large-scale trials of these therapies (Zellner and Eyer 2020). There is some evidence from animal data that the leukotriene antagonist montelukast is also beneficial in treating severe exposure (Hamamoto et al. 2017).

Phosgene is widely used in industry, manufactured by passing purified carbon monoxide and chlorine gas through a bed of activated charcoal. It is used in the manufacture of isocyanates, the precursors of polyurethanes and polycarbonates.

Patients exposed to phosgene may show few symptoms initially but should be treated as high-priority casualties. They should not exert themselves and need observation for at least 24 hours as exertion leads to rapid deterioration and even sudden death. Inhalation of high doses of phosgene (greater than 150 ppm, a potentially lethal dose) causes immediate tissue changes, although there is a clinical latent period. In WW1, phosgene-exposed individuals complained of cigarettes tasting foul before developing clinical effects.

On initial exposure to high concentrations of phosgene, a vagal reflex response may cause shallow respiration and reduced ventilation. This is quickly followed by the latent phase which may last a number of hours. During this phase, there is interstitial swelling in lung tissue, and there may be a rise in the haematocrit. As the pulmonary oedema develops, the casualty begins to develop tachypnoea, tachycardia, dyspnoea, and cyanosis. In very large exposures, sudden death occurs due to occlusion of the pulmonary circulation following intravascular haemolysis and thrombosis.

The aim of the treatment is to block the inevitable inflammatory cascade during the latent period. Casualties should receive non-steroidal anti-inflammatory drugs, steroids, bronchodilators (if symptomatic), and montelukast and *N*-acetyl cysteine by inhalation, which has also been shown to be effective in small studies.

Patients who develop pulmonary oedema will need ventilatory support, some requiring high inspired fractions of oxygen. Initially, continuous positive airway pressure (CPAP) may be effective, but with phosgene, early intubation and a ventilation strategy based on those employed in Adult Respiratory Distress Syndrome are recommended. Pressors and inotropic support may also be required. Other therapies that have been advocated include prostaglandin E1, a surfactant, antihistamines, asparaginase, calcium, atropine, anticoagulants, e-amino caproic acid, urease, hypothermia, and extra-corporeal oxygenation (Pauluhn 2021).

Asphyxiants: cyanide and hydrogen sulphide

Classification

The asphyxiant agents (particularly cyanide, cyanogen chloride, and hydrogen sulphide) were originally classified as "blood agents" but are in fact cellular asphyxiants or respiratory chain poisons. The blood agents are non-persistent agents that poison the intracellular mitochondria.

Hydrogen cyanide is colourless and has a smell of bitter almonds. It is extremely volatile and lighter than air, whereas cyanogen chloride is heavier than air and has a pungent pepper odour. Hydrogen sulphide is also heavier than air and smells of rotten eggs but at toxic levels quickly causes olfactory fatigue and loss of smell.

The asphyxiants are released as gases. Following inhalation, they rapidly enter the blood stream, where they are redistributed to the cells and exert their effect by causing cytotoxic anoxia within the mitochondria. Poisoning is caused by the impairment of oxidative phosphorylation, due to the inhibition of cytochrome oxidase that includes cytochromes a and a_3 (Complex IV) by the asphyxiant agents which have a greater affinity for the ferric ion found in the haem moiety of the oxidised form of this enzyme. The resulting chemical combination results in loss of the effectiveness of the enzyme, and tissue utilisation of oxygen is inhibited with rapid impairment of the vital functions.

Historical scenarios

Cyanide was first used on the battlefield by the French in 1915, when a release of 4000 tons occurred without much effect, probably due to the chemical's volatility. They were more successful when they introduced cyanogen chloride the following year. Cyanide was also allegedly used by the Japanese in the World War 11 and the Iraqis on the Kurds in the 1980s. However, due to its ease of production, it is more likely to be encountered as a terrorist weapon when released in a confined space. The terrorist "Mubtakar" ("invention") device is relatively easy to make and use, and the design was widely shared across terrorist publications. The doomsday cult Aum Shinrikyo attempted to release a hydrogen cyanide weapon in the Tokyo Shinjuku Subway Station in May 1995, and an Al-Qaeda plot, in 2003, to release hydrogen cyanide into the New York Subway was disrupted.

Cyanide and hydrogen sulphide gas detectors are common in industries where the products are used, and cyanide is regularly encountered in housefires, but the diagnosis of poisoning remains challenging as there is no rapid clinical test. Serum lactate greater than 8 mmol/l is 94% sensitive and 70% specific for cyanide toxicity when there is clinical suspicion of exposure. Whilst some labs can measure cyanide levels, they are rarely available in time to assist in clinical management. Similarly, blood and urine tests for hydrogen sulphide are unlikely to be clinically useful. Diagnosis relies on a good clinical history and an index of suspicion.

Symptoms and signs

Cyanide and hydrogen sulphide are absorbed through the respiratory tract and mucous membranes. For cyanide, signs and symptoms begin at blood concentrations of approximately 40 mol/l. After exposure, there is an initial deep inspiratory breath (gasp) followed by tachypnoea. At low dose, casualties complain of headache, dizziness, and shortness of breath. They are tachycardic. With further exposure, casualties develop seizures, bradycardia, hypotension, loss of consciousness, and death. Onset of symptoms is extremely rapid.

Hydrogen sulphide disassociates in blood to hydrosulphide and is distributed to the brain, liver, kidney, pancreas, and small intestines. Hydrogen sulphide also irritates the respiratory tract, causing rhinorrhoea, sore throat, wheeze, chest tightness, haemoptysis, and severe dyspnoea. The biological half-life of hydrogen sulphide in tissue is very short, and hydrosulphide is rapidly cleared, so the effects of exogenous hydrogen sulphide are brief once exposure is terminated. In high dose, it causes similar symptoms to cyanide with only a few breaths leading to unconsciousness, seizures, coma, respiratory paralysis, and death.

The body is well adapted to manage low-dose cyanide exposure. Cyanide detoxification mainly occurs via the rhodanese enzyme system, which is abundant in the liver. Thiosulfate reacts with cyanide, catalysed by rhodanese, to produce thiocyanate which is excreted in the urine.

Management

Initial management involves removal from exposure and early administration of high-flow oxygen. Inhalation patients who are still breathing five minutes after termination of exposure are likely to survive, whereas those who develop respiratory arrest after massive exposure are unlikely to receive treatment in time.

Active therapies include the use of sulphur donors (sodium thiosulfate), methaemoglobin formation using sodium nitrite, and cobalt donors that react with cyanide to produce cyanocobalamin (Hendry-Hofer et al. 2019).

In cyanide poisoning, sodium nitrite and sodium thiosulfate are usually combined, although this is unnecessary in hydrogen sulphide poisoning. Sodium nitrite is rapidly effective, whilst sodium thiosulfate is very slow acting but is generally safe. In sodium nitrite methaemoglobin therapy, the nitrite oxidises the red blood cell haem groups to form ferric methaemoglobin from ferrous haemoglobin. Cyanide has a high affinity for ferric iron, which binds any circulating free cyanide. Although the cyanide has a greater affinity for cytochromes a and a_3, the large amount of ferric iron in the blood counteracts this effect. The methaemoglobin will release the cyanide slowly to be detoxified by hepatic rhodanese. However, sodium nitrite should not be used in the management of cyanide poisoning due to fire smoke inhalation as the combination of methaemoglobinaemia and carboxyhaemoglobin (from carbon monoxide) can severely reduce the oxygen-carrying capacity of the blood and could lead to profound and dangerous hypoxia. Nitrites can also cause significant hypotension and even cardiovascular collapse.

Dicobalt edetate and hydroxocobalamin are cobalt donors which preferentially bind cyanide to form cobalt cyanides and cyanocobalamin, respectively; these are cleared by the kidneys. Whilst 1 mole of a cobalt salt should combine with 6 moles of cyanide, hydroxocobalamin will only bind 1 mole of cyanide per mole of hydroxocobalamin.

Adverse effects of hydroxocobalamin administration include transient hypertension (a benefit in hypotensive patients), reddish brown skin, mucous membrane and urine discoloration, and rare anaphylaxis and anaphylactoid reactions, whereas the adverse effects of dicobalt edetate include hypertension, tachycardia, retrosternal chest pain, sweating, facial and laryngeal oedema, vomiting, urticaria, and a feeling of impending doom.

Ideally, these antidotes should not be given to non-cyanide poisoned individuals. Finally, whilst studies have shown dicobalt edetate to be effective at preventing death in a mini-pig model, hydroxocobalamin failed to prevent death even at the minimum lethal dose of cyanide. Unfortunately, dicobalt edetate is no longer manufactured, so alternative new compounds are being evaluated as replacement antidotes.

Blistering agents/vesicants: sulphur mustard chlorine and lewisite

Classification

The blistering agents, otherwise known as vesicants, are a group of CW agents which include sulphur mustard (HD), nitrogen mustard (HN), lewisite (L), and the halogenated oximes such as phosgene oxime (CX). The latter is not covered here as the properties and effects of CX do not align with that of the other vesicants. The vesicants result in acute ocular, respiratory, and dermal effects placing casualties in the medical treatment chain with good survival rates but significant acute and chronic clinical morbidities.

Pharmacology and toxicology

Sulphur mustard is a colourless liquid, has a melting point of 14 °C, and gives off a colourless vapour that is heavier than air, with a volatility of 910 mg/m^3 at 25 °C, and in the presence of impurities may smell of garlic. Substitution of the sulphur group with nitrogen led, in 1935, to the synthesis of three nitrogen mustards,[1] all with the similar property of being strong alkylating agents, HN2 being the first chemotherapeutic compound. These highly reactive alkylating agents produce their effects by covalently linking an alkyl group (R-CH_2) to chemical species in nucleic acids and proteins. These compounds typically have two groups capable of reacting with, and breaking, DNA such that DNA replication enzymes are altered, including the activation of poly-adenosine diphosphate (ADP)-ribose polymerase (PARP), promoting cell senescence or death. Abnormal DNA processing has implications for the chronic development of cellular mutations and malignancies, but the exact mechanism of the acute clinical presentation following HD exposure is poorly understood; it may include guanine cross-linkage generating a cytotoxic cascade resulting in glutathione depletion and lipid peroxidation. Often, cells with high turnover are those most affected by vesicants, so at-risk areas include skin, intestine, and the haematopoietic system. Bone marrow dysplasias are reported as long-term outcomes in Iranian casualties and, more recently, lung cancers.

Dose, period of exposure, and environmental conditions dictate the toxicological effects of mustard exposure with mild acute effects (e.g. erythema) evident at vapour concentrations of 50–100 mg/min/m^3 or liquid exposures of 10–20 mcg/cm^2. In liquid exposures >100 mg/kg, death is likely to occur. Mustards can contaminate water and food being absorbed via inhalation, ingestion, and skin routes, particularly in regions that are moist, wet, or sweaty, thus affecting the eyes and the respiratory and skin systems.

Lewisite agents are seen as more toxic vesicants being dark oily liquids with rapid onset of effects. Synthesised in 1908 and weaponised during 1918, these agents were never used during WW1. Lewisite has been weaponised with mustard to reduce the freezing point of mustard, facilitating efficient ground dispersal and aerosol spraying. The combination (H/L) may have been deployed by Japan between 1937 and 1945, but this has never been corroborated. With the odour of geraniums, detected by smell at 20 mg/min/m^3, Lewisite has an LD_{50} of 30–50 mg/kg as a liquid and 1500 mg/min/m^3 as a vapour. As a rapid acting arsenical compound that is highly lipophilic, it penetrates mucosal surfaces to affect carbohydrate metabolism through the inhibition of pyruvate dehydrogenase and other enzymes while also reacting strongly with glutathione, inducing lipid peroxidation, cell membrane disruption, and cell death (DOI: https://doi.org/10.1016/j.ajpath.2016.06.012).

Environmental detection

In addition to the environmental clues that vesicants have been used, such as odour, detector papers can identify the presence in the liquid phase at the point of release. Ion mobility spectrometry, gas chromatography, and mass spectrometry are efficient detectors for the identification of vesicants from environmental samples in media such as wastewater, soil, and debris (Wooten et al. 2002). For volatile compounds, detection is possible via air-flow injection (Aldstadt et al. 1997) or commercially available Draeger TM tubes.

1 *N*-ethyl-2,2′di(chloroethyl)amine (HN1), *N*-methyl-2,2′di(chloroethyl)amine (HN2), and 2,2′,2″tri(chloroethyl)amine (HN3).

Historical scenarios

Sulphur mustard, first synthesised in 1822, was used initially at Leper (Ypres), Belgium, in 1917 and was responsible for 88% of chemical casualties sustained by the British. Most casualties exposed to HD had skin lesions, and the vast majority also has had eye involvement with three quarters presenting with airway symptoms as well. Use in warfare occurred more recently during the Iran–Iraq war, particularly in Halabja in 1988, and in 2016, the IHS Conflict Monitor reported that Daesh had used chemical weapons at least 52 times in Iraq and Syria since 2014, including both chlorine and sulphur mustard agents (*Islamic State Used Chemical Weapons at Least 19 Times around Mosul Since 2014, IHS Markit Says*, https://www.businesswire.com/news/home/20161122005494/en/Islamic-State-Used-Chemical-Weapons-at-Least-19-Times-around-Mosul-Since-2014-IHS-Markit-Says). US military and Kurdish militias were targeted (amongst others), suggesting that Daesh had sought to exploit asymmetric advantage by employing CW agents.

Symptoms and signs

Characteristically, sulphur mustard symptoms and signs do not emerge for several hours following an exposure developing to a peak after a number of days. Typically, early emergence of eye pain occurs within one hour that may be associated with nausea and vomiting and progressive respiratory tract complaints (such as hoarse voice, painful throat, and cough). Dysphonia was the earliest sign of HD exposure in Iranian casualties of the Iran/Iraq War. Skin signs and symptoms demonstrate an even clearer period of latency, much like moderate acute sunburn, with the delayed development of erythema (+/–oedema and pruritus) at 2–48 hours that progresses to the formation of small vesicles that subsequently combine to form larger fluid-filled blisters. Depending on the dose, location, environmental conditions, and casualty phenotypes, the extent of severe blistering varies, but generally systemic fluid shifts are less significant than those of thermal burn. The fluid-filled blisters are delicate, friable, and prone to rupture, but their contents are non-toxic/vesicant due to the widespread systemic uptake of vesicant. Such systemic absorption presents as gastrointestinal disturbance with pain, nausea, vomiting, and diarrhoea, contributing to a developing hypovolaemia in the absence of treatment alongside localised signs.

Much like miosis in NA exposure, the eye is extremely sensitive to HD exposures, presenting early as pain, a sensation of foreign body, excess lacrimation and conjunctivitis which progresses to photophobia, reduced acuity, and temporary blindness lasting 12 hours or so. Chronically and pathologically, there may be evidence of mustard gas keratitis, persistent and multiple corneal erosion/scarring with amyloid deposition.

The early respiratory tract symptoms of dry cough and painful throat result in a hesitancy to cough and diminished respiratory effort that combined with the lower airways direct damage due to mustard leads commonly to a productive bronchopneumonia.

Due to the highly reactive nature of sulphur mustard within biological tissues and fluid, bioanalytical detection of the parent molecule or its reactive metabolites is challenging, although species may be present in urine, on hair, skin, or clothing. Sulphur mustard metabolite thiodiglycol (TDG), and its oxidation product thiodiglycol sulfoxide (TDGO), can, however, be identified in urine using LC–MS and GC–MS (Xu et al. 2014, 2020; Nie et al. 2014).

Using modern techniques, the reactive species to which sulphur mustard has bound, namely DNA or proteins adducts, provides an alternative and more stable source of biological material for attribution. Sulphur mustard–DNA adducts such as N7-HETEG, Bis-G, and N3-HETEA may be present for weeks to months, whilst adducts with haemoglobin or albumin (e.g. a stable-hydroxyethylthioethyl [S-HETE] adduct) have half-life of detection equivalent to the clearance of the parent protein.

Clinically, Lewisite differs substantially from sulphur mustard, causing severe pain immediately at the time of exposure to eyes, skin, and airways. Local effects occur rapidly while the agent is rapidly absorbed to produce systemic dysfunction. The presence on skin generates a region of erythema with blister at the centre that extends outwards to become fluid-filled bullae containing toxic arsenic metabolites. Full thickness burns occur extending into the connective tissue associated with considerable vascular damage and the development of tissue necrosis. Skin pigmentation changes are less likely to be seen than with HD, and wounds tend to heal more rapidly with less chance of secondary infection. Counterintuitively, severe eye pain can be protective but does not avoid widespread conjunctival oedema, lid swelling, closure, and inflammation of cornea with iritis.

While the airway effects are similar to mustard, direct necrosis may lead to pseudomembranous formation and acute obstructions with concurrent generation of pulmonary oedema. In conjunction with systemic distribution of the agent, nausea and vomiting, abdominal pain, and bloody stools may be reported following a post-exposure latency period (15–30 min). With high doses “Lewisite shock” may occur from increased capillary permeability, resulting in intravascular fluid loss, hypovolemia, and potential organ hypoperfusion and failure.

Pathological hallmarks: Acute nuclear swelling within basal cells can be seen 3–6 hours following an exposure alongside liquefaction necrosis of epidermal basal cell keratinocytes. As wounds heal, transient hyperpigmentation is seen in healing deep mustard burns as melanocytes respond, reflective of post-inflammatory changes within the epidermis.

Management

Early decontamination by removal of clothing and application of an absorber of gross contamination, such as Fuller’s earth, or decontamination liquid, such as RSDL, are effective ways to reduce the effects of HD exposure. These should be applied as quickly as possible to avoid systemic absorption or oily spread of the agents across skin. Most mustard exposures will not result in death, but a spectrum of disease is seen from localised erythema through to systemic respiratory disease/oedema hypovolaemia fluid losses, infection, and organ failure. To date, there is no specific antidote for HD, and acute pharmacotherapy is supportive and should be tailored to the degree of clinical presentation. Skin lesions and eye symptoms are painful, so adequate analgesia is important alongside reassurance and psychological support as many will be anxious about blindness. Immediate and copious irrigation of the eyes may help to lessen ocular oedema and corneal damage alongside the use of lubricants and antibacterial dressings or ointments to lessen the risk of secondary infection. There may be a role for topical steroids.

Skin lesions are itchy and painful, so the use of topical agents, such as calamine lotion, or steroids, such as beclomethasone, can help alongside protection from secondary infection using silver sulfasalazine cream on lesions. Generally, small blisters do not require deroofing, while larger lesion evacuation remains controversial, it may help to accelerate healing and progression to sterile dressing. Should full thickness chemical burns exist, dermabrasion or laser debridement may help cellular healing and revascularisation. With an extensive burn, skin grafting may be required.

If acute airway signs are seen, oxygen therapy, steroids (e.g. nebulised dexamethasone), and adrenaline, in the presence of stridor, may be of benefit. Widespread respiratory tract inflammation is likely to progress through the acute phase into a chronic dysfunction, such as chronic obstructive pulmonary disease (COPD). This early lung inflammation risks secondary infection and may be managed with steroids (e.g. nebulised beclomethasone/oral prednisolone/IV/IM hydrocortisone) alongside bronchodilators and antibiotic prophylactic cover. Chronic structural lung changes occur in those previously exposed to HD, and they develop significant comorbidities, COPD, bronchiolitis obliterans, pulmonary fibrosis, and pulmonary hypertension, where interventions including recombinant tissue plasminogen activator (rt-PA), interferon gamma-1b, and phosphodiesterase-5 inhibitors have respective benefits (see Rafati-Rahimzadeh 2019).

Treatment of lewisite exposure can be undertaken with British Anti-Lewisite (BAL; dimercaprol) which reduces the amount of direct tissue damage but is unpleasant causing tachycardia and hypertension and requires an IM injection regimen of four hourly injections for two days with dose reductions out to 10 days. Alternative treatments including Succimer (dimercaptosuccinic acid (DMSA)) and Unithiol (2,3-dimercaptopropanol-sulfonic acid (DMPS); Dimaval) are water-soluble alternatives but less effective than BAL. Historically, topical BAL was available, and when applied rapidly this can prevent vesication, but this is no longer manufactured.

Other chemical warfare agents

3-Quinuclidinyl benzilate (BZ)

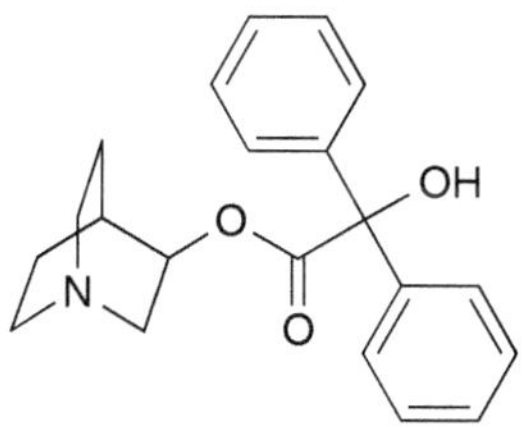

Pharmacology and toxicology

BZ is white crystalline powder with a bitter taste and no odour. It is an ester of benzilic acid and a derivative of quinuclidine and is stable in organic solvents, water, and dilute acids but not in alkaline solution. BZ has an environmental half-life of over a month and persists on surfaces.

The compound was developed by Hoffman-LaRoche as an anti-spasmodic agent for treating gastrointestinal diseases. It is an anticholinergic agent, similar to atropine, and primarily a muscarinic acetylcholine receptor antagonist. BZ is effective by oral, inhalational, and parenteral routes and, bound to plasma proteins, widely distributed throughout the peripheral and CNS. It easily crosses the BBB. The widespread muscarinic antagonism, and high-affinity binding, produces a broad range of toxic effects including complex psychological and other central effects (Blain 2009).

The effective inhalation dose (ED_{50}) of BZ, as a human incapacitant, is estimated to be 60 mg min/m^3, which is less than that causing severe acute toxicity. The lethal dose is reported as 40 times the incapacitating dose (Committee on Acute Exposure Guideline Level 2013).

The United States assessed BZ and several other glycolate anticholinergics as less-lethal incapacitating agents in a programme that included other psychoactive, psychotomimetic, and psychedelic drugs, such as lysergic acid diethylamide (LSD) and tetrahydrocannabinol, and dissociative drugs such as ketamine and phencyclidine. Potent opioids, such as fentanyl analogues, have also been considered. The concept of use for BZ, as a CW agent, focused on the adverse effects on the mental state of human volunteers. A key advantage was its effective airborne dispersal by heat-producing munitions, and BZ was weaponised for delivery in the M44 generator cluster and the M43 cluster bomb. Stocks were destroyed in 1989 when the United States downsized its CW programme.

Prior to, and during, the Gulf War, the United Kingdom claimed that Iraq had produced large amounts of a glycolate anticholinergic incapacitating agent, then known as "Agent 15." This was later considered to be chemically identical to BZ or very closely related to it. After the war, no evidence was found that Iraq had stockpiled or weaponised Agent 15. In 2013, it was reported that Agent 15 was used by the Syrian Government in Homs although intelligence reports had never included it in the Syrian chemical weapons programme.

BZ is listed as a Schedule 2 compound by the OPCW. The Czech Army uses an indicative gas detector tube PT-51 for personal alert of BZ exposure and their deployable chemical laboratory, PPCHL-90, for airborne detection in the field.

Symptoms and signs

There is a direct dose–response for severity and duration of toxic effects. Low exposure causes a dry mouth and anhidrosis, pupil dilatation and loss of accommodation, tachycardia, and slowing of gut motility. There is mild sedation and decreased cognitive agility. These effects increase in severity with increasing exposure doses, the most incapacitating being the CNS effects that include profound cognitive dysfunction, hallucinations, a state of delirium, motor incoordination, and loss of the capability to perform psychomotor tasks. The clinical signs include a marked mydriasis (affecting visual acuity), tachycardia, dermal vasodilation, and hyperthermia. At very high exposures the individual may become stuporose. The clinical effects can last for 2–3 days, with the eye effects persisting longer (Blain 2009).

The constellation of symptoms and signs in an exposed individual should suggest exposure to a potent anti-cholinergic agent, especially if anti-muscarinic effects predominate. Initially, physostigmine, a standard treatment for anticholinergic poisoning, especially the esters of glycolic acid such as atropine, was considered, but this drug has its own toxicity issues and is short acting. Acridine derivatives were evaluated, and tacrine (1,2,3,4-tetrahydro-9-aminoacridine) was found to be most effective (Fusek et al. 1974). Tacrine is an anticholinesterase drug and increases the concentration of acetylcholine at muscarinic receptors. It has a long duration of action (Fusek 1977). Unfortunately, its toxicity profile, especially hepatotoxicity, compromised its use. A 7-methoxy derivative of tacrine was found to be more effective and has an acceptable safety profile. It also potentially had a prophylactic protective effect against the toxic anticholinesterase effects of organophosphates (Fusek et al. 1986).

A hybrid molecule (7-methoxytacrine (7-MEOTA)–donepezil) combines two anticholinesterases with dual binding sites for greater efficacy and can be administered by oral or parenteral routes (Korabecny et al. 2014). This would now be the antidote of choice for BZ exposure.

Opiates and opioids

Opiates and opioids are different. Opiates are chemical substances extracted from opium in the exudate of the unripe seed capsule of *Papaver somniferum* (the poppy plant) and include several types of alkaloids such as morphine and codeine. An opioid is a synthetic compound with morphine-like activity. Purely synthetic opioids include pethidine (meperidine), methadone, dextropropoxyphene, tramadol, fentanyl, and a range of fentanyl derivatives including alfentanil, sufentanil, remifentanil, and carfentanil.

Fentanyl

Fentanyl is an opioid drug first synthesised from pethidine in the 1960s by Janssen Pharmaceuticals. It is widely used as a safe short-acting IV analgesic/anaesthetic with an efficacy potency 100 times that of morphine. In clinical practice its role is in short-term anaesthesia and pain relief. It has few cardiovascular adverse effects and does not cause histamine release.

Carfentanil

Carfentanil is a fentanyl analogue estimated to be some 10 000 times as potent as morphine (100 times fentanyl) and is effective in humans at a 1-μg dose. The lethal dose is between 2 and 20 μg (a grain of salt). Carfentanil is intended for large animal use only as its extreme potency makes it highly dangerous to humans. Unfortunately, carfentanil, synthesised in China, is now widely available through the illicit drug market (Heslop and Blain 2020).

Pharmacology of opioids

Opioids interact with specific neuronal receptors widely distributed in the brain, spinal cord, peripheral nervous system, and gastrointestinal tract. These "opioid" receptors are a group of inhibitory G protein-coupled receptors for endogenous neuropeptide agonists such as endorphins, enkephalins, and endomorphins (see section titled on "Bioregulators").

Opiates and opioids are receptor agonists, and their principal effects arise from their activity on μ receptors. Activation of the μ_1 receptor reduces conscious awareness, whereas activation of the μ_2 receptor reduces respiratory drive. It is the latter that compromises the safe use of these drugs.

Dubrovka theatre siege in Moscow

In October 2002, 50 armed Chechen rebels entered the Dubrovka Theatre and took around 900 people hostages. Chechen women, the so-called black widows, were positioned in the auditorium with explosives.

Fifty-seven hours into the siege, an elite Russian Army troop started an assault. In the auditorium, one survivor recalled seeing a grey mist descending on to the hostages. As this mist entered the auditorium, the Chechen gunmen fled, but the women remained in the auditorium with the hostages and were rendered unconscious by the gas, along with everyone else. First responder rescuers had access to antidotes to treat the hostages affected by the incapacitant gas, but these were outside and not rapidly administered. This oversight led to the death of some hostages, who had either stopped breathing or mechanically obstructed their airway. At least 129 civilians died during the storming of the theatre, and of those rescued, 245 were hospitalised, eight in a serious condition. In total, more than 750 hostages were rescued alive.

Following pressure from Western Governments, the Russian Health Minister revealed that "*a fentanyl-based substance was used to neutralise the terrorists.*" Clothing from the British members of the audience was analysed at Dstl in the United Kingdom. Traces of carfentanil and remifentanil were identified (Riches et al. 2012).

Many lessons were identified from this use of fentanyl analogues as incapacitants. One obvious difficulty was targeting with a safe but effective dose and controlling the duration of exposure, especially with the rapidity of effect by the inhalation route. In the auditorium, there was a steep concentration gradient for the aerosol depending on the location and so a difficulty managing the level and duration of exposure. Added to this was the inability of drowsy/unconscious victims to flee exposure and limit the dose received. The deployment was totally indiscriminate and significantly compromised the safety of those exposed, who varied from children through to the old and healthy to those with pre-existing morbidities.

Naloxone is a non-selective competitive opioid receptor antagonist that antagonises all opiate and opioid effects, but to varying degrees. Respiratory depression is reversed but sedation less completely. Naloxone has a short duration of action (some 20–30 min) and may need to be administered as repeated bolus doses or an infusion, as many opioids are longer acting. Naloxone is available for emergency clinical use as 400 μg, 2 mg, and 5 mg pre-filled syringes. A nasal spray (2 mg naloxone) is increasingly available for use by law enforcement officers and first responders. Naltrexone and nalmefene are longer acting opioid receptor antagonists.

Perfluoroisobutene (PFIB)

Toxicology

Teflon is an organofluorine polymer, and many similar polymers are widely used in industry because of their physical and chemical properties. Combustion products from burning of organofluorines include a mix of pyrolysis by-products implicated in the causation of polymer fume fever. One of these by-products is PFIB.

PFIB is a colourless toxic gas that causes a severe irritation of the respiratory tract that may lead to non-cardiac pulmonary oedema, similar to that with phosgene inhalation. PFIB is also an irritant to human skin, eyes, and other mucous membranes, and acute exposures cause marked irritation of conjunctivae, throat, and lungs with wheezing, cough, dyspnea, excessive secretions, and cyanosis (Patocka and Bajgar 1998).

The symptoms of pulmonary oedema may be delayed for several hours, but death can occur suddenly following high exposure, and even, brief inhalation can result in acute lung injury and death. PFIB is approximately 10 times more toxic than phosgene.

The severe human toxicity has resulted in PFIB being designated a potential CW agent and included in Schedule 2A of CWC.

Symptoms and signs

The primary mechanisms of action are oxidative injury and secondary hyperinflammatory responses. As with phosgene, the pulmonary oedema appears earlier and more intensive if the exposed individual exercises post-exposure. This may relate to increased pulmonary blood flow and increased permeability of the alveolar/capillary interface.

The typical latent period for low PFIB exposure is 1–4 hours before exertional dyspnoea is experienced, followed by dyspnoea at rest. Radiological findings parallel the clinical findings. Recovery does occur following moderate exposures, often with few sequelae, but high dose exposure is invariably fatal.

No specific antidotes are currently available, so symptomatic and supportive treatments, which include PEEP (positive end-expiratory pressure)/CPAP, oxygen, and antibiotic cover, are essential. Oral *n*-acetylcysteine (NAC), commonly used as a mucolytic in COPD, has shown some protection against inhalation of PFIB in rats when administered orally 4, 6, or 8 hours prior to exposure. The duration of protection was associated with the duration of elevated plasma cysteine, glutathione, and overall NAC levels.

Unfortunately, NAC as a post-exposure treatment is ineffective, but a potential therapeutic approach, in early-stage acute respiratory distress caused by PFIB, is a combination treatment of NAC with a natural surfactant (Curosurf) used in neonatal respiratory distress syndrome. Curosurf contains polar lipids, such as phosphatidylcholine, and specific low molecular weight hydrophobic proteins.

The treatment of PFIB-induced lung oedema is conventional with administration of diuretics, such as furosemide and torsemide. Anticholinergic drugs and steroids may have a therapeutic role.

There is concern that PFIB may break through certain respirator cannisters. The presence of PFIB in air can be detected using GC–MS (Timperley 2000).

Bioregulators

Bioregulators are compounds, often peptides, that occur naturally in organisms and have specific cellular effects. They are involved in all the functions of a living organism, from basic cellular processes to higher aspects of health such as emotional mood, level of consciousness, and sleep. Primarily, they act on cell receptors, initiate chemical transmitter release, or regulate physiological responses. So, they affect blood pressure, heart rate, temperature control, respiration, and immune responses as well as higher functions, essentially exerting homeostatic regulatory control on the body.

Several potent, rapid-acting bioregulators could be used to cause a serious acute illness in an individual, and as they regulate a wide range of physiological activities, exposure could present as an apparent natural death. Only a few might be used on a large scale and present significant challenges to the medical emergency response. The potential military or terrorist use of bioregulators is similar to that for toxins, but effective deployment and routes of exposure remain immature (Patocka and Merka 2004).

Advances in *in vitro* synthesis and production of significant quantities of potent peptides in pharmaceutical bioreactors could facilitate their use as chemical weapons. Synthetic derivatives or slightly modified forms of these compounds could drastically alter their toxic effects, and the discovery of novel bioregulators, especially bioregulators causing rapid incapacitation, enhances their threat potential. Some of these compounds may be many hundreds of times more potent than traditional CW agents. Novel bioregulators could offer significant future challenges as their toxic action can be rapid and physiologically specific, and more so, if the molecule also penetrated protective respiratory filters and vehicle ventilation equipment to produce militarily relevant incapacitation.

Endorphins and enkephalins

Endorphins are small peptides that activate opiate receptors and produce feelings of well-being and tolerance to pain. β-Endorphin is the most active and is about 20 times more potent than morphine. Concentrations *in vivo* are very low.

High activity, and specificity, made endorphins attractive compounds for clinical use, but they are active only if injected into the blood (or the cerebrospinal fluid). Most peptides are digested in the stomach or by peptidase enzymes, and their size and structure make them unable to penetrate the BBB. β-Endorphin (and α-endorphin and γ-endorphin) are also produced by blood macrophages and lymphocytes (Terenius 1992).

Met and leu enkephalins are endogenous pentapeptides and a component of the body's pain control mechanisms. They activate opioid receptors, having the highest affinity for the ∂ receptor, but are relatively weak analgesics. Enkephalins are found in many areas of the body but predominantly those in the CNS associated with nociception (e.g. in periaqueductal grey area and dorsal horn) (Przewlocki and Przewlocka 2001).

Neurokinins, including substance P

Neurokinins are found in the spinal cord, the sensorial nuclei of the brain stem, and in the ends of peripheral sensory nerve fibres. Neurokinin A (NKA), neurokinin B (NKB), and neurokinin-1 (NK1) also occur in cold-blooded animals and as a group are called tachykinins. Neurokinins play various roles in the regulation of cardiovascular system, pain pathways, and inflammatory reactions.

Substance P (NK1) is an 11-amino-acid polypeptide member of the tachykinins. It is active in doses of $<1\ \mu g$, causing a rapid drop in blood pressure and unconsciousness. Substance P is found in the gastrointestinal tract as well as the brain and responsible for excitatory effects on both the central and the peripheral neurons. It causes contraction of smooth muscle, constriction of bronchioles, and increased capillary permeability. Release from afferent nerves causes neurogenic inflammatory response, including mast cell degranulation (Sandberg and Iversen 1982).

NKA and NKB are more potent bronchoconstrictors than substance P.

Substance P in an aerosol with thiorphan (which inhibits enkephalinase degradation of endogenous enkephalins) was reported to be highly toxic and incapacitated at low air concentrations (Koch et al. 1999).

Endothelins

Endothelins are a group of four peptides called endothelin-1 (ET-1), endothelin-2 (ET-2), endothelin-3 (ET-3), and vasoactive intestinal contractor (VIC). Endothelins are highly potent endogenous vasoconstrictors and vasopressors and are produced in many cell types such as smooth muscle, neurons, melanocytes, and the parathyroids.

Individual ETs have different physiological roles in different target tissues. Secretion of ETs is stimulated by epinephrine, angiotensin II (AT II), arginine vasopressin, transforming growth factor beta, thrombin, interleukin 1, and hypoxia. ETs act to stimulate contraction of many smooth muscle tissues including blood vessels, uterus, bladder, and intestine. ET-1 is the most potent vasoconstrictor peptide yet discovered (Miller et al. 1993).

Interestingly, snake venom sarafotoxins (sarafotoxins S6a and S6b) are similar to endothelins and are vasoconstrictors and potent coronary artery constrictors, causing cardiac arrest in a few minutes (Kloog and Sokolovsky 1989).

Bradykinin

Bradykinin is a mediator of the kinin system, along with kallidin, and produced from an α-2-globulin precursor by kallikreins, trypsin, or plasmin. It is a vasoactive nonapeptide. Bradykinin causes hypotension by dilating blood vessels and increases vascular permeability. In bronchial smooth muscles, intestines, and uterus, bradykinin produces muscle contraction. Bradykinin also plays a role in pain pathways and inflammation and is one of the most potent known substances for inducing pain. Bradykinin antagonists are used for treating inflammation and pain in rheumatic arthritis, osteoarthritis, and pancreatitis. Bradykinin has a powerful influence in stimulating smooth muscle contraction, increasing blood flow, and permeability of capillaries (Cyr et al. 2001).

Angiotensin

Angiotensin is a decapeptide produced by the kidney that elevates blood pressure through direct arteriole vasoconstriction, a sympathomimetic effect, and aldosterone release.

The inhibition of angiotensin converting enzyme (ACE) results in a double hypotensive effect because the formation of blood pressure raising AT II and the degradation of the blood pressure lowering kinin are both inhibited. AT II agonists are used for treatment of shock and collapse as a normal blood pressure can be restored quickly. ACE inhibitors and AT II antagonists are used as antihypertensive agents (Mazzolai et al. 1998).

Neurotensin

Neurotensin is a 13-amino-acid polypeptide that causes hypotension and has actions on smooth muscle that include relaxation of the duodenum and contraction of the ileum and uterus. Neurotensin is also a CNS neurotransmitter and is involved with memory function, temperature control, movement, and pain. In Alzheimer's patients, there are deficits of neurotensin in regions involved with memory function. Neurotensin also regulates dopamine pathways (Moore and Black 1991).

Other Bioregulators

Other bioregulator candidates exist, such as the orexins and oxytocin. The orexins have a central role in the sleep/wake cycle as well as appetite control. Their importance is seen in narcolepsy when a dysfunction of the orexin receptor pathway results in sudden loss of muscle tone and somnolence. Oxytocin is primarily involved in uterine muscle contraction and stimulation of breast milk production. Its role in the establishment of the maternal bond with the newborn also reflects a more general neurohumoral action of promoting bonding between individuals and generating an emotion of trust.

Summary

Chemical weapons have been used for decades and more and can be classified into a small number of groups. Such weapons, despite being banned under international conventions, are used by both individuals, criminals, terrorist groups, and rogue states. It is inevitable that such agents will be used in some settings and it is essential that healthcare professionals have the basis of knowledge to recognise the nature of the agent and to put in place or escalate to appropriate specialists in order to reduce morbidity and mortality and mitigate the effects of exposure to such agents. This chapter provides the basic principles to appropriately equip individuals working in the forensic setting and will raise awareness of the huge range of signs and symptoms and potential lethality of such agents. It should be expected that new, or alternative agents will be used or become available in the future.

References

Aldstadt, J.H., Olson, D.C., and Martin, A.F. (1997). Determination of volatile arsenicals in ambient air by flow injection. *Anal Chim Acta* 338: 215–222.

Asia Pacific Journal (2019). "I was exposed to nerve agent on Okinawa" – US soldier sickened by chemical weapon leak at Chibana Ammunition Depot in 1969 breaks silence on what happened that day. https://apjjf.org/2019/20/Mitchell.html (accessed 17 December 2021).

Bao, Y., Liu, Q., Chen, J. et al. (2012). Quantification of nerve agent adducts with albumin in rat plasma using liquid chromatography-isotope dilution tandem mass spectrometry. *J Chromatogr A* 1229: 164–171.

BBC News (2011). Tokyo 1995 sarin attack: Aum Shinrikyo cult trials end. https://www.bbc.co.uk/news/world-asia-15815056 (accessed 17 December 2021).

Bellingcat (2017). The Khan Sheikhoun chemical attack, the evidence so far. https://www.bellingcat.com/news/mena/2017/04/05/khan-sheikhoun-chemical-attack-evidence-far/ (accessed 17 December 2021).

Blain, P.G. (2009). Human incapacitants. In: *Clinical Neurotoxicology: Syndromes, Substances, Environments* (ed. M.R. Dobbs), 660–673. Philadelphia, PA: Saunders Elsevier.

Brimijoin, S., Chen, V.P., Pang, Y.-P. et al. (2016). Physiological roles for butyrylcholinesterase: a BChE-ghrelin axis. *Chem Biol Interact* 259 (Pt B): 271–275.

Campbell, C. (2017). The mysterious death and life of Kim Jong Nam. *Time*. https://time.com/4688208/kim-jong-nam-north-korea-kuala-lumpur/ (accessed 13 June 2022).

Colović, M.B., Krstić, D.Z., Lazarević-Pašti, T.D. et al. (2013). Acetylcholinesterase inhibitors: pharmacology and toxicology. *Curr Neuropharmacol* 11 (3): 315–335.

Committee on Acute Exposure Guideline Levels; Committee on Toxicology; Board on Environmental Studies and Toxicology; Division on Earth and Life Studies; National Research Council (2013). *Acute Exposure Guideline Levels for Selected Airborne Chemicals*, vol. 14. Washington, DC: National Academies Press (US).

Costa, L.G. (2006). Current issues in organophosphate toxicology. *Clin Chim Acta* 366: 1–13.

Cyr, M., Eastlund, T., Blais, C. Jr. et al. (2001). Bradykinin metabolism and hypotensive transfusion reactions. *Transfusion* 41: 136–150.

de Araujo Furtado, M., Rossetti, F., Chanda, S., and Yourick, D. (2012). Exposure to nerve agents: from status epilepticus to neuroinflammation, brain damage, neurogenesis and epilepsy. *Neurotoxicology* 33: 1476–1490.

Eddleston, M., Buckley, N.A., Eyer, P., and Dawson, A.H. (2008). Management of acute organophosphorus pesticide poisoning. *Lancet* 371: 597–607.

Ellman, G.L., Courtney, K.D., Andres, V. Jr., and Featherstone, R.M. (1961). A new and rapid colorimetric determination of acetylcholinesterase activity. *Biochem Pharmacol* 7: 88–90.

Emmett, S. and Blain, P. (2020). Chemical terrorism. *Medicine* 48 (3): 182–184.

Fusek, J., Patocka, J., Bajgar, J. et al. (1974). Pharmacology of 1,2,3,4,-tetrahydro-9-aminoac ridine. *Activ Nerv Sup* 16: 226–228.

Fusek, J. (1977). Tacrin and its analogues, antidotes against psychotomimetics with anticholinergic effects. *Voj Zdrav Listy* 46: 21–27.

Fusek, J., Patocka, J., Bajgar, J. et al. (1986). Anticholinesterase effects of 9-amino-7-methoxy-1,2,3,4-tetrahydroacridine. *Activ Nerv Sup* 28: 327–328.

Grange-Messent, V., Bouchaud, C., Jamme, M. et al. (1999). Seizure-related opening of the blood-brain barrier produced by the anticholinesterase compound, soman: new ultrastructural observations. *Cell Mol Biol (Noisy-le-grand)* 45: 1–14.

Hamamoto, Y., Ano, S., Allard, B. et al. (2017). Montelukast reduces inhaled chlorine triggered airway hyperresponsiveness and airway inflammation in the mouse. *Br J Pharmacol* 174 (19): 3346–3358. https://doi.org/10.1111/bph.13953.

Haslam, J.D., Russell, P., Hill, S. et al. (2022). Chemical, biological, radiological, and nuclear mass casualty medicine: a review of lessons from the Salisbury and Amesbury Novichok nerve agent incidents. *Br J Anaesth* 128 (2): e200–e205. https://doi.org/10.1016/j.bja.2021.10.008.

Hendry-Hofer, T.B., Ng, P.C., Witeof, A.E. et al. (2019). A review on ingested cyanide: risks, clinical presentation, diagnostics, and treatment challenges. *J Med Toxicol* 15 (2): 128–133. https://doi.org/10.1007/s13181-018-0688-y.

Heslop, D.J. and Blain, P.G. (2020). *Threat Potential of Pharmaceutical Based Agents*. Intelligence and National Security.

Hulse, E.J., Haslam, J.D., Emmett, S.R., and Woolley, T. (2019). Organophosphorus nerve agent poisoning: managing the poisoned patient. *Br J Anaesth* 123 (4): 457–463.

Ikegaya, H., Iwase, H., Hatanaka, K. et al. (2001). Diagnosis of cyanide intoxication by measurement of cytochrome c oxidase activity. *Toxicol Lett* 119 (2): 117–123.

Jiri Bajgar, P.H., Zdarova, J.K., Kassa, J. et al. (2010). A comparison of tabun-inhibited rat brain acetylcholinesterase reactivation by three oximes (HI-6, obidoxime, and K048) in vivo detected by biochemical and histochemical techniques. *J Enzyme Inhib Med Chem* 25 (6): 790–797.

Jokanović, M. (2009). Medical treatment of acute poisoning with organophosphorus and carbamate pesticides. *Toxicol Lett* 190: 107–115.

Jokanović, M., Kosanović, M., Brkić, D., and Vukomanović, P. (2011). Organophosphate induced delayed polyneuropathy in man: an overview. *Clin Neurol Neurosurg* 113: 7–10.

King, A.M. and Aaron, C.K. (2015). Organophosphate and carbamate poisoning. *Emerg Med Clin North Am* 33 (1): 133–151. https://doi.org/10.1016/j.emc.2014.09.010.

Kloog, Y. and Sokolovsky, M. (1989). Similarities in mode and sites of action of sarafotoxins and endothelins. *Trends Pharmacol Sci* 10: 212–214.

Koch, B.L., Edvinsson, A.A., and Koskinen, L.O. (1999). Inhalation of substance P and thiorphan: acute toxicity and effects on respiration in conscious guinea pigs. *J Appl Toxicol* 19: 19–23.

Korabecny, J., Dolezal, R., Cabelova, P. et al. (2014). 7-MEOTA-donepezil like compounds as cholinesterase inhibitors: synthesis, pharmacological evaluation, molecular modelling and QSAR studies. *Eur J Med Chem* 82: 426–438.

Lakshmi, J., Mukhopadhyay, K., Ramaswamy, P., and Mahadevan, S. (2020). A systematic review on organophosphate pesticide and type II diabetes mellitus. *Curr Diabetes Rev* 16: 586–597.

Lenz, D.E., Yeung, D., Smith, J.R. et al. (2007). Stoichiometric and catalytic scavengers as protection against nerve agent toxicity: a mini review. *Toxicology* 233 (1–3): 31–39.

Li, J., Ren, F., Li, Y. et al. (2019). Chlorpyrifos induces metabolic disruption by altering levels of reproductive hormones. *J Agric Food Chem* 67 (38): 10553–10562.

Lotti, M. and Moretto, A. (2005). Organophosphate-induced delayed polyneuropathy. *Toxicol Rev* 24 (1): 37–49. https://doi.org/10.2165/00139709-200524010-00003.

Marrs, T.C. (2004). The role of diazepam in the treatment of nerveagent poisoning in a civilian population. *Toxicol Rev* 23 (3): 145–157. https://doi.org/10.2165/00139709-200423030-00002.

Mazzolai, L., Nussberger, J., Auber, J.F. et al. (1998). Blood pressure-independent cardiac hypertrophy induced by locally activated renin-angiotensin system. *Hypertension* 31: 1324–1330.

McDonough, J.H., McMonagle, J.D., and Capacio, B.R. (2021). Anticonvulsant effectiveness of scopolamine against soman-induced seizures in African green monkeys. *Drug Chem Toxicol* 1–8. https://doi.org/10.1080/01480545.2021.1916171.

Miller, R.C., Pelton, J.T., and Huggins, J.P. (1993). Endothelins – from receptors to medicine. *Trends Pharmacol Sci* 14: 54–60.

Mirzayanov, V.S. (1992). *State Secrets: An Insider's Chronicle of the Russian Chemical Weapons Program*. Outskirts Press.

Moore, M.R. and Black, P.M. (1991). Neuropeptides. *Neurosurg Rev* 14: 97–110.

Mutafova-Yambolieva, V.N., Yamboliev, I.A., and Mihailova, D.N. (1993). Comparative effects of the anticholinesterase drug galanthamine on the mechanical activity of isolated rat jejunum and ileum. *Gen Pharmacol* 24: 1253–1256.

Nepovimova, E. and Kuca, K. (2018). Chemical warfare agent NOVICHOK - mini-review of available data. *Food Chem Toxicol* 121: 343–350.

Nie, Z., Zhang, Y., Chen, J. et al. (2014). Monitoring urinary metabolites resulting from sulfur mustard exposure in rabbits, using highly sensitive isotope-dilution gas chromatography-mass spectrometry. *Anal Bioanal Chem* 406: 5203–5212.

Nohara, M. and Segawa, K. (1996). Ocular symptoms due to organophosphorus gas (Sarin) poisoning in Matsumoto. *Br J Ophthalmol* 80 (11): 1023.

Noort, D., Fidder, A., van der Riet-van Oeveren, D. et al. (2021). Verification of exposure to novichok nerve agents utilizing a semitargeted human butyrylcholinesterase nonapeptide assay. *Chem Res Toxicol* 34 (8): 1926–1932.

Nowak, T.V. and Harrington, B. (1985). Effect of cholinergic agonists on muscle from rodent proximal and distal small intestine. *Gastroenterology* 88: 1118–1125.

Ogane, N., Giacobini, E., and Messamore, E. (1992). Preferential inhibition of acetylcholinesterase molecular forms in rat brain. *Neurochem Res* 17: 489–495.

Okumura, T., Hisaoka, T., Yamada, A. et al. (2005). The Tokyo subway sarin attack--lessons learned. *Toxicol Appl Pharmacol* 207 (2 Suppl): 471–476. https://doi.org/10.1016/j.taap.2005.02.032.

OPCW (2021). The convention on the prohibition of the development, production, stockpiling and use of chemical weapons and on their destruction -chemical weapons convention. https://www.opcw.org/chemical-weapons-convention (accessed 17 December 2021).

Patocka, J. and Bajgar, J. (1998). Toxicology of perfluoroisobutene. *ASA Newsl* 5 (69): 16–18.

Patocka, J. and Merka, V. (2004). Bioregulators as agents of terrorism and warfare. *Ned Mil Geneesk T* 57: 12–15.

Patrick, W.J., Besley, G.T., and Smith, I.I. (1980). Histochemical diagnosis of Hirschsprung's disease and a comparison of the histochemical and biochemical activity of acetylcholinesterase in rectal mucosal biopsies. *J Clin Pathol* 33 (4): 336–343.

Pauluhn, J. (2021). Phosgene inhalation toxicity: update on mechanisms and mechanism-based treatment strategies. *Toxicology* 450: 152682. https://doi.org/10.1016/j.tox.2021.152682.

Przewlocki, R. and Przewlocka, B. (2001). Opioids in chronic pain. *Eur J Pharmacol* 429: 79–91.

Rafati-Rahimzadeh, M. (2019). Caspian. *J Intern Med* 10 (3): 241–264. https://doi.org/10.22088/cjim.10.3.241.

Riches, J.R., Read, R.W., Black, R.M. et al. (2012). Analysis of clothing and urine from Moscow theatre siege casualties reveals carfentanil and remifentanil use. *J Anal Toxicol* 36: 647–656.

Sandberg, B.E. and Iversen, L.L. (1982). Substance P. *J Med Chem* 25: 1009–1015.

Savelieva, E.I. (2021). Scopes of bioanalytical chromatography–mass spectrometry. *J Anal Chem* 76 (10): 1198–1210.

Singh, N., Golime, R., Acharya, J., and Palit, M. (2020). Quantitative proteomic changes after organophosphorous nerve agent exposure in the rat hippocampus. *ACS Chem Neurosci* 11 (17): 2638–2648.

Steinritz, D., Striepling, E., Rudolf, K.D. et al. (2016). Medical documentation, bioanalytical evidence of an accidental human exposure to sulfur mustard and general therapy recommendations. *Toxicol Lett* 244: 112–120.

Taylor, J.L., Mayer, R.T., and Himel, C.M. (1994). Conformers of acetylcholinesterase: a mechanism of allosteric control. *Mol Pharmacol* 45: 74–83.

Terenius, L. (1992). Opioid peptides, pain and stress. *Prog Brain Res* 92: 375–383.

Timperley, C.M. (2000, Chapter 29). Highly toxic fluorine compounds. In: *Fluorine Chemistry at the Millennium. Fascinated by Fluorine* (ed. E. Banks), 499–538. Elsevier.

US Department of State (2003). Saddam's chemical weapons campaign: Halabja, March 16, 1988. https://2001-2009.state.gov/r/pa/ei/rls/18714.htm (accessed 17 December 2021).

Wooten, J.V., Ashley, D.L., and Calafat, A.M. (2002). Quantitation of 2-chlorovinylarsonous acid in human urine by automated solid-phase microextraction--gas chromatography--mass spectrometry. *J Chromatogr B Anal Technol Biomed Life Sci* 772: 147–153.

Worek, F. and Thiermann, H. (2013). The value of novel oximes for treatment of poisoning by organophosphorus compounds. *Pharmacol Ther* 139: 249–259.

Worek, F., Thiermann, H., and Wille, T. (2016). Oximes in organophosphate poisoning: 60 years of hope and despair. *Chem Biol Interact* 259: 93–98.

Xu, H., Nie, Z., Zhang, Y. et al. (2014). Four sulfur mustard exposure cases. *Toxicol Rep* 1: 533–543.

Xu, F., Ashbrook, D.G., Gao, J. et al. (2020). Genome-wide transcriptome architecture in a mouse model of Gulf War Illness. *Brain Behav Immun* 89: 209–223.

Zellner, T. and Eyer, F. (2020). Choking agents and chlorine gas - history, pathophysiology, clinical effects and treatment. *Toxicol Lett* 320: 73–79. https://doi.org/10.1016/j.toxlet.2019.12.005.

Index

Note: Page numbers followed by "*f*", "*t*" and "*b*" indicate figures, tables and boxes respectively.

Current Practice in Forensic Medicine, Volume 3, First Edition. Edited by John A.M. Gall and J. Jason Payne-James.